Computational and Experimental Studies

Computational and Experimental Studies

Editors

Y. Villacampa
Universidad de Alicante, Spain

G.M. Carlomagno
University of Naples "Federico II", Italy

C.A. Brebbia
Wessex Institute, UK

S. Ivorra
University of Alicante, Spain

WIT*PRESS* Southampton, Boston

Editors:

Y. Villacampa
Universidad de Alicante, Spain

G.M. Carlomagno
University of Naples "Federico II", Italy

C.A. Brebbia
Wessex Institute, UK

S. Ivorra
University of Alicante, Spain

Published by

WIT Press
Ashurst Lodge, Ashurst, Southampton, SO40 7AA, UK
Tel: 44 (0) 238 029 3223; Fax: 44 (0) 238 029 2853
E-Mail: witpress@witpress.com
http://www.witpress.com

For USA, Canada and Mexico

Computational Mechanics International Inc
25 Bridge Street, Billerica, MA 01821, USA
Tel: 978 667 5841; Fax: 978 667 7582
E-Mail: infousa@witpress.com
http://www.witpress.com

British Library Cataloguing-in-Publication Data

A Catalogue record for this book is available
from the British Library

Library of Congress Catalog Card Number: 2018932127

ISBN: 978-1-78466-309-4
eISBN: 978-1-78466-310-0

The texts of the papers in this volume were set individually by the authors or under their supervision. Only minor corrections to the text may have been carried out by the publisher.

Printed in Great Britain by Lightning Source, UK.

The material contained herein is reprinted from special editions of the International Journal of Computational Methods and Experimental Measurements, Volume 6, Number 4, published by WIT Press.

Preface

The main scope of this book is to provide to the international technical and academic community information about the latest developments on the interaction and the complementary aspects of computational methods and experimental measurements. The main attention and relevance being committed to their reciprocal and advantageous integration

It is recognised that the constant progresses in computers efficiency and numerical techniques are producing a steady growth of computational simulations, which nowadays applies to an ever-widening range of engineering problems. Nonetheless, even if these simulations are continuously expanding and improving, there still exists the need for their validation, especially for the more complex cases, which can be only accomplished by performing dedicated experimental tests. Experimental techniques are becoming increasingly complex and sophisticated so that both their running as well as data collection can only be performed by computers. Finally, it must be emphasised that, for the majority of measurements, the data obtained must be processed numerically.

This volume contains a substantial number of excellent scientific papers, which present several advanced approaches to the application of Computational Methods and Experimental Measurements.

The Editors
Alicante

Contents

COMBINED EXPERIMENTAL AND NUMERICAL APPROACH TO MODEL, DESIGN AND OPTIMIZE THERMAL PROCESSES

YOGESH JALURIA
Board of Governors Professor and Distinguished Professor, Mechanical Engineering Department, Rutgers University Piscataway, New Jersey, USA.

ABSTRACT
This paper focuses on combined experimental and numerical approaches to model thermal processes and obtain accurate results on system behaviour and performance. Interest lies in obtaining repeatable and dependable inputs for choosing appropriate conditions and parameters for enhancing the efficiency and the desired output. These results can also form the basis for system design and optimization. Several fundamental and practical problems are considered and typical results presented to discuss the implications and applications of this methodology. Circumstances where experimental data are used to validate the model, provide greater physical insight and define the boundary conditions, thus allowing the numerical simulation to be carried out, are also presented. Results from a concurrent, or parallel, simulation and experimentation approach are also presented to indicate the usefulness of such a strategy. It is stressed that experimental data are indispensable in obtaining accurate and realistic results for complex practical problems involving thermal transport processes.
Keywords: combined approach, concurrent, experiment, inverse problem, numerical, thermal processes, thermal systems

1 INTRODUCTION

Thermal systems and processes are of interest in a wide range of applications, from manufacturing and transportation to thermal management of electronics, environmental control, and power generation. Because of the complexity of these systems, arising from material property variations, complicated domains, combined transport mechanisms, turbulent flow and other aspects, numerical modeling is needed to study, predict, design and optimize [1–3]. However, in many cases, experimental inputs are essential to completing the modeling effort and a combined experimental–numerical approach is valuable in obtaining accurate and dependable results. Experimental data are critical to the validation of the mathematical and numerical model and to the establishment of the accuracy and predictability of the numerical simulation. This is particularly important for complex transport processes that arise in most practical thermal systems [4].

Experiments are also needed for the determination of material properties that are crucial to any accurate simulation. In many important thermal processes, the boundary conditions are not known or well defined. Numerical determination of the boundary conditions may also be quite involved as is the case in conjugate problems. In such cases, experimental work can be used to provide the appropriate boundary conditions that may be applied for the simulation. An inverse problem must often be solved, using both the experimental data and the numerical model, to obtain the appropriate boundary conditions and solve the problem. Also, there are many problems in which experimentation is particularly suitable over given parametric ranges, while numerical simulation is more appropriate over other regions. For instance, turbulent flow and contact resistance are better treated experimentally than by analysis. Then, a

DOI: 10.2495/CMEM-V6-N4-625-634

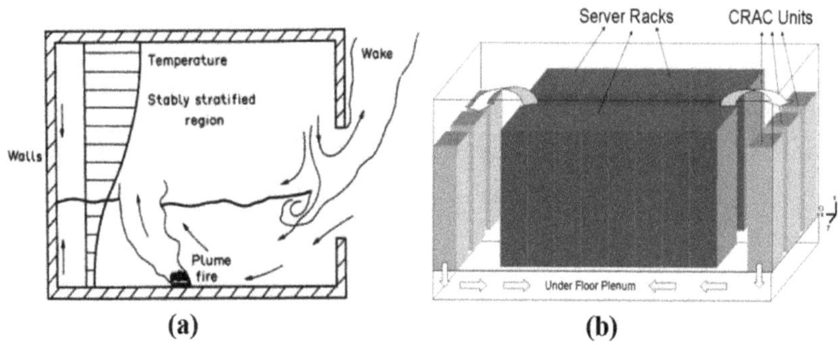

Figure 1. Examples of thermal systems: (a) Room with a fire; (b) Typical data center.

concurrent, parallel, experimental and numerical approach may be used to solve the problem more efficiently and accurately.

Of particular interest in this paper are the following aspects of a combined numerical and experimental approach:

- Validation
- Experimentally obtained boundary conditions
- Solution of inverse problems with experimental inputs
- Use of experimentation in model development
- Concurrent simulation and experimentation

All these are important in obtaining an accurate, efficient and realistic modeling and simulation of thermal processes and systems.

Figure 1 shows two common thermal systems in which experimentation and numerical simulation may be used to provide the inputs for design and for understanding the basic processes involved. The systems shown include a room with a fire, which has a stratified hot upper layer generated in the room due to the fire plume and flow exchange through an opening, and a typical data center, which involves electronic components, racks, servers and cooling arrangements [5]. Because of the various complexities mentioned earlier, an accurate study of the basic processes and of the system in these and other practical problems requires a strong coupling between experimentation and numerical simulation.

2 VALIDATION

An extremely important consideration in the modeling and simulation of thermal processes and systems is that of validation because of the simplifications used to treat various complexities. It is necessary to ensure that the numerical code performs satisfactorily and that the model is an accurate representation of the physical problem [6]. A consideration of the physical behavior of the results obtained is used to ensure that the results and trends are physically reasonable. Comparisons with available analytical and numerical results, particularly benchmark solutions, can then be used for validation of the mathematical and numerical models. Comparisons with experimental results are obviously desirable and it may become necessary to develop an experimental arrangement for providing data for validation. Figure 2 shows the validation of the mathematical and numerical model for the chemical vapor deposition (CVD) system shown.

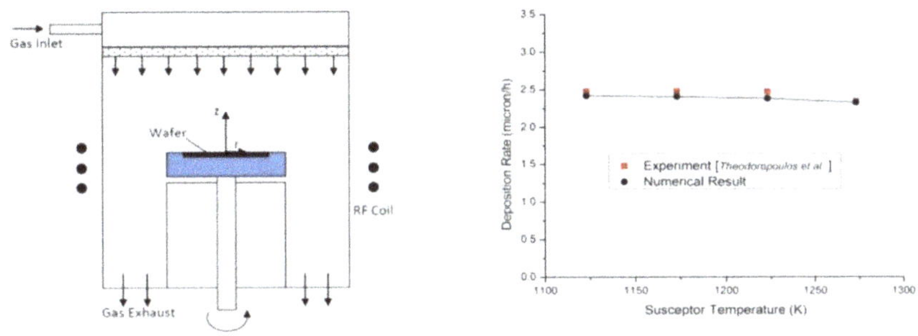

Figure 2. Schematic of a vertical rotating disk reactor along with numerical and experimental results on deposition rate.

The governing equations are the fluid flow and convective heat transfer equations with variable properties, along with chemical reactions and species equations. These may be given as:

$$\frac{D\rho}{Dt} + \rho \nabla . \bar{\bar{V}} = 0 \tag{1}$$

$$\rho \frac{D\bar{V}}{Dt} = \bar{F} + \nabla . \tau \tag{2}$$

$$\rho C_p \frac{DT}{Dt} = \nabla .(k \nabla T) + \dot{Q} + \beta T \frac{Dp}{Dt} + \mu \Phi \tag{3}$$

where ρ, C_p, k and β are the density, specific heat at constant pressure, thermal conductivity and coefficient of volumetric expansion, $\bar{V}$ the velocity vector, $\dot{Q}$ thermal energy source per unit volume, T the temperature, t the time, p the pressure, and $\bar{F}$ body force per unit volume. Also, D/Dt is the substantial or particle derivative, given in terms of the local derivatives in the flow. The stress tensor τ can be written in terms of the velocity if the fluid characteristics are known, yielding Navier–Stokes equations for common Newtonian fluids like air and water, often employed in cooling of electronic systems.

The species equations and chemical kinetics are given in terms of concentration ω, diffusion coefficient D, rate constant K and partial pressure p by

$$\frac{\partial(\rho u_j \omega_i)}{\partial x_j} = \frac{\partial}{\partial x_j}(\rho D_{ij} \frac{\partial \omega_i}{\partial x_j}) \tag{4}$$

$$K = \frac{K_o p_{SiH4}}{1 + K_1 p_{H2} + K_2 p_{SiH4}} \tag{5}$$

where the last equation is for silicon deposition. Usually, the problem is a very complicated one [7] and involves large number of chemical equations and species for the deposition of materials like gallium nitride, which is case in this example [8]. The figure shows good agreement between experimental and numerical results. Validation thus becomes critical for accurate and dependable results in practical processes and systems.

3 RESULTS AND DISCUSSION

A few results are presented to illustrate the application of combined experimental and numerical simulation to address the various aspects mentioned earlier. Only a brief discussion is given here and additional details may be obtained from the references given.

Experimentally obtained boundary conditions. In this case, the boundary condition is obtained experimentally because of the complicated nature of the analysis needed to determine it. Examples are the shape and dimensions of the dynamic meniscus in surface coating as the material plunges into a liquid [9]. This meniscus is determined experimentally and is then used as an input into the numerical model. Similarly, the temperature distribution at the surface of a CVD susceptor, or wafer, such as in the system shown in Fig. 2, depends on the heat flux, conduction heat transfer, convection at the surface, geometry, etc. It is more accurate and simpler to measure it experimentally, as shown in Fig. 3 for a vertical impinging CVD reactor [10]. With the measured temperature distribution provided as input to the numerical model, the results on flow, thermal field and deposition rate may be obtained for various operating conditions and design parameters. A few results are shown in Fig. 3 to indicate the dependence of deposition rate on inlet velocity and on the inflow concentration of the reactants. Similarly, other results may be obtained and the system can be optimized for high product quality at acceptable deposition rates.

Solution of inverse problems with experimental inputs. There are circumstances where experiment data can be obtained over only a limited region because of access, time or other limitations. In such cases, numerical modeling and experimentation may be used together to solve an inverse problem in order to define and quantify the boundary conditions and then proceed to the numerical simulation of the complete problem [11,12]. An example of this

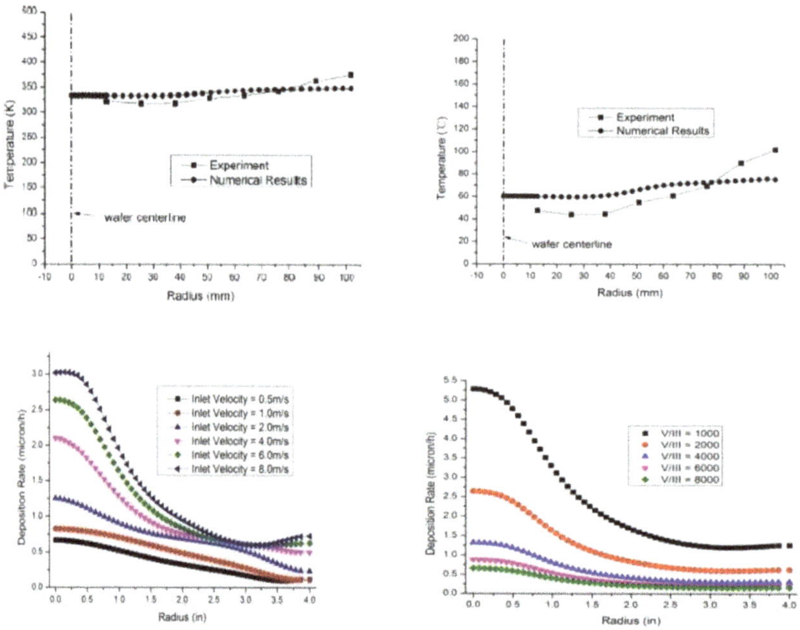

Figure 3. Measured susceptor temperature with no flow and no rotation (top left) and with 1 m/s inlet flow with rotation at 60 rpm (top right), along with calculated results on deposition rate at 600 rpm.

problem is given in Fig. 4, which shows a heated jet in cross-flow. If limited data taken downstream can be used to determine the location and conditions at the inlet, it would allow the determination of, for example, a polluting source as well as the impact on the environment.

The temperature data obtained downstream in the flow is used to solve the inverse problem to determine the inlet velocity and temperature of the jet. Figure 4(b) shows the improvement

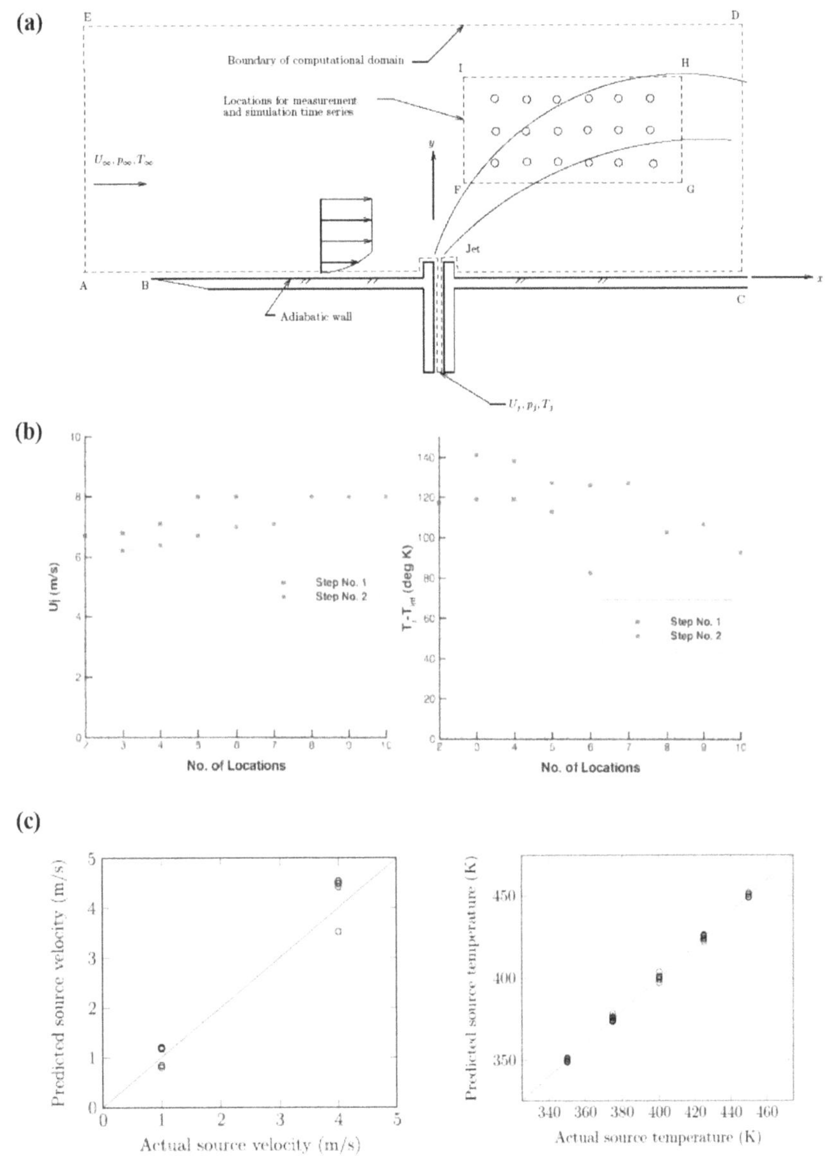

Figure 4. Inverse problem to determine jet inlet temperature and velocity from experimental data taken downstream. (a) Flow configuration; (b) Using experimental data to determine inlet conditions; (c) Accuracy of prediction.

in the determination as more data points are used. Step 1 refers to only one unknown, with the other variable known, and step 2 refers to the case of both velocity and temperature as unknown. Figure 4(c) shows a comparison between actual values and predicted ones from the inverse solution. A good agreement is seen. This approach was also used for finite heat sources in a channel and for determining the temperature distribution at the wall of a furnace [13].

Use of experimentation in model development. This is a particularly important application of the combined experiment and simulation approach. It is frequently used in practical thermal systems for developing a valid, accurate and physically realistic model. Figure 5 shows an example of this approach by considering a microchannel flow for heat removal from an electronic chip. The experimental system is sketched in Fig. 5(a), indicating a heater and a silicon block containing the microchannel. Several models were considered for simulating the thermal processes involved [14]. These included microchannel flow with imposed boundary conditions, microchannel with the heater and the entire system. The last two are referred to as Model II and III, respectively. The experimental results were compared with the numerical results, as shown in Fig. 5(b). At high flow rates, both the models gave results which were very close. But, at low flow rates, Model II did not perform satisfactorily, indicating the need to model the entire system, as given by Model III. Similarly, in other applications, such as casting, models may be developed, going from simple models to fairly elaborate models, and the experimental data may be used to choose the appropriate model [15].

Another example is shown in Fig. 6, where the comparison between experimental and computed results, for two different heat input conditions in the system sketched in Fig. 6(a), are used to choose the appropriate boundary condition as adiabatic and thus develop a more realistic model for the system shown.

Concurrent simulation and experimentation. In the solution of practical thermal convection problems, it is often found that numerical simulation is particularly suitable over a certain domain, whereas experimentation is more appropriate and accurate over other domains. Then, the two could be used *concurrently* or in parallel to obtain a more efficient approach to solving the problem.

Conventional engineering design and optimization are based on sequential use of computer simulation and experiment, with the experiments generally being used for validation or for providing selective inputs, as discussed earlier. However, the conventional methods fail to use

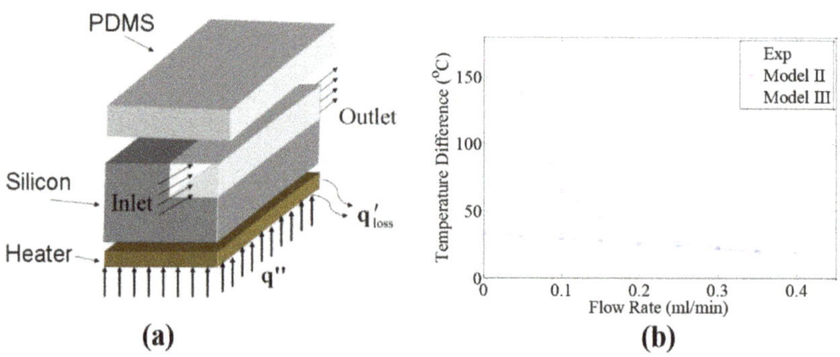

Figure 5. Sketch of a microchannel flow system for heat removal from an electronic chip and the experimental-numerical results on the temperature.

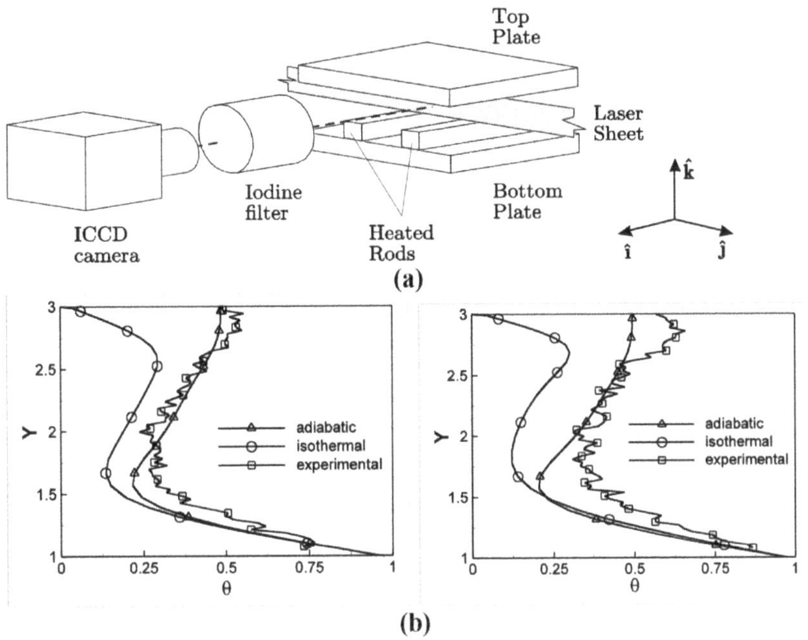

Figure 6. An experimental system for heat removal from two isolated heat sources that approximate electronic devices and comparison between experimental and numerical results to determine the correct boundary condition.

the advantages of using experiment and simulation concurrently in real time. Numerical simulation can easily accommodate changes in geometry, dimensions and material, whereas experiments can more conveniently study variations in the operating conditions such as flow rate, imposed pressure and heat input. Also, laminar and stable flows can be simulated conveniently and accurately, whereas transitional and turbulent flows are often more accurately investigated experimentally. By using concurrent numerical simulation and experimentation, the entire domain of interest can be studied for system design and optimization efficiently and accurately. This is the main motivation for this approach.

A simple physical system, consisting of multiple isolated heat sources, which approximate electronic components, located in a horizontal channel in a two-dimensional configuration, with or without a vortex generator to enhance heat transfer, is considered. Figure 7(a) and (b) shows the computed streamlines in the two cases. Numerical and experimental methods are used concurrently to study a wide range of design variables and operating conditions. The temperature and velocity distributions, the heat removal rates and pressure drop are calculated for laminar flows, as well as the beginning of oscillatory flow. Experiments are used for translational and turbulent flows. The first part of the simulation results deals with the determination of the critical flow conditions up to which numerical simulation can be used satisfactorily.

Figure 7(c) and (d) shows the results for a wide range of conditions with and without a vortex generator, respectively. The heat transfer from the first heat source facing the incoming flow is plotted against the Reynolds number Re based on channel height. The results at small values of Re, up to transition, are based on laminar flow calculations, whereas the results at

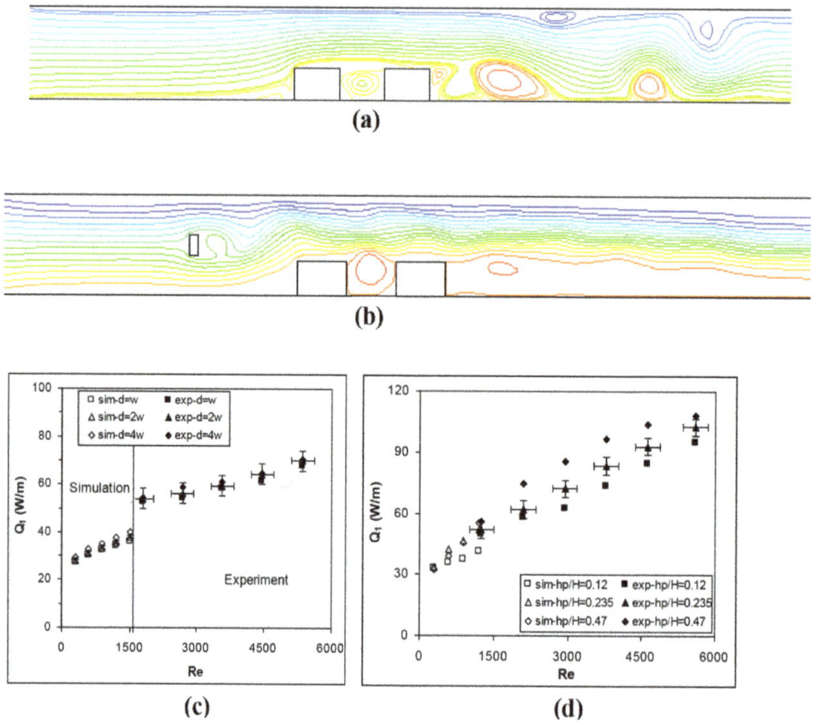

Figure 7. Concurrent experimentation and numerical simulation. (a) Computed streamlines for isolated heat sources in a channel; (b) Computed streamlines for isolated heat sources in a channel with a vortex generator; (c) and (d) Numerical and experimental results on heat transfer from the first source for (a) and (b).

larger Re are experimental ones shown with error bars. First, validation of the model is easily established. The results, which cover a wide range of Re and other parameters, are obtained efficiently by selectively using simulation and experimentation. Then the results are used for design and optimization of the system to maximize the heat transfer while keeping the pressure head within acceptable limits [16–18].

4 CONCLUSIONS

Experimentation is needed in various thermal processes and systems in order to provide the inputs needed for accurately defining the boundary conditions, simplifying the modeling and obtaining results over regions where simulation is inaccurate, inconvenient or inefficient. In addition, experimental data are needed for the validation of the models used. This paper presents various circumstances where the numerical simulation may be efficiently combined with experimentation, and indeed driven by experimental data, to obtain accurate, valid and realistic numerical predictions. Several examples of such problems are given and the difficulties with specifying the boundary conditions as well as with simulating the entire domain for design and optimization are outlined. Approaches for using experimental data driven simulation in such cases are discussed and results are presented for some simple and complex problems. It is shown that such approaches are critical to an accurate numerical simulation in many cases of practical interest.

5 ACKNOWLEDGEMENTS

The author acknowledges the support of the National Science Foundation, through several grants, for much of the work reported here. The author also acknowledges the interactions with Professors D. Knight, T. Rossmann and R. Bianchini, and the work done by several students.

REFERENCES

[1] Incropera, F.P., Convection heat transfer in electronic equipment cooling. *ASME Journal of Heat Transfer*, **110**, pp. 1097–1111, 1988.
https://doi.org/10.1115/1.3250613

[2] Paek, U.C., Free drawing and polymer coating of silica glass optical fibers. *ASME Journal of Heat Transfer*, **121**, pp. 774–788, 1999.
https://doi.org/10.1115/1.2826066

[3] Jaluria, Y., Challenges in the accurate numerical simulation of practical thermal processes and systems. *International Journal of Numerical Methods for Heat & Fluid Flow*, **23**, pp. 158–175, 2013.
https://doi.org/10.1108/09615531311289169

[4] Jaluria, Y., *Design and Optimization of Thermal Systems,* Second Edition, CRC Press, Boca Raton, FL, 2008.

[5] Joshi, Y. & Kumar, P., Eds., *Energy Efficient Thermal Management of Data Centers*, Springer, New York, 2012.

[6] Roache, P.J., *Verification and Validation in Computational Science and Engineering*, Hermosa Publishers, Albuquerque, New Mexico, 1998.

[7] Mahajan, R.L., Transport phenomena in chemical vapour deposition systems. *Advances in Heat Transfer*, **28**, pp. 339–425, 1996.
https://doi.org/10.1016/s0065-2717(08)70143-6

[8] Meng, J. & Jaluria, Y., Transient behaviour of thin film deposition: Coupling micro and macroscale transport. *Numerical Heat Transfer*, **68**, pp. 355–368, 2015.
https://doi.org/10.1080/10407782.2014.986373

[9] Ravinutala, S. & Polymeropoulos, C.E., Entrance meniscus in a pressurized optical fiber coating applicator. *International Journal of Experimental Heat Transfer, Fluid Mechanics,* **26**, pp. 573–580, 2002.
https://doi.org/10.1016/s0894-1777(02)00168-1

[10] Meng, J., Wong, S. & Jaluria, Y., Fabrication of GaN films in a chemical vapour deposition reactor. *Journal of Thermal Science and Engineering Applications*, **7**, pp. 021003, 2015.
https://doi.org/10.1115/1.4029353

[11] Darema, F., *Dynamic data driven application systems: A new paradigm for application simulations and measurements.* 4th International Conference on Computational Science, Springer-Verlag, Berlin, pp. 662–669, 2004.

[12] Rossmann, T., Knight, D.D. & Jaluria, Y., Data assimilation optimization for the evaluation of inverse mixing and convection flows, *Fluid Dynamics Research*, **47**, 2015.
https://doi.org/10.1088/0169-5983/47/5/051405

[13] VanderVeer, J. & Jaluria, Y., Solution of an inverse convection problem by a predictor-corrector approach. *International Journal of Heat and Mass Transfer*, **65**, pp. 123–130, 2013.
https://doi.org/10.1016/j.ijheatmasstransfer.2013.05.055

[14] Zhang, J., Jaluria, Y., Zhang, T. & Jia, L., Combined experimental and numerical study for multiple microchannel heat transfer system. *Numerical Heat Transfer,* **64**, pp. 293–305, 2013.
https://doi.org/10.1080/10407790.2013.791781

[15] Jaluria, Y., Thermal processing of materials: from basic research to engineering. *ASME Journal of Heat Transfer*, **125**, pp. 957–979, 2003.
https://doi.org/10.1115/1.1621889

[16] Icoz, T. & Jaluria, Y., Design of cooling systems for electronic equipment using both experimental and numerical inputs. *ASME Journal of Electronic Packaging*, **126**, pp. 465–471, 2005.
https://doi.org/10.1115/1.1827262

[17] Icoz, T. & Jaluria, Y., Design optimization of size and geometry of vortex promoter in a two-dimensional channel. *ASME Journal of Heat Transfer*, **128**, pp. 1081–1092, 2006.
https://doi.org/10.1115/1.2345433

[18] Zhao, H., Icoz, T., Jaluria, Y. & Knight, D., Application of data driven design optimization methodology to a multi-objective design optimization problem. *Journal of Engineering Design*, **18**, pp. 343–359, 2007.
https://doi.org/10.1080/09544820601010981

AN ELASTIC-VISCO-PLASTIC DEFORMATION MODEL OF Al–Li WITH APPLICATION TO FORGING

L. B. BORKOWSKI, J. A. SHARON, & A. STAROSELSKY
United Technologies Research Center, East Hartford, CT.

ABSTRACT

Recent alloy developments have produced a new generation of Al–Li alloys that provide not only weight savings, but also many property benefits such as excellent corrosion resistance, good spectrum fatigue crack growth performance, a good strength and toughness combination and compatibility with standard manufacturing techniques. The forging of such alloys would lead to mechanical properties that closely match the aircraft engine requirements including lower weight, improved performance and a longer life. As a result, detailed analyses need to be performed to determine which material properties are best suited for a specific structure and how to achieve the required mechanical and damage tolerant properties during material processing.

We developed an integrated physics-based model for prediction of microstructure evolution and material property prediction of third-generation Al–Li alloys. In order to develop such a model, an elastic-plastic crystal plasticity model is developed and incorporated in finite element software (ANSYS). The model accounts for microstructural evolution during non-isothermal, non-homogeneous deformation and is coupled with the damage kinetics. Our model bridges the gap between dislocation dynamics and continuum mechanics scales.

Model parameters have been calibrated against lab tests including micropillar in-situ simple compression tests of Al–Li alloy 2070. Numerical predictions are verified against the lab results including stress–strain curves and crystallographic texture evolution.

Keywords: crystallographic texture, light weight alloys, material characterization, material processing, micro-scale testing

1 INTRODUCTION

Aircraft weight is an important factor in fuel economy. Even a modest decrease in airplane weight reduces the fuel consumption and hence, allows increase in the maximum flight range, which in turn makes longer direct flights possible, avoiding extra take offs and landings, increasing engine reliability, reducing maintenance cost, not mentioning the additional fuel savings. As a rule-of-thumb, a 1% weight reduction in the gross weight of an empty aircraft corresponds to 0.25%–0.75% reduction in fuel consumption [1, 2]. Our overall goal is to achieve weight reductions possible through material swaps in aircraft engines.

The first component in a turbofan engine is a fan composed of fast rotating blades. Currently most of the blades of the fan are made of titanium. There is a drive to develop them from lighter materials, such as aluminum alloys. The current design of large fan blades in commercial turbines is generally too heavy and generates excessively high stresses on the component and the surrounding structures. Weight reduction afforded by a switch to Al–Li alloys would resolve both of these issues.

The fan blades are the subject of significant temperature variation, high inertia stresses and a centripetal force through its root. Furthermore, these components must last for at least 10,000 flights, displaying sufficient flexibility and damage tolerance to non-interruptive operation under severe environmental, thermal, and the occasional dynamic (bird impact) conditions. The development and characterization of forged Al–Li alloys are needed to enable the significant increase of engine reliability and reduced fuel consumption that would come with their deployment in engines. Given the anisotropy in properties and the relative

DOI: 10.2495/CMEM-V6-N4-635-646

immaturity of these alloys for turbine engine applications, much work still needs to be done to build the capability to predict forged Al–Li structural properties as a result of the material processing parameters.

Experience from the 1990's showed significant issues with Al–Li due to high planar anisotropy, unusual crack paths, and lack of thermal stability. However, recently, there has been a renewed interest in the new generation of Al–Li alloys that eliminated or lessened these manufacturing concerns. This new interest is being driven by the challenge to meet significantly higher performance requirements demanded by the next generation of developed commercial aircraft engines [1–3]. High damage tolerance is particularly important for airframe structures and hence of primary interest to airframe manufacturers.

Predictive methodologies for Al–Li need to fully couple material processing parameters (forging and heat treatment), forged part mechanical properties, and damage tolerance required by turbine engine service conditions. The developed and implemented microstructure-based crystal plasticity computational framework allows predicting the effect of local morphology on the mechanical behavior of the components. Such a model requires accurate measurements of single crystal elastic–plastic properties as well as dislocation density distributions and its incorporation in the part level polycrystalline model. Values of these model parameters are estimated from calibration experiments (compression, tensile) for both single crystal and polycrystalline specimens under a wide range of thermal mechanical conditions. The constitutive equations are implemented in finite element software as a user routine [4, 5] and used for the modelling of the microstructure and properties evolution during the forging procedure and for the final part design optimization. The predictive accuracy of the model in these domains is being quantified through validation testing.

2 CONSTITUTIVE MODEL

The overall polycrystalline plastic response is taken as a sum of responses of each of the many single crystals that comprise the representative volume element (RVE). Material behavior is modeled by FEM where each element quadrature point represents an RVE and where both compatibility and equilibrium are satisfied. The deformation of a crystal is taken as the combined contributions from an overall elastic distortion of the lattice and plastic deformation.

The governing variables in the constitutive model are taken as: (1) Cauchy stress $\mathbf{T}$, (2) total deformation gradient, $\mathbf{F}$ and (3) plastic deformation gradient $\mathbf{F}^p$ with $\det\mathbf{F}^p = 1$. Each crystal slip system is specified by a unit normal $\mathbf{n}_0^\alpha$ to the slip plane, and a unit vector $\mathbf{m}_0^\alpha$ denoting the slip direction. We define twelve octahedral {111}<110> slip systems to be operative. The elastic deformation gradient is defined by decomposition of the total deformation gradient as follows: $\mathbf{F} = \mathbf{F}^e\mathbf{F}^p$ where $\det\mathbf{F}^e > 0$. $\mathbf{F}^e$ describes the elastic distortion of the lattice and gives rise to the stress $\mathbf{T}$. For metallic materials the constitutive equation for the second Piola–Kirchhoff stress tensor is taken as a linear relation

$$\mathbf{T}^* = \mathbf{C}[\mathbf{E}^* - \mathbf{A}\cdot(\Theta - \Theta_0)]; \quad \mathbf{E}^* = \tfrac{1}{2}\left[\mathbf{F}^{eT}\mathbf{F}^e - \mathbf{1}\right] \tag{2}$$

where $\mathbf{A}$ is the second order thermal expansion tensor, $\mathbf{C}$ is the temperature-dependent anisotropic elasticity tensor of the fourth rank, Θ is the current temperature and Θ_0 is the reference temperature. Deformation gradients, stress and strain tensors are second order tensors and represented by (3 by 3) matrices.

The Cauchy stress tensor is calculated as follows:

$$\mathbf{T} = \frac{1}{\det(\mathbf{F}^e)}\mathbf{F}^e\mathbf{T}^*\mathbf{F}^{eT} \tag{3}$$

The evolution equation for the viscoplastic deformation gradient is given by the flow rule:

$$\dot{\mathbf{F}}^{\mathrm{p}} = \mathbf{L}^{\mathrm{p}}\mathbf{F}^{\mathrm{p}}; \quad \text{where} \quad \mathbf{L}^{\mathrm{p}} = \sum_{slip\ systems} \dot{\gamma}^{\alpha}\mathbf{S}^{\alpha} \quad \text{and} \quad \mathbf{S}^{\alpha} = \mathbf{m}^{\alpha} \otimes \mathbf{n}^{\alpha} \tag{4}$$

The shear rate along each slip system, $\dot{\gamma}^{\alpha}$, is given in terms of the resolved shear stress (RSS), which is the projection of tensor $\mathbf{T}^{*}$ on a slip system $\alpha\left(\tau^{\alpha}=\mathbf{T}^{*}:\mathbf{S}^{\alpha}=\mathbf{T}^{*}:(\mathbf{m}_{0}^{\alpha}\otimes\mathbf{n}_{0}^{\alpha})=\left(\mathbf{T}^{*}\mathbf{n}_{0}^{\alpha}\right)\cdot\mathbf{m}_{0}^{\alpha}\right)$, slip system resistance and equilibrium back stress. Evolution of crystallographic texture is explicitly defined by the elastic part of the deformation gradient.

$$\mathbf{m}_{\mathbf{t}}^{\alpha} = \mathbf{F}^{\mathrm{e}}\mathbf{m}_{\mathbf{0}}^{\alpha}; \quad \mathbf{n}_{\mathbf{t}}^{\alpha} = \mathbf{F}^{\mathrm{e}-\mathrm{T}}\mathbf{n}_{\mathbf{0}}^{\alpha} \tag{5}$$

Particular expression for shear rates is [4]:

$$\left\{\dot{\gamma}^{c}\right\}^{\alpha} = \dot{\gamma}_{0}\left(\frac{\rho_{m}^{\alpha}}{\rho_{0}}\right)\left|\frac{\tau^{\alpha}-\omega^{\alpha}}{s^{\alpha}}\right|^{n} \operatorname{sgn}\left(\tau^{\alpha}-\omega^{\alpha}\right)\exp\left(-\frac{Q}{k\Theta}\right) \tag{6}$$

where τ^{α} is the RSS, ω^{α} is a back stress and s^{α} is the deformation resistance of α-th slip system. Arrhenius term allows accounting for the temperature changes. Next, latent hardening evolution has been described by modifying Asaro rule $\dot{s}^{\alpha} = h_{0}\left(1-\frac{s^{\alpha}}{s^{*}}\right)^{p}\sum_{\beta} h^{\alpha\beta}\left|\dot{\gamma}^{\beta}\right|$, with hardening matrix $h^{\alpha\beta} = \left\{q+(1-q)\delta^{\alpha\beta}\right\}$ for temperature dependent h_{0} and s^{*}. The back stress has a limiting saturation value, $\omega_{\infty} = \frac{c_{1}}{c_{2}}$, corresponding to the end of the primary creep stage, which evolves according to the following relationship [4–6]: $\dot{\omega}^{\alpha} = c_{1}\dot{\gamma}^{\alpha} - c_{2}\left|\dot{\gamma}^{\alpha}\right|\omega^{\alpha}$. The back stress requires two additional coefficients, c_1 and c_2, that are explicit functions of temperature. It is important to note that hardening terms indirectly account for rafting process and microstructure evolution during the first stage of viscoplastic deformation. Results reported in this paper have been obtained by using the hardening expressions shown above. We postulate that dislocation generation rate is proportional to the entropy production. Using concepts from chemical kinetics we have chosen to represent the evolution as two body interactions. We assume that dislocation immobilization takes place when two corresponding dislocation loops interact with each other [4, 7]. For the sake of simplicity, prediction of the texture evolution during rolling operation was obtained with $\rho_{m}^{\alpha}/\rho_{0} = 1$, which affects rate of relaxation but does not affect texture evolution.

3 ELASTIC MODULI OF AA2070 FROM SINGLE CRYSTAL MICROPILLAR COMPRESSION

3.1 Experimental

Single crystal data for AA2070 is sparse so micromechanical testing was employed to extract single crystal properties from small scale test structures fabricated within the individual grains of a polycrystalline billet. Specifically, micron size compression pillars were made with focus ion beam (FIB) machining. Locally milling test structures within a single grain of a polycrystalline coupon is an established methodology and has been successfully demonstrated in Al by Ng *et al.* [8] and Kunz *et al.* [9]. The material employed for this investigation came from a large polycrystalline H-beam forging of AA2070. The material was deformation processed then solutionized, quenched and aged. A small coupon approximately 25 mm x 25 mm x 5 mm was extracted from the artificially aged forging and then ground such that top and bottom faces were parallel. This coupon was then mounted to a metal puck that served as a holder for polishing as well as fixturing in a scanning electron microscope (SEM) and nanoindenter. Once mounted, the sample was polished and then orientation mapped using Quanta FEG 650 scanning electron microscope (SEM) (FEI; Hillsboro, OR) outfitted with a NordlysNano Electron Backscatter Diffraction (EBSD) detector (Oxford Instruments; Concord, MA). The orientation imaging microscopy with EBSD elucidated the crystallographic orientation and location of large grains suitable for micropillar fabrication with a FIB. The orientation mapping was performed at approximately 2,000X magnification with the electron beam set at 30 keV and a spot size of 5. The electron backscatter diffraction patterns were collected with 1 x 1 binning and step size of 0.5μm.

A Helios Nanolab 600 dual beam microscope (FEI; Hillsboro, OR) was then employed to machine microcompression pillars into the candidate grains with a 30 keV Ga FIB. The pillar fabrication approach followed the two step annular milling method of Volkert [10]. The target pillar geometry was a diameter of 1.5 μm, a height of 4 μm, and a taper of less than 2°, which follows the guidelines of Zhang *et al.* [11] for accurate micro-compression experiments.

After fabrication, the pillars were tested using a G200 nanoindenter (Keysight Technologies; Santa Rosa CA) outfitted with a flat diamond punch. To acquire load-deflection data for determining modulus, a series of quasi-static load–unload compression steps were performed to a total plastic strain not exceeding 10%. All experiments were conducted at room temperature. Post compression SEM imaging was also performed.

3.2 Results

Figure 1 presents the orientation map for three different regions scanned on the AA2070. The color corresponds to the out of plane crystallographic direction and the black line segments denote boundaries for which the misorientation angle exceeds 1.5°. It can be seen that the AA2070 is comprised of elongated grains. Large grains with widths on the order of 20 μm were selected for pillar fabrication as they are sizeable enough to accommodate the pillar footprint. Grains denoted 'A', 'B' and 'C' were selected for pillar fabrication. The pillars were milled in regions of the grain away from any boundaries. Next to each grain, a representative pillar is displayed. Two pillars in grain A were tested. This grain has the [18 14 21] orientation. Two pillars were also tested from grain B which was of the [1 1 4] orientation. A single pillar from grain C was tested. This pillar was of the [7 1 15] orientation.

Figure 2 presents a compilation of the compressive engineering stress-strain curves for each tested pillar. The nanoindenter outputs load-displacement. To compute the strain, the approach of Frick *et al.* [12] was followed with the displacement adjusted by the Sneddon correction [13] to account for the elastic displacement of the base as well as any elastic deformation of the tip. For this correction, the polycrystalline AA2070 properties were taken to be elastic modulus, $E = 77.1$ GPa and a Poisson ratio, $\nu = 0.31$. As for the diamond punch, $E = 1050$ GPa and $\nu = 0.2$ was assumed following the property specification of the manufacturer (Micro Star Technologies; Huntsville TX).

Table 1 summaries the modulus measurements determined from the unloading curves of the compression tests. Often the first, and sometimes the second, loading step resulted in a low (~50–60 GPa) modulus. As some pillars did not have a perfectly flat top, this artifact is assumed to stem from the indenter tip becoming fully seated on the pillar in the initial

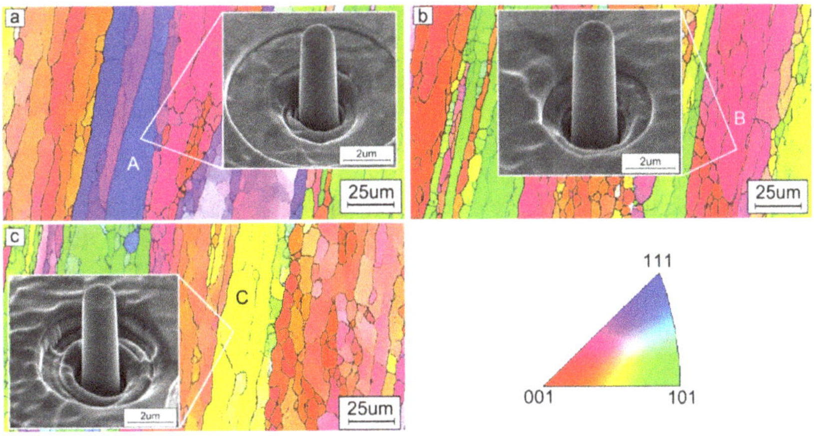

Figure 1: Out-of-plane orientation maps for three different regions (a), (b) and (c) of the AA2070 coupon. Labelled grains were those selected for pillar fabrication and the inset shows a representative pillar machined within that grain.

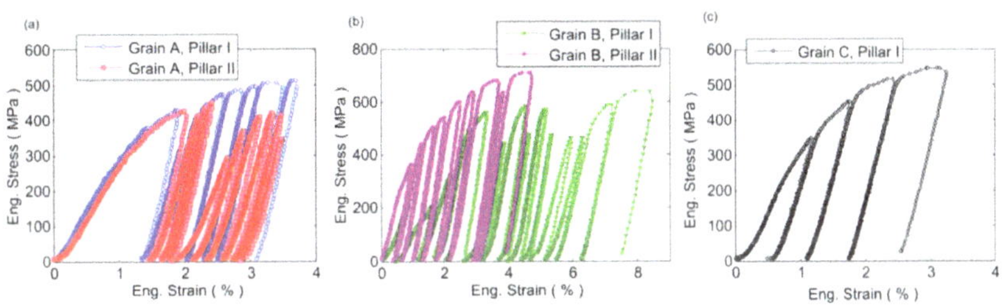

Figure 2: Compressive engineering stress–strain for (a) grian A, (b) grain B and (c) grain C.

compression cycle. These initially low modulus measurements were not considered in the reported average elastic modulus measurement. In addition, some of the modulus measurements from the latter compression cycles were discounted. For example, consider pillar II from grain A (Fig. 2a). The 5th loading step, which has a peak stress of 300 MPa at 2.5% strain, shows a deviation from the consistent behavior observed on the previous cycles and the modulus from this cycle was 63 GPa which is a 22% drop compared to the previous

Table 1: Single crystal elastic modulus measurements from microcompression pillars

Grain	Pillar ID	Avg. Modulus (GPa)	Std. Dev. (GPa)	Modulus of pure Al* (GPa)	Difference
A	I	84.1	2.2	75.4	11.5%
A	II	80.0	0.6		6%
B	I	78.6	3.4	67.3	16.8%
B	II	72.0	3.9		6.0%
C	I	72.2	3.3	68.0	6.2%

**calculated from the stiffness/compliance constants of pure Al as reported in Smithells Metals Handbook.*

cycles which returned a modulus value of 80.5 GPa. It is assumed that on this loading cycle some change has occurred in loading alignment or pillar geometry. This as well as the subsequent cycles should be discounted from the average modulus.

Figure 3 highlights the state of the pillars after deformation. As the AA2070 is a face-centered-cubic system, room temperature slip will occur on the 111-type planes in the 110-directions. The 111-planes with respect to the pillar orientation are highlighted. In Fig. 3(a), a few faint slip traces are discernable and they are aligned with the slip planes orientated nearly along the axis of the pillar. For grain B (Fig. 3(b)), no slip traces were readily evident however the increase in pillar diameter and decrease in height did match the total plastic strain. Figure 3(c) highlights the deformation of the pillar in grain C and the traces of slip on the 111-planes are clearly observed.

3.3 Discussion

From the measurement of several polycrystalline aluminum alloys containing various concentrations of lithium, Noble *et al.* [14] reports that as a general trend, modulus increases by ~6% for each wt.% of lithium present in the alloy. AA2070 contains from 1.0 to 1.4 wt.% lithium thus compared to pure Al, the modulus should be 6 to 8.4% higher. This polycrystalline trend appears to translate to single crystals. The measurements of this work find the modulus of AA2070 along single crystallographic directions is about 6% higher compared to pure aluminum (see Table 1). Pillar I from Grain B was found to have an exceptionally high modulus compared to the trend noted for the other pillars. It is speculated that this measurement could be influenced by the reduced sample size. Micropillar compression has been widely utilized to glean insights on the plasticity and deformation mechanisms in systems with micron and submicron length scales [15, 16]. For this work it is the elastic behavior that is of interest; nevertheless the reduced volume of the test specimen could impact the data. AA2070 is a heat treatable system that garners strength from both solutes and precipitates. Typically, the precipitates of aluminum alloys are reported to be stiffer than the matrix metal [14] ranging from 90 to 175 GPa. If a pillar had a high volume fraction of stiff precipitates, their presence could elevate the measured modulus. Additional analysis beyond the scope of this paper is required to ascertain the exact influence of the precipitates. As a first step, follow up experiments are being planned to serial section pillars to quantify the volume fraction of precipitates. The presented tests and results aid in determining the single crystal stiffness matrix **C** (Eq. 2) and single crystal yield parameters used in the constitutive model.

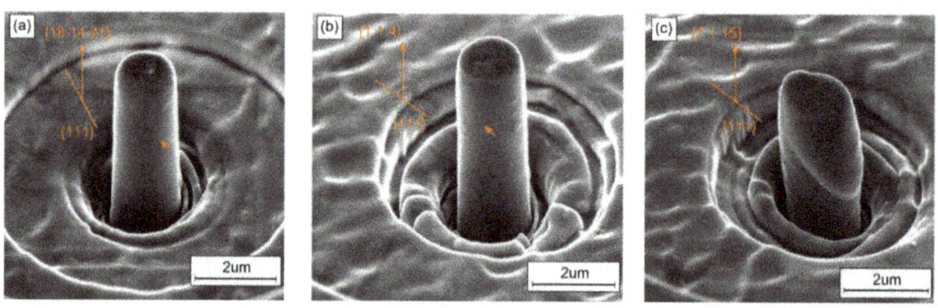

Figure 3: Representative secondary electron images of pillars after compression for (a) grain A, (b) grain B and (c) grain C. Faint slip traces are denoted with arrows and the corresponding 111-type slip plane is shown with respect to the compression axis.

4 TEXTURE PREDICTION UNDER FORGING CONDITIONS

Multiscale, physics-based models can serve as valuable tools in coupling processing and manufacturing operations to local material properties. This, in turn, allows such models to support the tailoring of manufacturing processes and parameters to yield desirable structural properties as a function of microstructural features such as crystallographic texture [17]. With this goal in mind, the finite element-based crystal plasticity model presented in Section 2 was called upon to simulate the forging process of an Al-Li H-beam. The H-beam forging is approximately 1 m long by 0.25 m wide and contains regions of three different thicknesses along its length (A: 2.54 cm, B: 5.08 cm, and C: 10.16 cm). The crystal plasticity simulation utilizes calibrated parameters from the micro-, single crystal tests presented in Section 3 as well as typical values for an FCC material at the forging temperature. The slip-related model parameters used for the analyses presented in this section are the following: $\dot{\gamma}_0$ = 0.001 s^{-1}, n=10, s_0=16 MPa. Based on the observed constitutive behavior of Al–Li at the forging temperature, ~450°C, hardening was not included in the model, therefore perfect plasticity was assumed. The prescribed slip parameters govern the deformation on the 12 octahedral slip systems, {111}<110>, that are active in the FCC material. Researchers have shown that additional, non-octahedral slip systems can be active in aluminum alloys at elevated temperatures such as during hot rolling or forging [18, 19], however it has been demonstrated that deformation on the non-octahedral slip systems contributes relatively little to the final forging texture, even under triaxial forging scenarios [20].

The forging process begins with a cast ingot containing a collection of grains with no preferential orientation (i.e., randomly oriented). For the forging simulation, the initial, random microstructure is applied to a cuboid FEM mesh of 125 fully-integrated elements. Therefore the model contains 1000 integration points onto which 1000 grains with orientations sampled from a random distribution were mapped. For each integration point (i.e., grain) in the FEM mesh, the crystal plasticity model described in Section 2 is called via a usermat subroutine by ANSYS in order to solve for the deformation on each of the 12 octahedral slip systems. The aggregate of the contributions of each of the slip systems within each integration point produces the macroscale constitutive behavior observed in the material stress-strain curves. Based on the local deformation at the integration points, the corresponding grain will experience lattice rotation, thereby evolving the texture of the material as a whole.

Rolling and forging processes are typically simulated using plane strain compression boundary conditions based on the constraints applied to the material preform during its formation into the final shape. Plane strain compression of the FEM microstructure containing 1000 randomly oriented grains results in the typical Copper-type texture presented in Fig. 4. Similarly, the pole figure corresponding to a location along the mid-thickness and mid-width of the

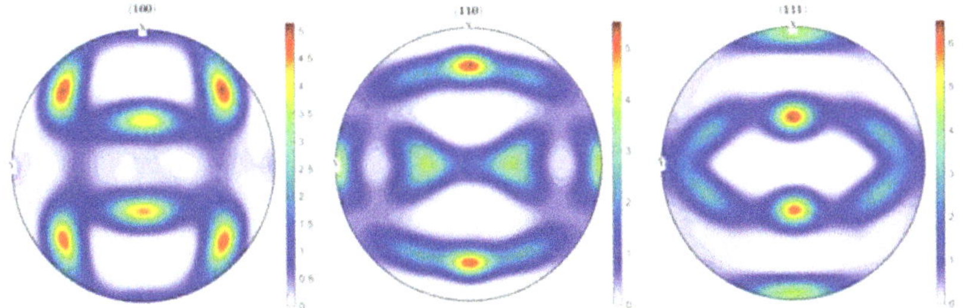

Figure 4: Copper-type texture resulting from plane strain compression of Al-Li microstructure.

10.16 cm H-beam section is presented in Fig. 5. Comparing Figs 4 and 5, one can observe that some of the texture components are predicted correctly by the plane strain compression simulation, while others are missed entirely. Although plane strain boundary conditions are sufficient to simulate simple rolling or forging processes, it is evident from this comparison that other mechanisms are contributing to a more complicated texture profile in the case of the forged H-beam. In particular, as annotated in Fig. 5, an unusually high intensity of the Goss texture component is evident in the experimental pole figure. In addition, the pole figure in Fig. 5 appears rotated; however, this may be a result of specimen placement in the microscope.

Traditionally, the Goss texture component is associated with recrystallization. Therefore a detailed investigation was carried out on the EBSD data to determine if the presence of substantial recrystallization is responsible for the observed high-intensity Goss texture. A grain orientation spread (GOS) analysis was conducted on EBSD data taken from the 10.16 cm region of the H-beam. The GOS value represents an average deviation of the orientation of each point in the grain with respect to the average grain orientation. Using this approach, grains with low GOS values (e.g., ≤2°) can be considered recrystallized, while grains with higher values (e.g. ≥5°) are determined to be deformed. The results of this analysis are shown in Fig. 6. By separating the deformed grains with GOS values ≥5° from the recrystallized grains with values ≤2°, it is possible to visualize the extent and location of recrystallization in this particular microstructure. As is evident from Fig. 6, the area fraction of recrystallized grains is small (4.30%) compared to the area fraction of deformed grains (70.63%). From these results, it is clear that recrystallization cannot be the primary contributor to the high-intensity Goss texture. Based on the observation that the high-intensity Goss texture can be attributed to the deformed grains, further analysis is needed to determine the specific grains responsible for this texture. Separating the grains primarily responsible for the Goss, Brass (Bs) and Copper (Cu) texture components, we can easily observe from Fig. 7 that very long, clearly deformed grains have a Goss texture. In order to accurately predict the microstructural texture of the forged H-beam, it is therefore imperative to determine the source of the deformation-related Goss texture.

Biaxial and triaxial forging of FCC materials at elevated temperatures has been shown to alter the area fractions of various texture components [20]. For example, compression along the ST and LT directions increase the relative number of grains with the Goss orientation. It is with this knowledge that further details regarding the forging process were pursued. Although specific details of the forging operation are proprietary, we were able to deduce the general steps involved in producing the H-beam. These steps, beginning with a cast ingot,

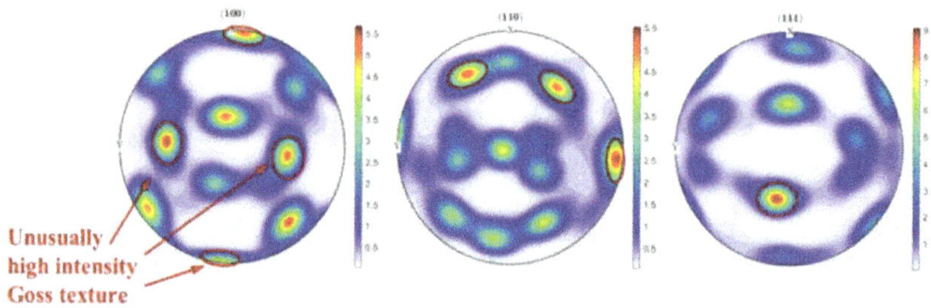

Figure 5: H-beam forging texture in Region C (10.16 cm thickness) – pole figure with regions of high Goss texture intensity highlighted.

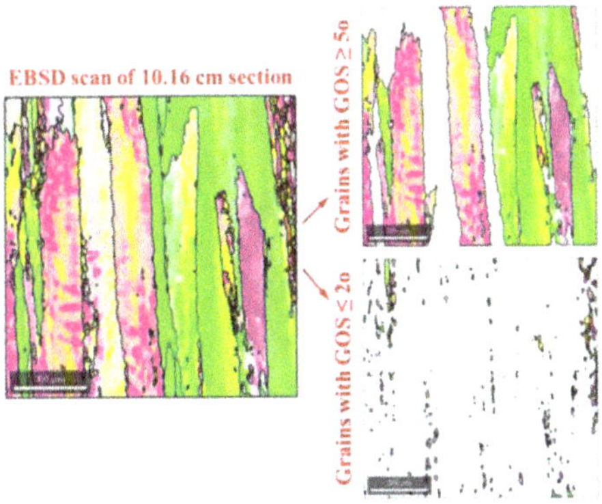

Figure 6: GOS analysis results for EBSD scan of H-beam forging.

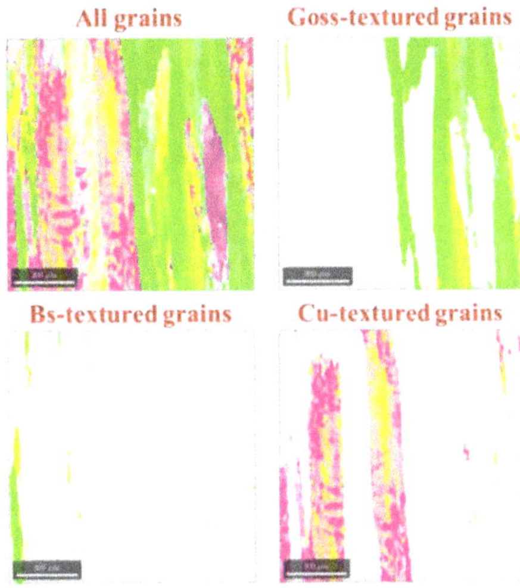

Figure 7: Grains in EBSD image separated by texture components.

include: (1) A upset, (2) B upset, (3) draw and step preform and (4) closed-die forge H-beam. These forging steps are followed by heat treatment and aging. The three steps responsible for producing the texture observed in the experimental micrographs are illustrated in Fig. 8. In order to simulate these forging steps, a three step simulation approach was utilized where the final texture from each step was applied as the initial texture of the following step, beginning with a random texture to represent the cast ingot microstructure.

Steps 1 and 2 (AB upset) were simulated as simple compression deformation, while Step 3 was approximated as plane strain compression due to the rolling/drawing-like deformation behavior. It should be noted that the A and B upsets include compression in the LT and ST directions, respectively. These two steps, in particular the A upset, intensify the Goss texture. This is in agreement with the findings of Ref. [20].

The three-step forging process was simulated using the calibrated crystal plasticity FEM model and a comparison between the simulated and forged textures is presented in Fig. 9. One can observe that all of the major texture components are represented in the pole figure of the simulated microstructure. In addition, the rotation of the pole figure is accurately predicted. It was initially assumed that the rotation observed in the experimental pole figure was a result of specimen rotation prior to EBSD imaging. However, by predicting the pole figure rotation, the simulation was able to provide otherwise-hidden insight into the deformation

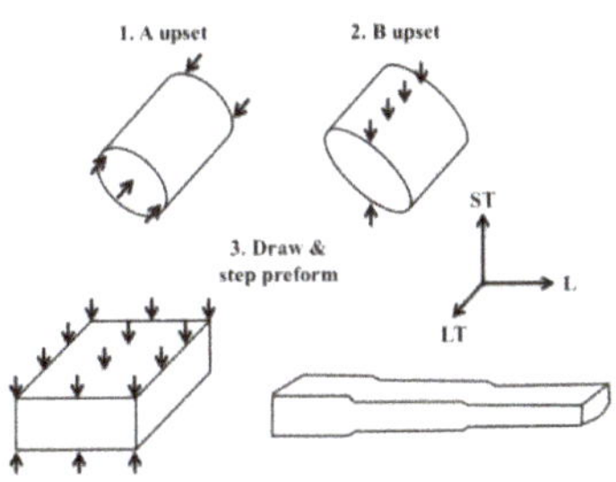

Figure 8: H-beam forging steps prior to closed-die forging into final shape.

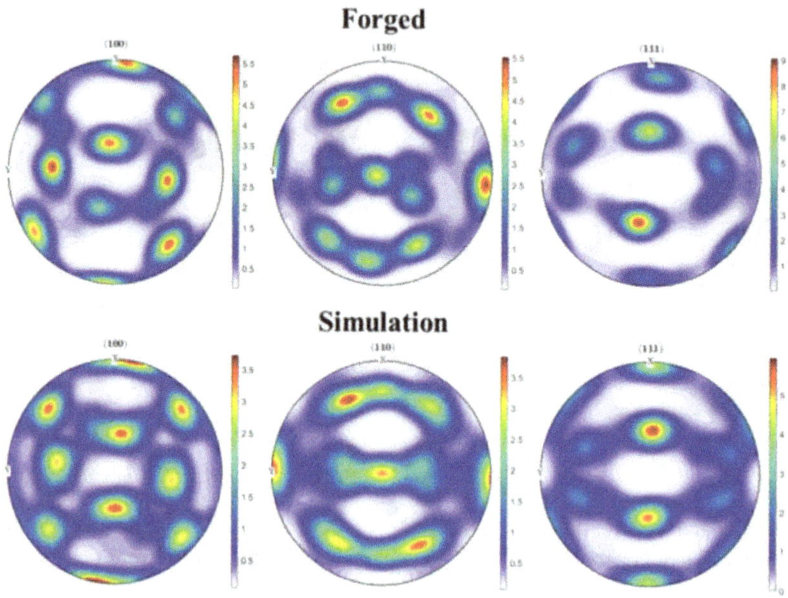

Figure 9: Pole figures of forged and simulated microstructures.

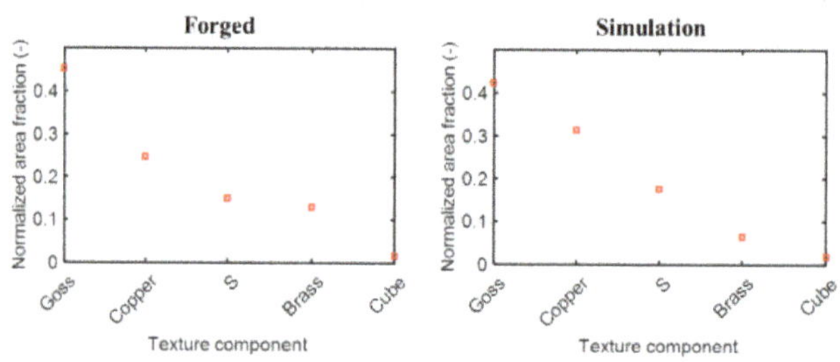

Figure 10: Texture components of forged and simulated microstructures.

behavior occurring during sequential forging operations. Analyzing the texture component area fractions, as presented in Fig. 10, allows additional quantification of the simulated forging operation. The area fractions of the Goss, Copper, S, Brass and Cube texture components are plotted due to their prevalence in rolled and forged FCC materials. The normalized area fraction of the Goss, S, and cube texture components are accurately predicted; however, the copper and brass components are over- and under-predicted, respectively. Since access to the exact strains imparted on the preform during each of the three steps is restricted due to their proprietary nature, the texture component area fractions of the simulated microstructure represent the forged microstructure accurately given the available information.

5 CONCLUSIONS

The goal of the developed model and forging simulation framework was to provide a tool to accurately predict the crystallographic texture of an H-beam forging. Based on experimental analysis of the EBSD data using pole figures, GOS analysis, and quantification of texture component volume fraction, it was determined that typical methods of simulating rolling or forging operations were insufficient to produce the observed forged texture. Considering previous effort in texture analysis of multi-axial forging operations provided guidance for accurately determining the sequential forging steps involved in producing the H-beam. These inferred steps were then verified with conversations with the forging company. In order to model the forging process, elastic properties were calibrated using micropillar tests on single crystals within the polycrystalline microstructure. These tests yielded crucial model parameters that are difficult to obtain from macroscale tests. Simulating the three-step forging process that precedes the final closed-die forging of the H-beam, we were able to accurately replicate the pole figure for material in the mid-width, mid-thickness location of the 10.16 cm section. Quantitative analysis of the texture component area fractions provided further validation of the developed procedure however exact replication of the area fractions was not possible since we lacked the necessary (proprietary) details of each forging step. Regardless, the developed model was demonstrated capable of accurately predicting the microstructural evolution and final texture of the H-beam, thereby satisfying a primary goal of this effort. The ability to accurately predict texture evolution during complex manufacturing processes including rolling, forging, and drawing permits the tailoring of microstructures in order to optimize directional material properties such as yield. Since one of the drawbacks of Al–Li alloys is its yield and strength anisotropy, models such as the one presented in this work can be called upon to mitigate this drawback and provide a tool to design manufacturing processes to yield more desirable material properties. This tool can therefore facilitate the design of Al–Li structures with directional strength properties tailored to specific load scenarios similar to how composite structures are designed.

ACKNOWLEDGEMENTS

The authors are grateful for support and funding from Lightweight Innovations for Tomorrow (LIFT), operated by the American Lightweight Materials Manufacturing Innovation Institute (ALMMII).

REFERENCES

[1] Encyclopedia of Energy Engineering and Technology 1, Barney L. Capehart CRC Press. ISBN 978-0-8493-3653-9, (2007).

[2] Lee, J.J, *Historical and future trends in aircraft performance, cost and emissions*, MS Thesis, Massachusetts Institute of Technology, Cambridge, 2000.

[3] Gupta, R.K., Nayan, N., Nagasireesha, G. & Sharma, S.C., Development and characterization of Al-Li alloys. *Materials Science and Engineering*, **A420**, pp. 228–234, 2006. https://doi.org/10.1016/j.msea.2006.01.045

[4] Staroselsky, A. & Cassenti, B.N. Combined rate-independent plasticity and creep model for single crystal. *Mechanics of Materials,* **42**(10), pp. 945–959, 2010. https://doi.org/10.1016/j.mechmat.2010.07.005

[5] Staroselsky, A., *Crystal plasticity due to slip and twinning*, PhD Thesis, MIT, 1997.

[6] Stouffer, D.C. & Dame, L.T., *Inelastic deformation of metals*. John Wiley & Sons, Inc, New York, 1996.

[7] Staroselsky, A. & Cassenti, B.N., Mechanisms for tertiary creep of single crystal superalloy. *Mechanics of Time-Dependent Materials,* **12**(4), pp. 275–289, 2008. https://doi.org/10.1007/s11043-008-9065-6

[8] Ng, K.S. & Ngan, A.H.W., Stochastic nature of plasticity of aluminum micro-pillars. *Acta Materialia,* **56**(8), pp. 1712–1720, 2008. https://doi.org/10.1016/j.actamat.2007.12.016
[9] Kunz, A., Pathak, S. & Greer, J.R., Size effects in Al nanopillars: Single crystalline vs. bicrystalline. *Acta Materialia,* **59**(11), pp. 4416–4424, 2011. https://doi.org/10.1016/j.actamat.2011.03.065
[10] Volkert, C.A. & Lilleodden, E.T., Size effects in the deformation of sub-micron Au columns. *Philosophical Magazine*, **86**(33–35), pp. 5567–5579, 2006. https://doi.org/10.1080/14786430600567739
[11] Zhang, H., Schuster, B.E., Wei, Q. & Ramesh, K.T., The design of accurate micro-compression experiments. *Scripta Materialia*, **54**(2), pp. 181–186, 2006. https://doi.org/10.1016/j.scriptamat.2005.06.043
[12] Frick, C.P., Clark, B.G., Orso, S., Schneider, A.S. & Arzt, E., Size effect on strength and strain hardening of small-scale [1 1 1] nickel compression pillars. *Materials Science and Engineering*, **A 489**(1–2), pp. 319–329, 2008. https://doi.org/10.1016/j.msea.2007.12.038
[13] Sneddon, I.N., The relation between load and penetration in the axisymmetric boussinesq problem for a punch of arbitrary profile. *International Journal of Engineering Science*, **3**(1), pp. 47–57, 1965. https://doi.org/10.1016/0020-7225(65)90019-4
[14] Noble, B., Harris, S.J. & Dinsdale, K., The elastic modulus of aluminium-lithium alloys. *Journal of Materials Science* **17**(2), pp. 461–468, 1982. https://doi.org/10.1007/bf00591481
[15] Uchic, M.D., Shade, P.A. & Dimiduk, D.M., Plasticity of micrometer-scale single crystals in compression. *Annual Review of Materials Research*, **39**, pp. 361-386, 2009. https://doi.org/10.1146/annurev-matsci-082908-145422
[16] Greer, J.R., & De Hosson, J.T.M., Plasticity in small-sized metallic systems: Intrinsic versus extrinsic size effect. *Progress in Materials Science*, **56**(6), pp. 654–724, 2011. https://doi.org/10.1016/j.pmatsci.2011.01.005
[17] McDowell, D. L., A perspective on trends in multiscale plasticity. *International Journal of Plasticity*, **26**(9), pp. 1280–1309, 2010. https://doi.org/10.1016/j.ijplas.2010.02.008
[18] Caillard, D. & Martin, J.L., Glide of dislocations in non-octahedral planes of fcc metals: a review. *International Journal of Materials Research*, **100**(10), pp. 1403–1410, 2009. https://doi.org/10.3139/146.110190
[19] Contrepois, Q., Maurice, C. & Driver, J. H., Hot rolling textures of Al–Cu–Li and Al–Zn–Mg–Cu aeronautical alloys: experiments and simulations to high strains. *Materials Science and Engineering: A*, **527**(27), pp. 7305–7312, 2010. https://doi.org/10.1016/j.msea.2010.07.095
[20] Ringeval, S., Piot, D., Desrayaud, C. & Driver, J. H., Texture and microtexture development in an Al–3Mg–Sc (Zr) alloy deformed by triaxial forging. *Acta Materialia,* **54**(11), pp. 3095–3105, 2006. https://doi.org/10.1016/j.actamat.2006.02.047

EFFECT OF MOULD TYPE AND SOLIDIFICATION TIME ON BIFILM DEFECTS AND MECHANICAL PROPERTIES OF AL–7SI–0.3MG ALLOY CASTINGS

MAHMOUD AHMED EL-SAYED[1] & KHAMIS ESSA[2]
[1]Department of Industrial and Management Engineering, Arab Academy for Science, Technology and Maritime Transport, Egypt.
[2]School of Mechanical Engineering, University of Birmingham, UK.

ABSTRACT

The properties of light alloy castings are strongly affected by their inclusion content, particularly double oxide film defects (bifilms), which not only decrease the tensile and fatigue properties, but also increase their scatter. Recent research has suggested that oxide film defects may alter with time, as the air inside the bifilm would react with the surrounding melt, while the hydrogen dissolved in the melt could diffuse into the bifilm cavity to form hydrogen porosity. The mechanical properties of the casting were shown to be significantly dependent upon the new morphology of its entrained bifilms. In this work, the Weibull moduli of the tensile properties of three Al castings, all expected to contain oxide films of, approximately, the same amount were compared. The first casting was poured into a resin-bonded sand mould while the second and third castings were poured into ceramic moulds with the mould for the third casting being preheated prior to pouring. The results of mechanical property analysis and electron microscopy examination suggested a considerable influence of the type of the mould and the solidification time on the morphology of bifilms and by implication, on the reliability and reproducibility of the tensile properties.

Keywords: bifilms, castings, mechanical properties, oxide film defects

1 INTRODUCTION

During the casting of aluminium alloys, the melt surface is exposed to air, which results in the formation of a surface oxide film. As a result of surface disturbance, such as during metal transfer and pouring, the oxidised melt surface can be folded over onto itself and trap a layer of the mould atmosphere, creating a bifilm defect [1, 2]. This defect can be incorporated into the bulk liquid by entrainment.

Double oxide film defects lower mechanical properties, but also lead to variable mechanical properties, contributing to the scatter of properties associated with shape castings [2, 3]. Apart from their effect on mechanical properties, double oxide film defects have also been held to contribute to the formation of other defects, such as hydrogen porosity and intermetallic phases [2]. Their effect on the mechanical properties of the final casting depends upon several factors, such as their initial size, subsequent changes in size and volume due to melt conditions such as velocity and bulk turbulence, solidification time and hydrogen content of the melt.

Once a double oxide film defect has become incorporated into the melt, it would be expected to contain an atmosphere that would be predominantly air, which would be expected to react with the surrounding liquid metal. It was suggested by Nyahumwa *et al.* [4], Raiszadeh and Griffiths [5], El-Sayed *et al.* [6, 7] and Griffiths *et al.* [8, 9] that the internal atmosphere of bifilms would react with the surrounding liquid Al, forming different Al oxides and AlN, and therefore was gradually consumed over time. Also, hydrogen was found to diffuse into the bifilm as the melt was kept longer in the liquid state. The consumption of the bifilm interior atmosphere as well as the H ingress into the defect was suggested by El-Sayed *et al.* to

DOI: 10.2495/CMEM-V6-N4-647-655

result in an alteration of the shape and size of the bifilm with a direct effect on the mechanical properties of the casting [11, 12]. This change in morphology with time suggested that the defect could perhaps become less detrimental to mechanical properties.

The aim of this work was to study the effect of the type of the mould and the solidification time of the casting on entrained double oxide films, and the corresponding change in mechanical properties of aluminium alloy castings. Understanding these issues could lead to the development of techniques by which the effect of double oxide film defects in aluminium castings might be reduced or eliminated.

2 EXPERIMENTAL WORK

Three casting experiments were carried out to obtain tensile test bars with approximately similar bifilm content but with different bifilm ages and/or H content. In each experiment, two top-pouring moulds with a rectangular mould cavity of 120 x 100 x 20 mm, each producing 10 test bars, were used. In the first experiment, in which castings with relatively high H content were expected, resin bonded sand moulds were prepared one day before the casting experiment. In the second experiment, two ceramic shell moulds were produced via investment casting technique. The moulds were then dried prior to pouring, which was carried out while the moulds were at room temperature. The third experiment was exactly same as the second one except that the ceramic moulds were preheated to 800°C prior to pouring. Ceramic moulds were used in the second and third experiments instead of sand moulds to prevent the solidifying casting from picking up H from the solvent of the resin used as a binder for the sand moulds.

In each experiment, about 7 kg of 2L99 alloy (Al–7Si–0.3Mg) was melted in an induction furnace and then was held at 800°C under a vacuum of about 80 mbar for one hour, a procedure intended partially degas the molten metal [12] as well as to remove most, or all, previously introduced oxide films from the melt. The liquid metal was then poured from a height of about 1 meter into sand/ceramic moulds, which was expected to be sufficient for the creation and entrainment of new double oxide film defects, and their introduction into the melt.

For each experiment, two Leco samples for solid-state H determination were taken; one from the melt after vacuum treatment and before pouring, and the second sample was taken from the running bar of the castings. Both samples were analyzed to determine the hydrogen content of the castings. After solidification, each of the castings was machined into 10 tensile test bars with a gauge length of 90 mm and of a rectangular cross-section in the gauge length of 10 x 7 mm. They were pulled to fracture using a tensile testing machine model MTS 810, with a strain rate of 1 mm min^{-1}. Tensile results were evaluated using a Weibull statistical analysis approach to assess the influence of the different casting conditions on the variability of the mechanical properties of the castings. Finally, SEM with EDX analysis was used to investigate the fracture surfaces of the test bars.

For each of the three experiments, an extra mould was cast to measure the solidification time of the casting in each experiment. In each experiment a type K-Inconel-sheathed thermocouples (of 1 mm diameter and 500 mm length and with a maximum allowable temperature of 1100ºC) was inserted into the mould cavity. The thermocouples were connected to a data logger, which was used to record the change of temperature at the thermocouple tip with time at a rate of 1 Hz. For each casting, the temperature recording was terminated when the casting temperature reached 610ºC, the liquidus temperature of the alloy.

3 RESULTS AND DISCUSSION

This solidification time measurement experiment was carried out to determine the time taken for the aluminium castings poured into the ceramic and the sand moulds to solidify after casting. It was found that the solicitation times of the castings poured into the sand mould (experiment 1) and un-preheated ceramic mould (experiment 2) were about 5 and 10 seconds, respectively. However, in the third experiment, where the liquid Al was decanted into a pre-heated ceramic mould, the melt had taken about 230 seconds to reach the liquidus temperature (about 610°C). Such significant difference in the solidification times would be due to the mould preheating prior to pouring. This could be helpful in determining more accurately the age of a double oxide film, or the time spent by the defect in the liquid metal before freezing.

The two-parameter Weibull distribution was used to analyze the scatter in the mechanical properties of the Al–5Mg alloy castings produced under different casting conditions, as it was suggested to be more appropriate than a normal distribution [13, 14]. The Weibull modulus (the slope of the line fitted to the log-log Weibull cumulative distribution data) is a single value that shows the spread of properties; a higher Weibull modulus reveals less variability among the studied property.

Weibull plots of the UTS and % elongation of the test bars cut from the castings from the three experiments are shown in Figs 1 and 2, respectively. The values of the correlation coefficients (R^2) suggested that the data points expressing both the UTS and % elongation values were linearly distributed. It was noted that for both the UTS and % elongation, the Weibull moduli (the slope of the trend line) of the castings from experiment 3, where a preheated ceramic mould was used, were the highest among all castings.

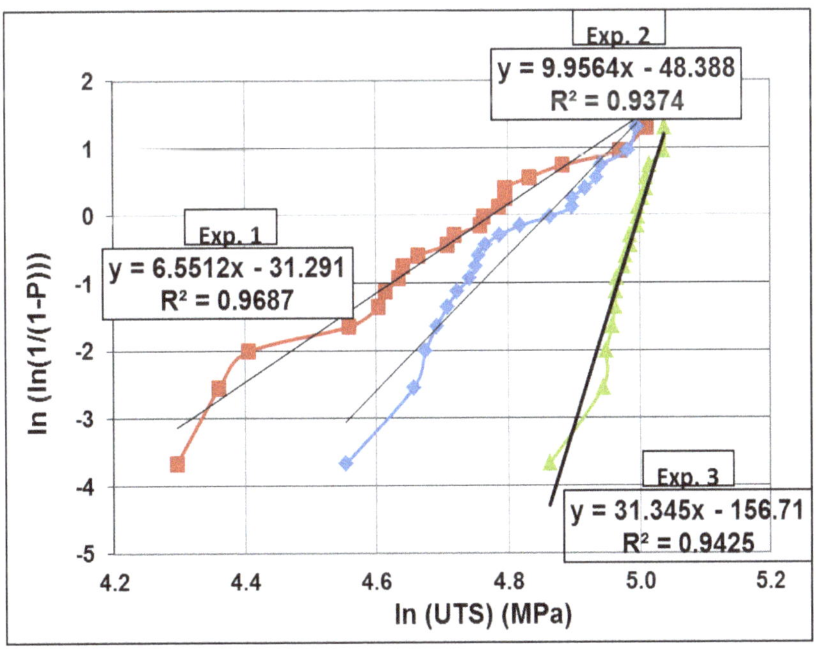

Figure 1: Weibull distribution of ultimate tensile strength of Al–7Si–0.3Mg alloy specimens from different experiments.

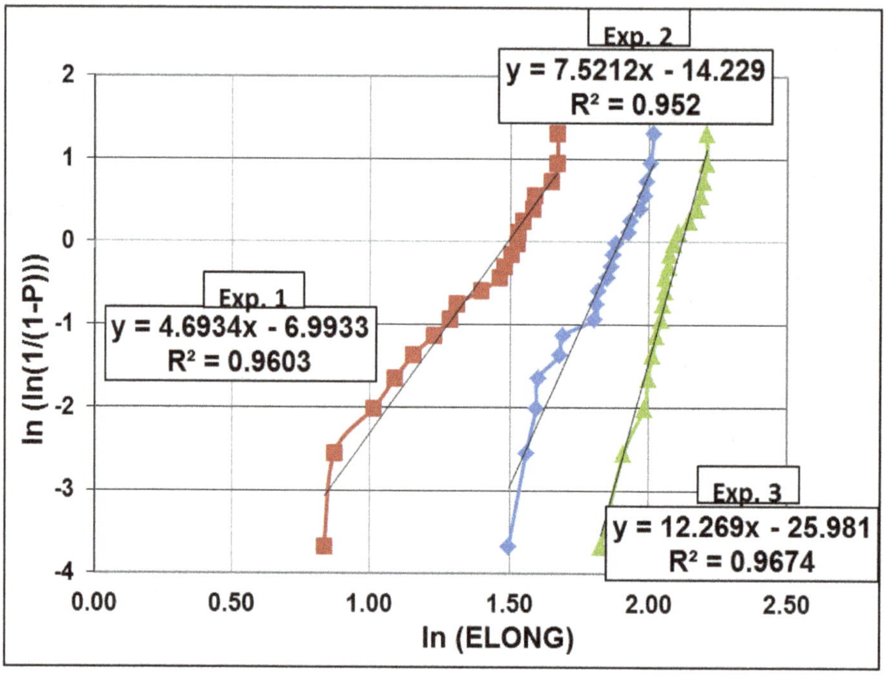

Figure 2: Weibull distribution of % elongation of Al–7Si–0.3Mg alloy specimens from different experiments.

Table 1 shows the results of the Weibull analysis of both of the UTS and percentage elongation values obtained from the different experiments, together with the position parameter and the results of H measurements from different experiments.

The H content of the Leco sample, taken from the melt before pouring, was 0.131 cm^3/100g Al. From the results presented in Table 1 It was found that, the H content of the solidified casting poured into the sand mould (experiment 1) was measured to be 0.235 cm^3/100g Al, almost twice the H content of the liquid casting, probably due to H pick-up from the resin-bonded sand moulds. However, the castings poured into ceramic moulds had H content of 0.139 and 0.152 cm^3/100g, respectively for experiments 2 and 3. In the second experiment (when the melt was poured into an un-preheated ceramic mould), the H content was almost the same as that of the melt before pouring, which was expected to be due to the almost

Table 1: Results of the H measurement and Weibull analysis for the test bars of the castings from different experiments.

Casting Conditions			H Content of Solidified Casting (cm^3/100g Al)	UTS (MPa)		% Elongation	
Exp. No.	Mould Material	Preheating		Weibull Modulus	Position Parameter	Weibull Modulus	Position Parameter
1	Sand	No	0.235	6.6	118.9	4.7	4.4
2	Ceramic	No	0.139	10.0	129.0	7.5	6.6
3	Ceramic	Yes	0.152	31.3	149.4	12.3	8.3

instantaneous solidification. In the third experiment (when the melt was poured into a pre-heated ceramic mould) the H content was slightly higher, perhaps due to the H pick-up by the casting from the surrounding atmosphere during solidification (which took about 4 minutes).

Oxide film defects were found on all the fracture surfaces of test bars investigated from both experiments 1 and 2. Figures 3(a) and (b) show examples of SEM images of oxide films found on the fracture surfaces of test bars from experiments 1 and 2 respectively. The corresponding EDX analysis suggested the presence of $MgAl_2O_4$ spinel on the surfaces. However, the size of oxide film in the specimen from experiment 1 was relatively larger than that in experiment 2. The oxide film in Fig. 3(a) is about 1.5 x 0.5 mm, while that in Fig. 3(b) is about 0.3 x 0.3 mm), perhaps due to the larger H content of the formers.

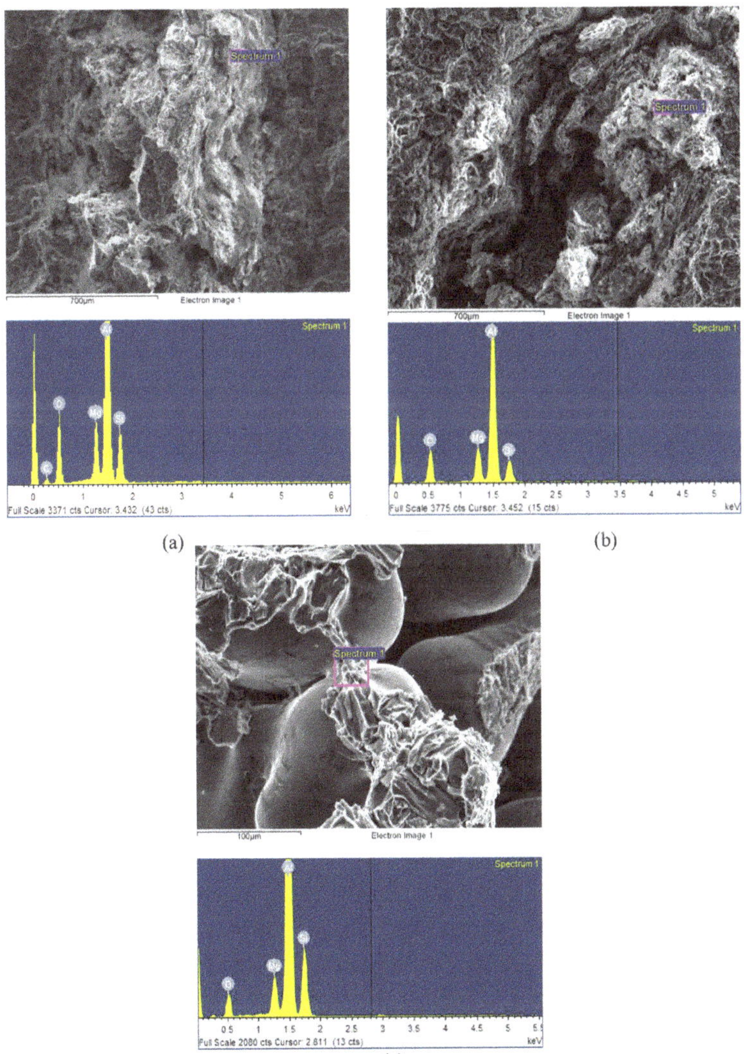

(c)

Figure 3: (a), (b) and (c) SEM images and corresponding EDX analysis of bifilms found at the fracture surfaces of test bars from experiments 1, 2 and 3, respectively.

Figure 3(c) shows an SEM image of a pore found on the fracture surfaces of a specimen from 2L99 alloy casting that was poured into preheated a ceramic mould and took about four minutes to solidify (experiment 3). Oxide films detected in this experiment were always found covering or trapped within the dendrites inside the pores. The EDX analysis suggested the existence of spinel, indicated by the Mg peak.

No other oxide films were found on the fracture surfaces of the test bars from all castings from experiment 3. The SEM images of the pores on the fracture surfaces of specimens from this experiment showed oxide layers with different forms inside these pores. The EDX analysis of these oxides indicated the presence of $MgAl_2O_4$, indicating that the origins of the pores were primarily oxide film defects that might have been changed their shapes and sizes during the solidification time to form pores. This might suggest a role of the time spent by the casting in the liquid state before solidification in eliminating bifilms in Al castings, or at least altering their morphology.

It has been suggested by Campbell [2] that, after entrainment and due to internal turbulence in the bulk liquid, the entrained film might become compacted in a convoluted form. Afterwards, and during solidification the bifilm might unfurl and its shape might then be changed to a planar crack or a pore. In addition, the results by Griffiths and Raiszadeh [15] and El-Sayed and Griffiths [12] suggested that both the solidification time of the casting and the H content of the melt can play a role in controlling the variability of the mechanical properties of Al castings.

In this experiment the tensile properties of three sets of 2L99 alloy castings with different H contents and/or solidification times were compared. The H content of the melt before pouring was 0.131 $cm^3/100$ g Al. The castings from experiments 1 and 2, which were solidified almost instantaneously, had practically not allowed any time during solidification for the unfurling mechanism to take place and therefore the main difference between the two castings was the mould material. For the castings from experiment 1, poured into sand moulds made one day before the experiment, the high temperature during pouring would probably cause the solvent (of the resin bonding the sand grains) to evaporate and release hydrogen, which was absorbed by the liquid metal. This had significantly increased the H content of the solidified casting to 0.235 $cm^3/100$ g. In contrast, for the casting poured into un-preheated, but dried, ceramic mould did not cause such H pick up from mould walls which caused the H content of the solidified casting to be much lower 0.139 $cm^3/100$ g Al, almost the same as the melt before pouring.

Dispinar and Campbell [16–19] suggested that hydrogen in excess of the solubility limit of the Al melt might diffuse into the bifilm gap to form H porosity in the Al castings. The same conclusion was reached by Raiszadeh and Griffiths [5].

Double oxide film defects should have been present in both castings from experiments 1 and 2 because of the poor running system design that was deliberately used, as was shown by the SEM investigation of the fracture surfaces, shown in Figs 3(a) and (b), respectively. However, the relatively lower H content of the casting in experiment 2 would have significantly reduced the amount of H diffused into the bifilms, decreasing their sizes and therefore enhancing the quality and reproducibility of the mechanical properties of the casting. Compared to experiment 1, the casting from experiment 2 had the Weibull moduli of the UTS and % elongation improved by about 52% and 60%, respectively, and the position parameters of both properties to increase by about 8% and 50%, respectively.

When the melt was poured into a preheated ceramic mould, the H content of the final casting was 0.152 $cm^3/100$g Al. The slight increase in the H content, compared to experiment 2, could be perhaps due to the H pick-up by the casting from the surrounding atmosphere during solidification (which took about 4 minutes). The SEM investigation of the fracture surfaces of specimens from this experiment did not detect any oxide films lying on the surfaces. Instead, all of the oxide films were found within pores, as was confirmed by the EDX analysis that detected the presence of spinel inside pores, as shown in Fig. 3 (c). Also, the sizes of oxide films were significantly smaller than those in experiment 2, despite the nearly similar H contents of both castings. Finally, there was a noticeable improvement in the mechanical properties. Compared to experiment 2, the casting from experiment 3 had the Weibull moduli of the UTS and % elongation to improve by about 213% and 64%, respectively, and the position parameters of both properties to increase by about 16% and 26%, respectively.

El-Sayed *et al.* [11] had suggested that during solidification the bifilms experienced two competing mechanisms. The first mechanism is the consumption of air inside the bifilms, and the other mechanism is the diffusion of H into the defects. Both mechanisms were expected to influence the morphology of the bifilm defects and by implication their effect on the mechanical properties of the casting. The solidification times of the castings in experiments 2 and 3 were about 10 second and 230 seconds respectively. Therefore, it could be suggested that during solidification of the casting in the preheated ceramic mould (experiment 3) some morphological changes occurred to the double oxide film defects. This could be perhaps due to the consumption of their interior atmospheres while a little amount of H was picked-up from the ceramic mould and/or the surrounding atmosphere (causing it to slightly increased from 0.131 to 0.152 cm^3/100 g Al), which might have caused them to be less harmful to the mechanical properties.

To summarize, the Weibull analysis of the tensile properties of 2L99 casting specimens from different experiments, containing double oxide film defects with different ages (time taken during solidification) and/or H contents suggested the occurrence of two competing mechanisms that controlled the mechanical properties. The consumption of air inside the bifilms due to reaction with the surrounding molten metal may lead to improvements in mechanical properties, but this may be accompanied by hydrogen passing into the bifilms, which would have a deleterious effect on properties. Also, the melt was found to pick H from either the walls of sand moulds and surrounding air during solidification.

Therefore, if the H content of the melt could be decreased (by degassing) prior to pouring into a ceramic mould and then the filled mould could be held in a furnace with a dry atmosphere that contains the minimum possible water vapour content, this would allow a casting producer to benefit from the holding treatment in reducing the bifilm atmospheres (by reactions with the surrounding melt), while preventing the formation of H porosity. This would decrease the size of bifilm defects and in turn lead to castings with higher and more reproducible mechanical properties.

4 CONCLUSIONS

1. Oxide film defects were detected on all the fracture surfaces of the test bars from different experiments either lying on all the surfaces or inside pores, which demonstrated a role for such defects in influencing the failure of Al castings.
2. Increasing the H content of a 2L99 casting from 0.139 to 0.235 cm^3/100 g Al increased the amount of H diffused into the bifilms, increasing their sizes and caused the Weibull moduli of both the UTS % elongation to decrease by about 35%.
3. Preheating of a ceramic mould to 800°C increases the solidification time of the casting which might allow the interior atmosphere of its entrained bifilms to react with the

surrounding melt, reducing their sizes and accordingly their harmful effect on mechanical properties.

4. Not only the amount and age of oxide film defects but also the H content of the melt can determine the reliability and the reproducibility of the mechanical properties of Al alloy castings.

REFERENCES

[1] Campbell, J., *Castings*, 2nd. ed: Butterworth-Heinemann, 2003.

[2] Campbell, J., Entrainment defects. *Materials Science and Technology*, **22**, pp. 127–145, 2006.
https://doi.org/10.1179/174328406x74248

[3] Basuny, F.H., Ghazy, M., Kandeil, A.-R.Y. & El-Sayed, M.A., Effect of casting conditions on the fracture strength of Al-5 Mg alloy castings. *Advances in Materials Science and Engineering*, **2016**, pp. 1–8, 2016.
https://doi.org/10.1155/2016/6496348

[4] Nyahumwa, C., Green, N.R. & Campbell, J., Effect of mold-filling turbulence on fatigue properties of cast aluminum alloys. *AFS Trans*, **106**, pp. 215–223, 1998

[5] Raiszadeh, R. & Griffiths, W.D., A method to study the history of a double oxide film defect in liquid aluminum alloys. *Metallurgical and Materials Transactions B*, **37**(6), pp. 865–871, 2006.
https://doi.org/10.1007/bf02735007

[6] El-Sayed, M.A., Salem, H.A.G., Kandeil, A.Y. & Griffiths, W.D., Determination of the lifetime of a double-oxide film in al castings. *Metallurgical and Materials Transactions B,* **45**(4), pp. 1398–1406, 2014.
https://doi.org/10.1007/s11663-014-0035-x

[7] El-Sayed, M.A., Salem, H.A.G., Kandeil, A.Y. & Griffiths, W.D., A study of the behaviour of double oxide films in Al alloy melts. *Materials Science Forum,* **765**, pp. 260–265, 2013.
https://doi.org/10.4028/www.scientific.net/msf.765.260

[8] Griffiths, W.D., Caden, A. & El-Sayed, M., An investigation into double oxide film defects in aluminium alloys. *Materials Science Forum,* **783–786**, pp. 142–147, 2014.
https://doi.org/10.4028/www.scientific.net/msf.783-786.142

[9] Griffiths, W.D., Caden, A. & El-Sayed, M.A., The behaviour of entrainment defects in aluminium alloy castings. *Proceedings of the 2013 international symposium on liquid metal processing and casting,* John Wiley & Sons, Inc., pp. 187–192, 2013.

[10] El-Sayed, M., Hassanin, .H & Essa, K., Effect of casting practice on the reliability of Al cast alloys. *International Journal of Cast Metals Research*, **29**(6), pp. 350–354, 2016.
https://doi.org/10.1080/13640461.2016.1145966

[11] El-Sayed, M.A., Salem, H.A.G., Kandeil, A.Y. & Griffiths, W.D., Effect of holding time before solidification on double-oxide film defects and mechanical properties of aluminum alloys. *Metallurgical and Materials Transactions B*, **42**(6), pp. 1104–1109, 2011.
https://doi.org/10.1007/s11663-011-9571-9

[12] El-Sayed, M. & Griffiths, W., Hydrogen, bifilms and mechanical properties of Al castings. *International Journal of Cast Metals Research*, **27**(5), pp. 282–287, 2014.
https://doi.org/10.1179/1743133614y.0000000113

[13] Weibull, W., A statistical distribution function of wide applicability. *Journal of Applied Mechanics*, **13**, pp. 293–297, 1951.
[14] Green, N.R. & Campbell, J., Statistical distributions of fracture strengths of cast Al-7Si-Mg alloy. *Materials Science and Engineering: A*, **173**(1–2), pp. 261–266, 1993. https://doi.org/10.1016/0921-5093(93)90226-5
[15] Griffiths, W.D. & Raiszadeh, R., Hydrogen, porosity and oxide film defects in liquid Al. *Journal of Materials Science*, **44**(13), pp. 3402–3407, 2009. https://doi.org/10.1007/s10853-009-3450-7
[16] Dispinar, D., Akhtar, S., Nordmark, A., Di Sabatino, M. & Arnberg, L., Degassing, hydrogen and porosity phenomena in A356. *Materials Science and Engineering: A.*, **527**(16–17), pp. 3719–3725, 2010. https://doi.org/10.1016/j.msea.2010.01.088
[17] Dispinar, D. & Campbell, J., Critical assessment of reduced pressure test. Part 1: porosity phenomena. *International Journal of Cast Metals Research,* **17**, pp. 280–286, 2004. https://doi.org/10.1179/136404604225020696
[18] Dispinar, D. & Campbell, J., Use of bifilm index as an assessment of liquid metal quality. *International Journal of Cast Metals Research,* **19**(1), pp. 5–17, 2006. https://doi.org/10.1179/136404606225023300
[19] Dispinar, D. & Campbell, J., Effect of casting conditions on aluminium metal quality. *Journal of Materials Processing Technology*, **182**(1–3), pp. 405–410, 2007. https://doi.org/10.1016/j.jmatprotec.2006.08.021

COMPUTATIONAL APPROACH TO IMPROVE BEARINGS BY RESIDUAL STRESSES BASED ON THEIR REQUIRED BEARING FATIGUE LIFE

F. PAPE[1], O. MAISS[2], B. DENKENA[2] & G. POLL[1]
[1]Institute for Machine Design and Tribology, Leibniz Universitaet Hannover, Germany.
[2]Institute of Production Engineering and Machine Tools, Leibniz Universitaet Hannover, Germany.

ABSTRACT
In drive systems and component technology a high reliability is very important for machines. Machine element dimensions are calculated for reliability. The properties for these elements are based on conventional manufacturing techniques. Very high stresses are applied on bearings in their operating time. To improve the endurance life, residual stresses can be induced into the subsurface zone. In contrast to a conventional grinding process, the mechanical surface modification process deep rolling is able to induce very high compressive residual stresses. A computational approach was developed to establish an appropriate residual stress depth profile matching the applied loads. Thus, the costs of manufacturing can be chosen in accordance to the required properties. The method to determine the residual stresses is based on an iterative reverse calculation of an existing bearing fatigue life model of Ioannides *et al.* The model originates from the approach of Lundberg and Palmgren (1947) including a stress fatigue limit τ_u. For the term τ_i, the fatigue criterion of Dang-Van is applied. The equation accounts for the maximum orthogonal shear stress and the local hydrostatic pressure p_{hyd}, corrected for residual and hoop stress. The inputs into the computational model are the stresses on the surface, which are simulated based on the load and geometry of the contact between roller and bearing surface. As an output the required residual stress profile underneath the bearings raceway is given to achieve a bearing fatigue life as required for the given application. In order to verify the model, the bearing fatigue life was experimentally determined for a given residual stress profile by experiments.
Keywords: bearing fatigue life, inverse computational model, residual stresses.

1 INTRODUCTION

Roller bearings are typically made of 100Cr6 bearing steel (German standard corresponding to AISI52100). The investigated bearings of type NU206 have cylindrical rolling elements (rollers), which are held by a polyamide cage, Fig. 1 shows the bearing's components. In roller bearings, the contact between raceway and rolling elements is exposed to cyclic load due to the rotational motion. Alternating stresses are induced in varying depths beneath the surface of the contacting partners. The depth of the highest load induced stresses can be calculated according to the Hertzian contact theory including the approaches of Karas [1] and Föppl [2]. The normal loads onto the surface cause changes of the microstructure and may trigger crack formation beneath the surface of the raceways in the depth of highest load induced stresses.

According to Voskamp, bearing fatigue life for roller bearings can be separated into three phases [3]:

- The first phase, the so-called shakedown-phase. Within the first 1000 revolutions plastic micro deformation occurs on the raceway. These deformations lead to work-hardening of the material and a resulting decrease of the retained austenite in the martensitic matrix and a moderate change in compressive residual stresses. The influence depends on the level of the load.
- The second phase, the steady state phase, features no plastic deformations and no changes of the microstructure. The duration of the stationary phase depends on the applied stresses.

DOI: 10.2495/CMEM-V6-N4-656-666

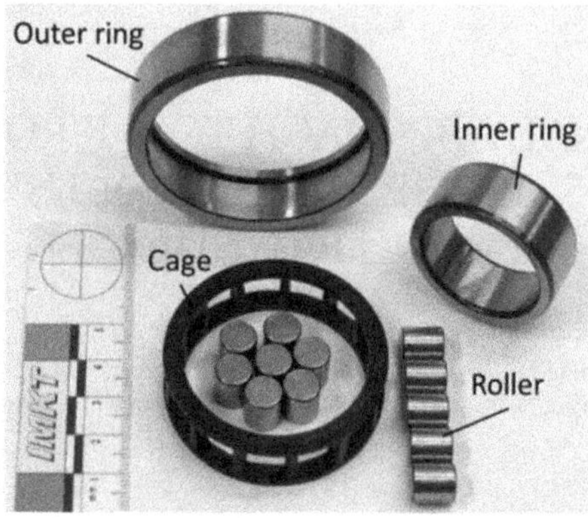

Figure 1: Components of the roller bearing.

For loads above the so-called 'Shakedown Limit' the third phase starts immediately, whereas for very low loads the bearing remains in the second phase forever and no failure due to fatigue takes place.

- The third phase, the instability phase. The inner boundary layers, conditioned in the shakedown phase, lose the ability to carry the load elastically. A local change of the microstructure takes place. Due to the phase transformation and the complex residual stress state, the material is modified. The microplastic deformations transform into macroplastic deformations. With persistent cyclic load on the inner boundary layer, cracks grow to the surface and finally pittings occur on the raceway. The depth of these pittings is equal to the depth of the highest load induced stresses. In practical applications, this failure type (Fig. 2) on the surface is one of the main failure modes of bearings (under ideal lubrication conditions and avoidance of lubricant contamination).

To increase the duration of the steady state phase, residual stresses can be induced to the inner boundary zone in the depth of the highest load induced stresses. Thus, the shakedown phase can be forestalled. To achieve pre-induced residual stresses, bearing surfaces can be reinforced with a short period of overload in the shakedown-phase, as well as in the manufacturing process. To show this, Hacke subjected NU206 bearings to a two stage test [4]. In a first step, the tests were run at a higher load for 1.5 million revolutions. The test was then continued at normal load. The tests were performed under full film lubrication with a specific film thickness of $\lambda \geq 2$. It could be shown that the bearing fatigue life increased drastically. The manufacturing process can be used to induce residual stresses not only to the near

Figure 2: Damaged surface of a bearing inner ring due to fatigue.

surface zone but to the subsurface region, too. The mechanical surface modification process deep rolling is one effective possibility to induce the compressive residual stresses. Within the process, a hard ball is pressed on a surface of a rotating part. A hydraulic pressure of up to p_w = 600 bar can be used for the process. Neubauer *et al.* investigated the influence of manufacturing processes and induced residual stresses on bearing fatigue life. A computational model was set up and compared to bearing fatigue life tests [5, 6]. For the application of bearings, it is of interest to define a desired bearing fatigue life and calculate the demanded surface properties. Thus, the manufacturing process can be selected to fulfil these specifications. The aim of this paper is to introduce an inverse computational model for this application. Therefore, the calculation of bearings fatigue life will be described shortly. As a second step, the inverse computational model will be presented. From the results of the calculation, a manufacturing process will be chosen to manufacture roller bearing inner rings. At the end, the resulting bearings fatigue life will be determined experimentally.

2 CALCULATION OF BEARING FATIGUE LIFE

The influence of the residual stresses on the bearing fatigue life can be understood by taking a deeper look on the computational approaches. Roller bearings are highly stressed due to the cyclic contact load. To calculate the bearing fatigue life, the theoretical approaches are based on the weakest link concept by Weibull [7]. Weibull developed the following equation:

$$\log\left(\frac{1}{S}\right) = -\int_V n(\sigma)\, dV \tag{1}$$

S: Probability of survival
n(*σ*): Material dependent scale parameter
V: Damage risk volume [mm^3]

For homogeneous materials the following correlation has to be chosen [7]:

$$n(\sigma) = \left(\frac{\sigma - \sigma_u}{\sigma_o}\right)^m, \tag{2}$$

with σ_u, σ_o and m as experimentally detected constants, σ is the applied stress.

Lundberg and Palmgren developed a model for the calculation of bearing fatigue life, based on Weibull [8]. They expected a failure underneath the bearings surface for high orthogonal stresses and for imperfections of the material. In contrast to Weibull, it is expected that cracks due to fatigue occur on a weak point below the surface, e.g. an inclusion. Such cracks may extend further to the surface and finally lead to pittings. The probability of survival can be calculated as a function of the number of revolutions, the maximal orthogonal stresses, the stressed volume and the depth of the highest load induced stress.

$$\ln\left(\frac{1}{S}\right) \propto \frac{\tau_o^c \cdot N^e \cdot V}{z_o^h} \tag{3}$$

N: Number of load cycles
τ_o: Maximal orthogonal shear stress [MPa]
z_o: Depth of the max. orthogonal shear stress [mm]
c: Exponent for stress criterion
h: Factor in the bearing life equation
e: Exponent in the bearing life equation (standardized Weibull slope)

In the 1980s Ioannides and Harris introduced an advanced model for the calculation of bearing fatigue life based on Lundberg and Palmgren [9]. The key point for the model is a statistical connection between the probability of surviving and a stress dependent fatigue criterion σ_i or τ_i, respectively, for a volume element of the bearings surface. The probability of surviving for a volume element of a bearing is described by the following equation:

$$\ln \frac{1}{\Delta S_i} = A_i \cdot H(\sigma_i - \sigma_u) \cdot (\sigma_i - \sigma_u)^c \cdot \Delta V_i \,. \tag{4}$$

A_i is an independent random variable and H(x) a heaviside function with

$$H(x) = \begin{cases} 1\, for\, x \geq 0 \\ 0\, for\, x \leq 0 \end{cases} . \tag{5}$$

The resulting stress field is used as input for the fatigue life model of Ioannides *et al.* [10]. The equation is:

$$\ln\left(\frac{1}{S}\right) \approx N^e \int_V \frac{(\tau_i - \tau_u)^c}{z'^h} dV \tag{6}$$

τ_i: Fatigue stress criterion
τ_u: Shear stress fatigue limit (Pa)
z': Stress weighted depth (depth from the surface) (m)
h: Factor in the bearing life equation

In order to account for the depth of the maximum shear stress the weighted depth z' is introduced:

$$z_i' = \frac{\int_0^{z_{max}} (\sigma_i - \sigma_u) \cdot z_i \, dz}{\int_0^{z_{max}} (\sigma_i - \sigma_u) dz} \,. \tag{7}$$

The model of Ioannides *et al.* [10] includes a stress fatigue limit τ_u and a fatigue stress criterion. For the term τ_i as fatigue stress criterion the criterion of Dang-Van *et al.* [11] is selected:

$$\tau_i = \tau_{Omax} - k_{hyd} \cdot p'_{hyd} \tag{8}$$

τ_{Omax}: Max. shear stress (Pa)

k_{hyd}: Weighting factor for hydrostatic stresses

p'_{hyd}: Hydrostatic stress including compressive stresses

$$p_{hyd}' = p_{hyd} - \left[\frac{(\sigma_{Rx} + \sigma_{Ry} + \sigma_{Rz})}{3} + \frac{\sigma_{hoop}}{3} \right] . \tag{9}$$

The Dang-Van criterion includes the maximum orthogonal shear stress and the local hydrostatic pressure, p'_{hyd}, corrected for residual and hoop stress. With increased compressive residual stresses the hydrostatic pressure is increased, thus the critical fatigue stress is reduced. The critical fatigue stress is highest in the zone of the highest load induced stresses beneath the surface. Concludingly, by applying residual stresses in this depth, the computational approach predicts that the bearing fatigue life is influenced positively.

2.1 Computational method to determine the influence of intrinsic stresses on bearing fatigue life

In order to assess the influence of the magnitude and depth of residual stresses on bearing fatigue life, Neubauer previously developed a computational method [5, 12]. A finite element model of a bearing inner ring was set up to compute the resulting three-dimensional stress field from the superposition of load stresses and residual stresses. Figure 3 depicts the flow chart of the computational model. The input variables are the external load and the residual stress depth profile. The external load is transformed into a Hertzian pressure distribution. The residual stress depth profiles are defined as initial stresses in the model. One result of the finite element (FE) model are the orthogonal shear stresses beneath the surface. The stresses are required for the forward calculation model based on Ioannides *et al.* [10], which is computed in MathWorks® Matlab. The method allows determining the resulting bearing fatigue life based on the initial residual stress state of the bearings surface.

3 INVERSE CALCULATION OF DESIRED RESIDUAL STRESSES FOR GIVEN BEARING FATIGUE LIFE TIME

The inverse model is based on an iterative reverse calculation of the bearing fatigue life model of Ioannides *et al.* [10]. It is set up to determine the required residual stress state for reaching a specified bearing fatigue life. The inputs into the computational model are the stresses on the surface, which are simulated based on the load and geometry of the contact between roller and bearing surface. The load and basic rating life L_{10} (bearing life that 90% of a sufficiently large group of identical bearings reach or exceed before fatigue appears). Output is the required residual stress state in the bearing surface to achieve a bearing fatigue life as required for a given application. The flow chart of the inverse model is shown in Fig. 4.

The calculation is performed in Matlab, the flow chart is shown in Fig. 5. For the model, the Dang-Van criterion as well as the approach of Ioannides and Harris are used. Due to the

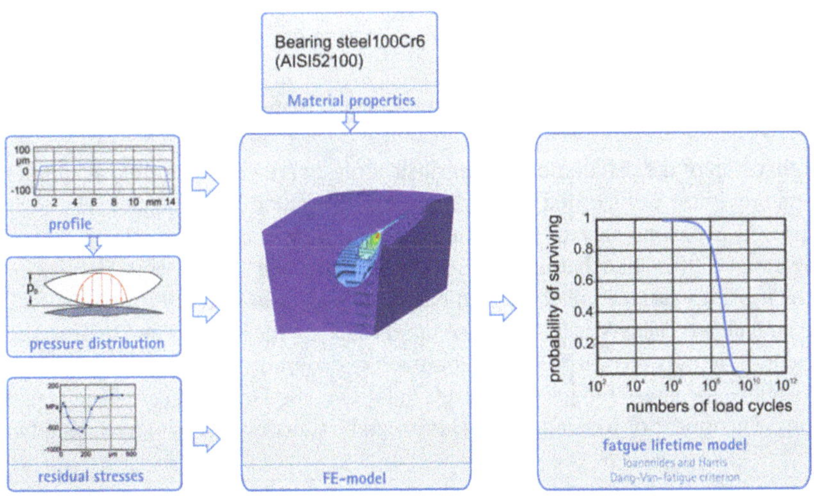

Figure 3: Flow chart of the forward computational approach based on Neubauer [5] to calculate the influence of residual stresses on bearing fatigue life.

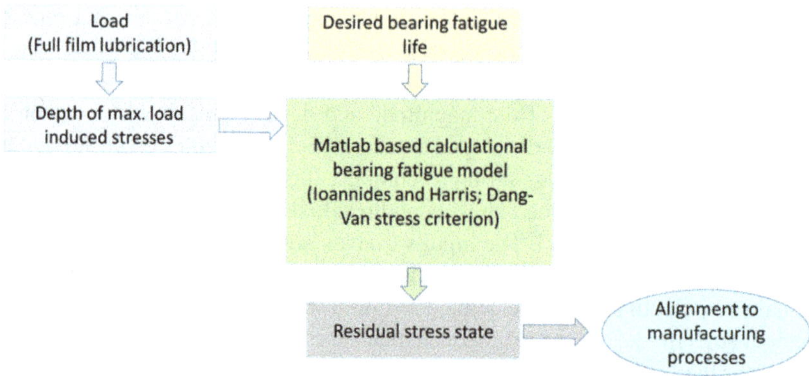

Figure 4: Flow chart for the inverse computational model to calculate the residual stress state desired to reach pre-defined bearing fatigue life.

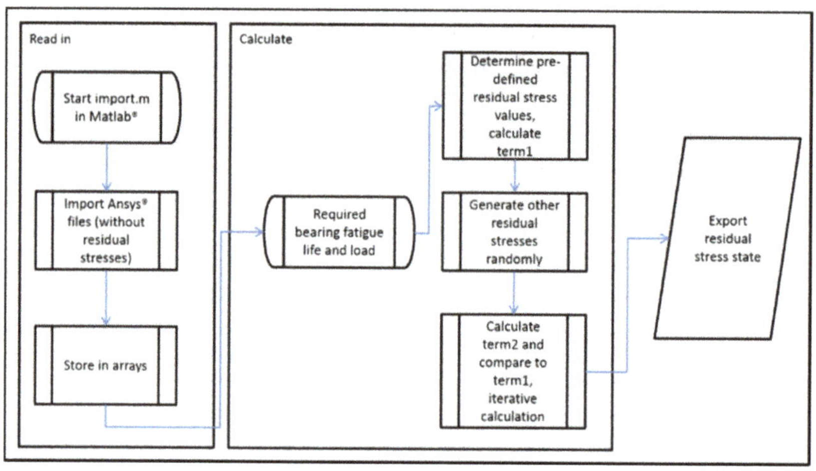

Figure 5: Flow chart of the Matlab program to define the residual stress state.

approach including a heaviside function the intrinsic stresses cannot be calculated directly, see eqn (9).

In a first step of the calculation, the intrinsic stresses are generated for one point randomly based on previous results of the forward calculation method by Neubauer. The early calculations proved a dependency of the residual stress formed in higher depth and achievable bearing fatigue life. Distribution functions of the residual stress depth profiles originating from the bearings surface are calculated based on the Matlab curve fitting tool.

To calculate the Dang-Van criterion the input based on the shear stress, hydrostatic pressure and residual stresses are used. As the approach of Ioannides and Harris considers that only stresses above the stress fatigue limit result in fatigue, the Dang-Van criterion is used for calculations. The model of Ioannides and Harris can be transformed by introducing two terms.

$$ln\left(\frac{1}{S}\right) = N^e \sum_{i=1}^{n} A_i H(\sigma_i - \sigma_u)(\sigma_i - \sigma_u)^c \left[\frac{\sum_{i=0}^{n-1}(\sigma_i - \sigma_u)}{\sum_{i=0}^{n-1}(\sigma_i - \sigma_u)\cdot z_i}\right]^h \Delta V_i, \tag{10}$$

$$term1 = \frac{ln\left(\frac{1}{S}\right)}{\tilde{A} \cdot N^e}, \tag{11}$$

$$term2 = \sum_{i=1}^{n} \frac{(\sigma_i - \sigma_u)^c}{z_i^{'h}} \Delta V_i \qquad \text{(12) for H = 1.}$$

For the generation of the residual stresses both equations are calculated and balanced. The iterations have to be continued, until both terms converge. After finishing the iteration the resulting residual stress state is calculated and plotted. Figure 6 shows calculated residual stress states for pre-defined bearing fatigue life under a contact pressure of 3,000 MPa. The graphs illustrate the trend that for increasing bearing fatigue life the residual stresses have to be increased and induced into higher depths. For increased bearing rating life the value and depth of the compressive residual stresses have to be adjusted. Figure 7 depicts the calculated residual stress state for a contact pressure of 2,300 MPa and a pre-defined bearing rating life L_{10} = 1,750 h. Due to the lower contact pressure the depth of highest load induced stress is in

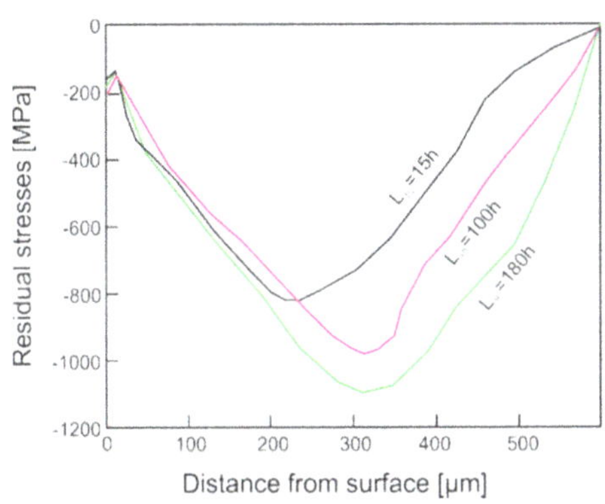

Figure 6: Calculated optimized residual stress state for pre-defined bearing fatigue life (contact pressure of 3000 MPa).

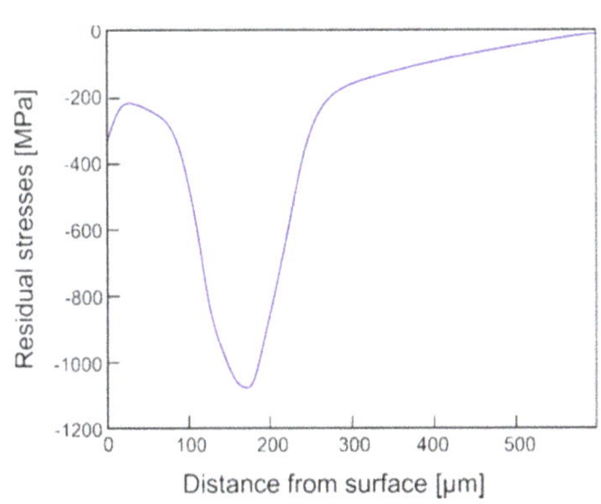

Figure 7: Pre-defined optimized residual stress state for a contact pressure of 2300 MPa; L_{10} = 1750 h.

lower depths compared to the graphs in Fig. 6. The compressive residual stresses should be highest in a depth of approx. 150 µm.

4 MACHINING OF ROLLER BEARINGS WITH SPECIFIC RESIDUAL STRESS DESIGN BY HARD TURNING AND DEEP ROLLING

To proof the presented model, machining tests were conducted. For the tests roller bearing inner rings type NU206 were machined by hard turning and deep rolling on a high precision lathe Hembrug Microturn 100. Within the experiments, the influence of process parameters on the residual stress depth profile was analysed. The residual stresses were measured with an X-ray diffractometer Seiffert XRD 3000 P using CrK*α* radiation. The experimental procedure and results are described in detail by Denkena *et al.* [13, 14].

The deep rolling process influences the residual stress state significantly. Figure 8 summarizes the experimental results. An increasing rolling pressure p_w leads to higher amplitude and larger depth of the compressive stresses. The size of the rolling tool, d_k, affects the resulting depth of the residual stress profile. The effect of the overlap factor u can be compared with the effect of the shake down phase in bearings. An increased overlap factor results in an increased number of rolling processes, which leads to higher compressive residual stresses. From the results, a regression model is generated to calculate the resulting subsurface properties by the processes hard turning and deep rolling. To achieve the residual stress state depicted in Fig. 7 the process parameters summarized in Table 1 were chosen. The achieved residual stress state is shown in Fig. 9.

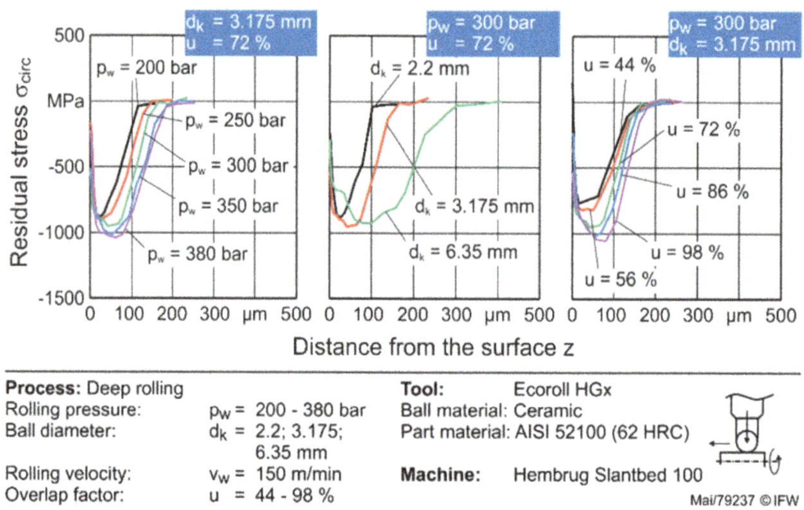

Figure 8: Effect of machining parameters in hard turning and deep rolling on residual stress state.

Table 1: Chosen process parameters to induce residual stress depth profile.

Hard turning			**Deep rolling**		
cutting speed v_c	m/min	200	rolling pressure p_w	bar	300
feed f	mm	0.05	rolling speed v_w	m/min	150
depth of cut a_p	mm	0.1	ball diameter d_k	mm	6.35
cutting edge radius $r_β$	µm	25	overlap factor u	%	98

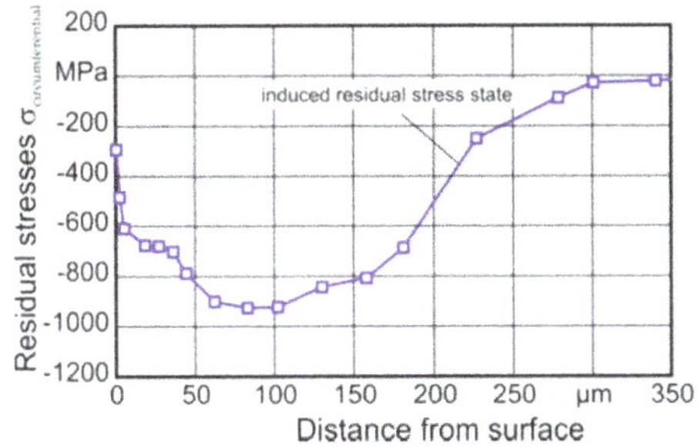

Figure 9: Pre-induced residual stress state.

5 EXPERIMENTAL EVALUATION OF THE INFLUENCE OF RESIDUAL STRESSES ON BEARING FATIGUE LIFE

To verify that with the chosen residual stress state resulting from the computational model the requested bearing fatigue life is achieved, experiments were conducted. Bearings without and with beneficial residual stress states were taken for the tests. Standard NU206 bearing inner rings, featuring no significant residual stresses in higher depth, were compared to bearings with a beneficial residual stress state induced by the manufacturing process. By increasing the pressure for the deep rolling process, the magnitude of the residual stresses can be enhanced whereas by increasing the diameter of the rolling tool the stress field is extended into greater depths. Modified bearing inner rings and standard parts for the rollers and outer rings were subjected to the tests.

The evaluation of the experimental bearing endurance time was performed at six identical 4-bearing endurance test benches. Each of the test rigs allows for running four equal roller bearings of type NU206 C3 at the same time. Thus, higher statistic coverage can be achieved faster in the experiments. All bearings are exposed to the same radial load in the test rig. The applied load is measured with a load cell during testing. The tests were performed under full film lubrication with a specific film thickness of $\lambda \geq 2$. The bearings are supplied with tempered oil (at 60°C). A synthetic oil was chosen with a kinematic viscosity of $\eta_{40} = 68\ mm^2/s$. Bearing failures due to pitting are detected with a condition monitoring system. The radial load was chosen with C/P = 4 equal to a maximum Hertzian contact pressure of $p_{max} = 2{,}300$ MPa in the bearing. The rotational speed was 4,050 min^{-1}. The tests were performed according to the sudden death test principle. When the fatigue life of one of the bearings is reached and a pitting has formed, the other three bearings are suspended. Figure 10 shows a schematic of the test rig.

After failure, the bearings were disassembled and analysed. For the statistical evaluation, the maximum-likelihood method is applied. Thus, the bearings without failure are accounted for as "suspended".

The standard honed bearings reached an L_{10} of 274 h, and an L_{50} of about 7,000 h. In comparison, for bearings with pre-induced residual stresses on the inner rings, an L_{10} of 1,924 h and an L_{50} of 8,446 h could be observed. The experimentally validated bearing life is close to the calculated L_{10} of 1,750 h (Fig. 7). For these bearings, the residual stress state remained nearly constant over the test duration. After an ultra-long running time of 378 Mio revolutions, the residual stresses start to decrease in the zone of the highest load induced stresses. Thus, in accordance with Voskamp [3], the breakdown phase is forestalled and the duration of the steady state phase could be increased.

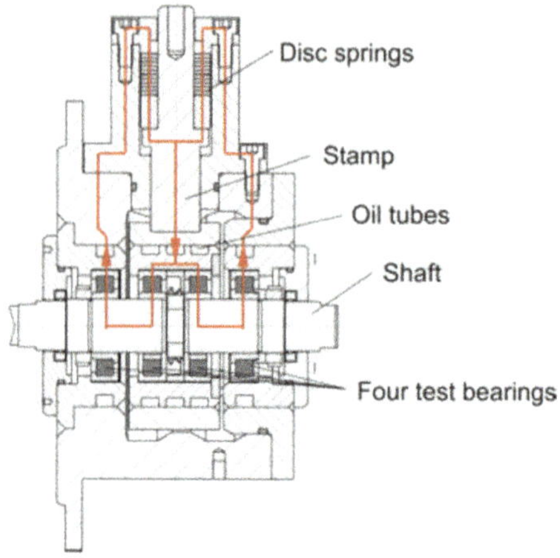

Figure 10: Schematic of the test rig to evaluate the computational approach.

6 CONCLUSION

Applying residual stresses on bearing surfaces offers the possibility to increase the bearing fatigue life drastically. To equip these bearings with the desired surface properties, a calculation procedure was established. The heaviside function of the bearing fatigue life model by Ioannides and Harris requires an iterative approach to determine the optimum residual stresses based on the Dang-Van criterion. The calculated residual stress state was compared to measurements showing a good agreement. The inverse computational model allows for choosing the proper manufacturing processes particularly deep rolling processes delivering the desired stress state. Thus, the manufacturing effort can be adjusted according to the desired bearing properties. The bearings can be equipped with improved properties and machines designed regarding improved resource efficiency. Especially, for applications with complicated and expensive replacement of failed bearings (such as windmills and aerospace), such approaches are valuable.

ACKNOWLEDGMENT

The authors thank the DFG (German Research Foundation) for supporting this project in the context of the research program 'Resource efficient Machine Elements (SPP1551)'.

REFERENCES

[1] Karas, F., Werkstoffanstrengung beim druck achsenparalleler Walzen nach den gebräuchlichen Festigkeitshypothesen. *Forschung auf dem Gebiet des Ingenieurwesens*, **11**(6), pp. 334–339, 1940. https://doi.org/10.1007/BF02583244

[2] Föppl, L., Der Spannungszustand und die Anstrengung des Werkstoffes bei der Berührung zweier Körper. *Forschung auf dem Gebiet des Ingenieurwesens*, **7**(5), p. 209, 1936. https://doi.org/10.1007/BF02585815

[3] Voskamp, A., Microstructural changes during rolling contact fatigue - metal fatigue in the subsurface region of deep groove ball bearing inner rings, Technische Universität Delft, Thesis, 1996.

[4] Hacke, B., Radnai, B. & Hinkelmann, K., Berücksichtigung von Betriebszuständen, Sonderereignissen und Überlasten bei der Berechnung der Wälzlager-Lebensdauer in Windenergieanlagen und Großgetrieben. Abschlussbericht FVA Forschungsheft Nr. 967, AiF-Nr. 15227 N, 2011.
[5] Neubauer, T., Betriebs- und Lebensdauerverhalten hartgedrehter und festgewalzter Zylinderrollenlager, Leibniz Universität Hannover, IMKT. Doctoral Thesis, 2015.
[6] Denkena, B., Poll, G., Maiß, O., Pape, F. & Neubauer, T., Enhanced boundary zone rolling contact fatigue strength through hybrid machining by hard turn-rolling. *Bearing World Journal, Volume 1, Proceedings of the 1st Bearing World Conference 12-13 April 2016*, Hanover/Germany, ISBN 978-3-8163-0705-1, pp. 87–102, 2016.
[7] Weibull, W., A Statistical Theory of the Strength of Materials. *Generalstabens litografiska anstalts förlag,* 1939. (Ingeniörsvetenskapsakademiens handlingar)
[8] Lundberg, G. & Palmgren, A., Dynamic Capacity of Rolling Bearings. Generalstabens Litografiska Anstalts Förl, 1947 (Acta polytechnica. Mechanical engineering series).
[9] Ioannides, E. & Harris, T.A., A new fatigue life model for rolling bearings. *Journal of Tribology*, **107**(3), pp. 367–377, 1985.
https://doi.org/10.1115/1.3261081
[10] Ioannides, E., Bergling, G. & Gabelli, A., An analytical formulation for the life of rolling bearings. In: Acta Polytechnica Scandinavica, Mechanical Engineering Series No. 137, 1999.
[11] Dang Van, K., Griveau, B. & Message, O., On a new mulitiaxial fatigue criterion: theory and application. In: M.W. Brown & K.J. Miller (eds), *Mechanical Engineering Publications*, London, pp. 479–496, 1989.
[12] Pape, F., Neubauer, T., Maiss, O., Denkena, B. & Poll, G., Influence of manufacturing processes and related residual stresses on bearing fatigue life, TFC - *Tribology Frontiers Conference,* STLE, 14. November 2016.
[13] Denkena, B. Grove, T. & Maiß, O., Influence of the cutting edge radius on surface integrity in hard turning of roller bearing inner rings. *Production Engineering*, Vol. 9(3), pp. 299–305, 2015.
https://doi.org/10.1007/s11740-015-0615-x
[14] Denkena, B., Grove, T. & Maiß, O., Influence of hard turned roller bearings surface on surface integrity after deep rolling. *Proceedia CIRP*, **45**, pp. 359–362, 2016.
https://doi.org/10.1016/j.procir.2016.02.047

APPLICATION OF AN EVOLUTIONARY ALGORITHM TO REDUCE THE COST OF STRENGTHENING OF TIMBER BEAMS

FRANCISCO B. VARONA, DAVID BRU-ORTS, FRANCISCO J. BAEZA & SALVADOR IVORRA
Department of Civil Engineering, University of Alicante, Spain.

ABSTRACT
The present paper describes the application of an evolutionary algorithm to the optimum design of the reinforcement of timber beams using FRP laminates and sheets. The objective function is the material cost of the strengthening and is subjected to ten constraints derived from the ultimate limit states for flexural and shear behaviour as well as the serviceability limit states. A genetic algorithm is used and the optimization problem is transformed into an unconstrained one by means of an adaptive penalty function. The design variables are the CFRP and GFRP mechanical properties and dimensions and they are encoded in a binary chromosome: type of composite material (CFRP or GFRP), reinforcement mechanical properties and geometric configuration. The search space for the minimum cost consists of 65 billion possible solutions. The crossover operator switches randomly between a fenotype crossover and flat crossover. An adaptive mutation scheme has been as well as an elitism criterion. The algorithm has been used for obtaining optimum designs in several specific load and geometry cases of glued laminated timber beams. The objective is finding whether there are specific reinforcement configurations more feasible for a certain loading situations: short or long beams and lower or higher loading increments. Five cases have been analysed. In the first three cases the length of the beams has constant values of 2, 2.5 and 3 m, whereas the value of loading was variable. In the latter case, the value of the load was fixed and the length of the beam was variable. The analysis of the results shows that the GFRP reinforcement is more efficient than CFRP for designs governed by shear failure, whereas CFRP is more effective in the case of flexural failure and deflection controlled strengthening of timber beams.
Keywords: adaptive operators, FRP strengthening, genetic algorithm, structural optimization, timber structures

1 INTRODUCTION

Although the use of wood as a building material is ancient, nowadays it still presents good structural qualities compared to steel and concrete materials. It stress ratio to specific weight makes it suitable for one way floors and, in particular, wood beams without knots or cracks have high values of compressive and tensile strength. However, elastic modulus and shear strength have very low values compared with steel materials (Triantafillou [1]). Nowadays, building reutilization and the current technical standards lead to an increase in service loads and to a decrease of the maximum deflection limits. In order to improve the structural safety and reliability of structural wood systems in existing buildings, composite-based reinforcement techniques are demanded. In this regard, fibre reinforced polymers (FRP) are an adequate alternative to improve structural capacity (Bru *et al.* [2]). This study is aimed at reaching and optimum design of the CFRP and GFRP reinforcement of wood beams by means of a genetic algorithm. Genetic algorithms (GAs) have been thoroughly used as an alternative to traditional structural optimization methods. GAs were introduced in the 70's by Holland [3]. Concerning the optimum design of the reinforcement for retrofitting of concrete structures, GAs have proven to be a robust technique, as shown in Perera and Varona [4]. Contrary to mathematical programming based optimization methods, GAs are of a probabilistic nature. They are able to handle groups of design points simultaneously and it is also possible to handle continuous and discrete variables. Their application does not need any

DOI: 10.2495/CMEM-V6-N4-667-678

explicit algebraic formulation of the objective functions and constraint functions and since they do not depend on gradient information, it is not necessary to perform a sensitivity analysis of these functions. Those are their principal advantages over mathematical programming based optimization. GAs are based on the Darwinian principle of evolution, in which the most feasible individuals within a generation are most likely to survive and pass their genetic material on to the next generation. Among their principal advantages, GAs are easily implemented and programmed and, given enough computation time, are always capable of reaching the global optimum. This can be justified because, as noted above, they are able to analyse simultaneously a wide range of possible solutions randomly generated, thus being less sensible to converge to local optima. Other heuristic optimization techniques, such as ant colony optimization and particle swarm optimization have not yet proven as robust as GAs for structural optimization.

2 MECHANICAL PROPERTIES OF MATERIALS

The study of the mechanical properties of the materials used in this research was based on the standard ASTM D3039/D3039M for glass fibre reinforced polymer (GFRP) laminates (Bru *et al.* [5]). These properties were obtained for bidirectional laminate (0/90º fibre orientation) and for different values of dosage. In this research, weight dosage values of 1/1 were selected, for woven type E glass fibre with 440 g/m² weight (213/217) and an epoxy resin supplied by SICOMIN Composites, type SR 5550 and SD 5503. The geometrical and mechanical properties for GFRP laminates are given in Table 1.

In Table 1, e_r is the equivalent thickness of the laminate; V_f and V_m are the volume fraction of the glass fibre and the matrix, respectively; E is the elastic modulus of the equivalent laminate; and $f_{u,f}$ is the tensile strength of the equivalent laminate. Other important input used in the genetic algorithm is the cost of the materials. For Sicomin epoxy resin, 34.5 €/m² has been considered. For glass fibre fabric, the following values have been considered: ULE630, 1.81 €/m²; BXE300, 1.62 €/m²; BXE315, 2.29 €/m²; BXE446, 1.98 €/m²; BXE600, 2.33 €/m²; BXE800, 2.92 €/m²; BE440, 1.67 €/m²; BE850, 2.74 €/m².

In the case of carbon fibre reinforced polymers (CFRP), Sika-CarboDur laminates were selected. The elastic modulus was 170 GPa and the tensile strength $f_{u,f}$ was 2300 MPa. The laminates were supplied with a thickness of 1.2 mm (case 50 and 80 mm wide laminates) and

Table 1: Mechanical properties of GFRP laminates.

Material	Fibre weight g/m²	Orient. º	e_r mm	V_f %	V_m %	E N/mm²	$f_{u,f}$ N/mm²
GF-ULE	630	0	0.797	37.2	62.8	27823	426
GF-BXE	300	+-45	0.718	17.7	82.3	14721	225
GF-BXE	315	+-45	0.724	18.6	81.4	15317	234
GF-BXE	446	+-45	0.765	26.4	73.6	20518	314
GF-BXE	600	+-45	0.793	35.5	64.5	26632	407
GF-BXE	800	+-45	0.815	47.3	52.7	34573	529
GF-BE	440	0/90	0.671	13.0	87.0	11545	176
GF-BE	600	0/90	0.718	17.7	82.3	14721	225
GF-BE	850	0/90	0.760	25.1	74.9	19684	301

1.4 mm (for 100 and 120 mm wide laminates). These correspond to laminate models 512E (31.00 €/m), 812E (47.10 €/m), 1014E (69.10 €/m) and 1214E (83.20 €/m), respectively.

The elastic properties of the timber beams were obtained from the Spanish standard, CTE-DB-SE-M, whereas the plastic properties were based on previous research [6]. In this way, average value of ultimate compression strain for timber beams is taken equal to 0.01. Currently the software is only implemented with homogenous and nonhomogeneous glulam properties, but different kind of wood and reinforcement materials will be available shortly.

3 STRUCTURE OF THE GENETIC ALGORITHM

This section presents the formulation of the structural strengthening optimization problem, its transformation into an unconstrained one, the definition of the genetic algorithm and its operators and the treatment of the different constraints.

3.1 Formulation of the structural optimization problem

The general formulation of an optimization problem is to minimize an objective function subjected to a set of constraint functions, as shown in eqns (1), (2). In this case the objective function, noted as $f(x)$, was the cost of the reinforcement materials:

$$\text{Minimize } f(x) = C_{CFRP} + C_{GFRP}. \tag{1}$$

$$\text{subjected to } g_i(x) \leq 0, \quad i = 1,2,\ldots,10. \tag{2}$$

where C_{CFRP} is the cost of the CFRP laminates, C_{GFRP} is the cost of the GFRP and $g_i(x)$ are the constraints considered for this problem.

The first seven constraints deal with ultimate limit states (ULS) of the retrofitted design. Constraint $g_1(x)$ corresponds to the bending capacity of the wood beam at the point of maximum moment along the reinforced zone. Constraint $g_2(x)$ corresponds to the bending capacity at the point of maximum moment along the unreinforced zone. Constraints $g_3(x)$ and $g_4(x)$ correspond, respectively, to the tension failure of the CFRP laminate and the GFRP sheet used for flexural reinforcement at the point of maximum moment. And additional constraint $g_5(x)$ handles the ultimate stress of the CFRP before delamination occurs at the end of the laminate. Constraint $g_6(x)$ corresponds to the shear capacity of the reinforced wood beam at the point of maximum shear force near the supports. Constraint $g_7(x)$ corresponds to the shear capacity of the timber beam along the zone without shear reinforcement.

Constraint $g_8(x)$ assesses the serviceability limit state (SLS) of vertical deformations. Constraints $g_9(x)$ and $g_{10}(x)$ refer to geometric limitations concerning the available beam width for installing the CFRP laminates (next to one another) and the available beam depth for installing the GFRP sheets, respectively. Figure 1 shows the locations where some of the previous design constraints are checked.

3.2 Formulation of an unconstrained optimization problem

The constrained optimization problem was transformed into an unconstrained one. This was done by creating a penalty function $P(x,t)$, which multiplied the original objective function $f(x)$. The penalty function depends on the degree of violation of each of the constraints as

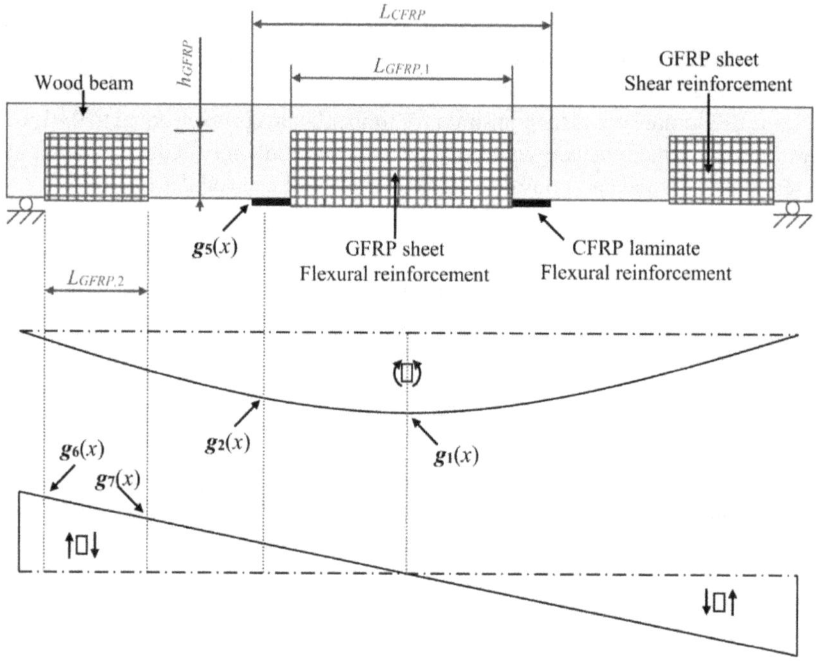

Figure 1: Assessment points of the constraints in the structural optimization problem.

well as the generation number t. The unconstrained optimization problem was thus rewritten in eqn. (3):

$$\text{Minimize } \Phi(x,t) = f(x)P(x,t). \tag{3}$$

The genetic algorithm starts with a randomly generated population of 100 individuals, each encoding one possible solution. This population becomes the first generation (t =1), which consists of j = 1, 2,..., 100 individuals. Each one of these consists of a binary string (Fig. 2) that encodes the reinforcement: type and number of CFRP laminates and length of the CFRP reinforced zone (L_{CFRP}); type and number of layers of GFRP sheets, fibre orientation, length and depth of reinforcement ($L_{GFRP,1}$, $L_{GFRP,2}$ and h_{GFRP}, see Fig. 1). The aim of the penalty function $P(x,t)$ is the following: when a given individual-chromosome encodes an inexpensive solution which does not satisfy one or several constraints, then its cost gets penalized and the genetic algorithm is less likely to select that individual for the genetic operators.

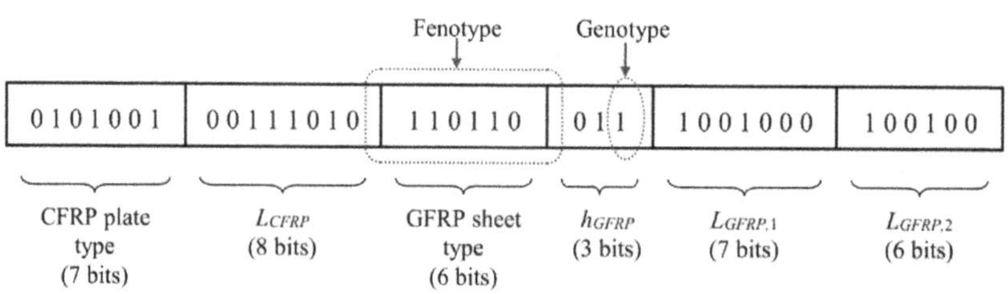

Figure 2: Example of chromosome that encodes a possible solution.

For this optimization problem it was found that an adaptative penalty function $P(x,t)$ of the type proposed by Gen and Cheng [7] and Perera and Varona [4] worked to satisfactory results. The adaptative penalty function is calculated through eqns (4) and (5):

$$P(x,t) = 1 + \frac{1}{m}\sum_{i=1}^{m}\left(\frac{\Delta b_i}{b_i(t)}\right)^k. \tag{4}$$

$$\Delta b_i = \max\{0\,;\, g_i(x) - b_i(t)\} \qquad \text{with} \quad b_i(t) = \frac{0.01}{t+1}. \tag{5}$$

where a value of k = 2 has been used in eqn (4). The parameter $b_i(t)$ is a penalty threshold, which, thanks to eqn (5), gets lower as the number of generations progresses.

Once function $\Phi(x,t)$ is calculated for each individual, a fitness parameter S_j is defined for each *j*-th individual through eqn (6):

$$S_j = 1 + \frac{\Phi_{\min,t} - \Phi_j(x,t)}{\Phi_{\max,t}}. \tag{6}$$

where $\Phi_{\min,t}$ and $\Phi_{\max,t}$ are, respectively, the minimum and maximum values of function $\Phi(x,t)$ within the *t*-th generation.

3.3 The genetic operators

A genetic algorithm has three basic operators: selection, crossover and mutation. The first one, selection, depends on the fitness values. The probability of selection $p_{S,j}$ for the *j*-th individual is defined in eqn (7):

$$p_{S,j} = \frac{S_j}{\sum_{j=1}^{100} S_j}. \tag{7}$$

The higher the value of $p_{S,j}$ the more likely will be the *j*-th individual eligible for reproduction and preservation of its genetic material (i.e. reinforcement type and dimensions) onto generation t + 1. Therefore, the selected individuals become parents for generation t+1.

The reproduction operator is performed on a pair of parents and works combining the genetic material of the parents, creating two new individuals (possible optimum solutions) which are designated as children. This operator is also known as crossover operator. For this particular genetic algorithm a reproduction probability of 70% has been used, which means that in some cases (30% approximately) the genetic material of the parents will not be combined and the pair of children will be identical to the pair of parents. In the rest of cases, the combination of genetic material is made randomly through two different methods: fenotype crossover and flat crossover (Radcliffe *et al.* [8]).

The last operator, mutation of genetic material, is not likely to produce more feasible children than their parents, but it must be used to avoid convergence to local optima. For the first 10 generations mutation probability varies from 20% to 15%; between the 11-th generation and the 50-th generation, mutation decreases from 15% to 5%. The mutation operator is performed at genotype level (see Fig. 2), whereas the crossover operator could work at a fenotype level.

From the 51-th generation onwards, mutation is set at a minimum of 5% but adaptative mutation scheme was added, making the GA able to detect whether there is little genetic diversity within a given t-th generation. This could happen in the event of converging to a local optimum. The genetic diversity D_t of generation t is defined as the quotient of the maximum value of the penalized objective function within the population $\Phi_{max,t}$ and its mean value $\Phi_{mean,t}$. Depending on D_t mutation probability will be corrected as shown in eqn (8):

$$\text{mutation probability} = \begin{cases} 13\%, & \text{if } D_t = 1 \\ 10\%, & \text{if } 1 < D_t \leq 1.05 \\ 8\%, & \text{if } 1.05 < D_t \leq 1.1 \end{cases} . \tag{8}$$

Finally, an elitism criterion has been applied, which means that generation t+1 will be formed by the best individuals from the group that includes the parents selected from generation t and their children (i.e. if a child is not a better solution for the optimization problem than a given parent, it does not enter generation t+1).

4 DESIGN CONSTRAINTS

To perform this research, two standards have been used to study bending and shear behaviour of reinforced timber beams: the Spanish standard CTE-DB SE-M and the Italian standard CNR DT201. Besides, for analysing the delamination failure, several previous research works were addressed as well. These design recommendations are based on limit-states design principles and allow plastic behaviour in compression. This approach sets acceptable levels of safety against the occurrence of both serviceability limit states and ultimate limit states. To assess these limit states, certain values for the design loading and the design strength of the materials must be assumed. Load factors and strength reduction factors stated are the ones in Spanish standard (which is harmonized to the European standard Eurocode 5). For example, eqn (9) shows the bending strength f_{wd} for timber beams:

$$f_{wd} = k_{mod} \cdot k_{sys} \cdot k_h \cdot \frac{f_{wk}}{\gamma_m}. \tag{9}$$

The algorithm automatically selects the characteristic strength f_{wk} of the wood according to its strength class. The software also selects the correct load factor k_{mod} according the load duration, the load sharing factor k_{sys}, the height factor k_h and the material partial safety factor γ_m. Other factors related to the type of load (dead load, variable load, etc.) are also used in the design application.

In the assessment of constraints, $g_1(x)$ and $g_2(x)$, which correspond to the bending capacity, the following assumptions have been considered: wood is considered an isotropic material, having tree planes of symmetry; shear effect is neglected and plane sections remain plane; the stress-strain relationships assumed for wood are elastic-plastic (in compression) and linear-elastic (in tension).

In the case of GFRP and CFRP, linear-elastic behaviour is considered for both compression and tension stress. For the case of a FRP reinforced timber beam in bending, Fig. 3 shows the stress and strain distributions of the composite cross-section. The ultimate bending moment of the section M_u has been obtained assuming strain compatibility and equilibrium of internal forces and moments. In Fig. 3, e_{rG} and e_{rC} are the laminate thickness for GFRP and

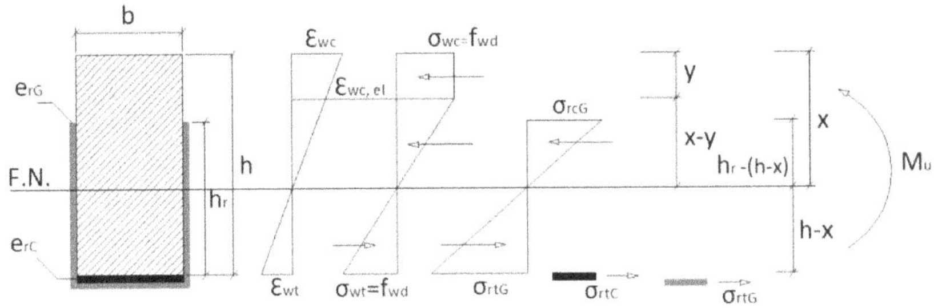

Figure 3: composite FRP-strengthened section in bending.

CFRP, respectively; σ_{rcG} is the compression stress in the GFRP; σ_{rtG} and σ_{rtC} are the tensile stresses in the GFRP and in the CFRP laminate, respectively.

After calculating the flexural capacity M_u of the strengthened timber beam and the factored value of the maximum bending moment at mid-span $M_{Ed,\max}$, constraint $g_1(x)$ can be obtained through eqn (10):

$$M_{Ed,\max} \leq M_u \quad \rightarrow \quad \frac{M_{Ed,\max}}{M_u} \leq 1 \quad \rightarrow \quad g_1(x) = \left(\frac{M_{Ed,\max}}{M_u} - 1 \right) \leq 0. \tag{10}$$

An analogous approach must be followed with constraints $g_2(x)$ to $g_{10}(x)$.

Constraints $g_3(x)$ and $g_4(x)$ correspond to the tension failure of laminate. In this case, the ultimate tensile strain of wood is equal to the maximum value of the laminate strain. Therefore, the stress level of the laminates will be around 14% of its ultimate stress. Thus, these constraints are not likely to be active during the optimization process. However, in future improvements of the algorithm, voids and defects in the wood, could change this hypothesis.

Constraints $g_5(x)$, $g_6(x)$ and $g_7(x)$ correspond to the shear capacity of wood beams. In these cases, two different analyses have been performed. On the one hand, the value of resisting shear force V_{Rd} was calculated by transforming the FRP reinforcement to equivalent wood, according to Triantafilou *et al.* [1], and using the formulation of Collignon-Jourawski; the ultimate strength was then calculated according to the Spanish standard. The equivalent wood thickness of the GFRP is notated as $e_{h,rG}$ and is equal to the GFRP thickness e_{rG} times the quotient of the laminate elastic modulus and the wood elastic modulus. In this case, it has been assumed that no other failure mechanism such as FRP shear failure occurs prior to wood shear failure. On the other hand, the second possible failure mechanism is due to debonding of the CFRP laminate. In this case, the maximum laminated normal stress $\sigma_{rtC,\max}$ was calculated according to Juvandes *et al.* [9], as shown in eqn (11):

$$\sigma_{rtC,max} = \frac{c_1 \cdot k_b \cdot k_c \cdot \sqrt{E_{rC} \cdot e_{rC} \cdot f_{wtm,p}} \cdot k_\mu}{e_{rC}}. \tag{11}$$

where c_1, k_c and k_μ are experimental factors equal to 0.76, 1, and 1, respectively; k_b is a geometrical function; E_{rC} is the elastic modulus of the CFRP laminate; and $f_{wtm,p}$ is the ultimate pull-off stress, equal to 2.5 MPa. Currently, the algorithm considers the maximum anchor length, according to the properties of the reinforcement. Moreover, this check is performed only for CFRP laminates, due to the lack of information about experimental test for debonding in wrap beams with GFRP.

In addition to the ultimate limit states, the serviceability of a member under service loads should satisfy the provisions of Spanish standard CTE-DB-E-M. This is verified through constraint $g_8(x)$ for the following three cases: structural integrity, user comfort and functionality. The calculation method of these three different provisions is indicated in the Spanish standard.

Finally, constraint $g_9(x)$ corresponds to the ratio between the total width of the CFRP plates and that of the timber beam cross-section. In the same way, constraint $g_{10}(x)$ corresponds to the ratio between the perimeter of GFRP and the roll width of fiberglass fabric.

5 APPLICATION OF THE GENETIC ALGORITHM

This section shows the application of the genetic algorithm to evaluate the efficiency of different schemes of reinforcing to simply supported timber beams.

The first example consists of a 3 m span beam, with a 0.08×0.16 cross section (width×depth). It is a glulam GL24h timber beam subjected to a fixed distributed dead load of 0.825 kN/m, plus a distributed live load. The unreinforced timber beam is able to support a live load of 2.475 kN/m according to CTE-DB-SE-M. The live load is increased at 10% intervals and the optimum reinforcing scheme for each of them is obtained with the genetic algorithm (10 runs are performed for each case). Table 2 shows the optimum designs. The notation used in Table 2 is the following: L_C is the length of the CFRP laminate, which is centred at mid span; H_G is the depth of the GFRP reinforcing, on the lateral faces of the timber beam; $L_{G,A}$ is the length of the GFRP from the beam ends, used for reinforcing in shear; $L_{G,C}$ is the length of the GFRP located at mid span (analogous to L_C); finally, the cost is given as a unit cost, comparing the total cost of reinforcing materials to that of the timber beam (hence the units, in €/€). As it was introduced in Table 1, BXE stands for bidirectional fibres (±45°) whereas ULE stands for fibres with single orientation.

When SLS are active for aesthetic reasons, the algorithm converges to an expensive solution installing CFRP laminates, which is the most effective way of increasing the inertia of the composite cross section and thus reducing the maximum vertical displacement. However, when SLS is not active, Table 2 shows that GFRP reinforcing for flexural bending is the less expensive scheme. In this case it also becomes clear that shear reinforcement is not needed

Table 2: Optimum solutions for Example 1: 3 m simply supported beam.

		CFRP		**GFRP**					
	Δq (%)	Type	L_C (mm)	Type	n. layers	H_G (mm)	$L_{G,A}$ (mm)	$L_{G,C}$ (mm)	**Cost** (€/€)
SLS + ULS	10	1×512E	450	ULE630	2	90	0	480	0.85
	20	1×512E	800	ULE630	3	90	0	730	1.71
	30	1×512E	1120	ULE630	3	90	0	1180	2.58
	40	1×512E	1250	ULE630	3	160	0	1320	3.53
ULS	10	-	-	BXE300	1	90	0	370	0.14
	20	-	-	BXE446	1	90	0	820	0.32
	30	-	-	ULE630	1	90	0	1080	0.42
	40	-	-	ULE630	2	90	0	1280	1.00
	50	-	-	ULE630	2	90	0	1420	1.11
	60	-	-	ULE630	3	90	0	1560	1.83

(because $L_{G,A}$ is null in all cases). Finally, for live load increments over 60% the algorithm did not find any feasible solutions: in the event of retrofitting a timber beam with the aforementioned characteristics for a live load increment over 60% its original value, strengthening solutions different from FRP materials should be taken into account

Tables 3 and 4 show the solutions obtained when applying the GA to simply supported timber beams of 2.5 and 2 m span, respectively. In the case of the 2.5 m beam, the original distributed live load for the unreinforced design was 4.125 kN/m. In the case of the 2 m beam, the original distributed live load was 6.49 kN/m.

The purpose of analysing shorter beams is forcing the algorithm to deal with designs which can be partially governed by shear failure. In the case of the 2 m span beam, there were no differences found between the optimum designs for ULS and SLS combined and those corresponding to ULS only; vertical deformations did not control the strengthening solution and they would not constitute critical constraints in short beams.

A remarkable result is that of the 2.5 m span beam retrofitted for an increase of 50% its original live load, with both ULS and SLS constraints active: this was found to be the most expensive design, needing a wider CFRP laminate as well as GFRP reinforcements for both bending and shear.

Finally, the smallest possible live load increment corresponded to the shortest beam (Table 4). FRP strengthening solutions would not be adequate for a retrofitting demanding an increase of over 30% the original distributed live load.

Next examples are of beams subjected to fixed values of dead and live distributed loads, but with increasing lengths. In example 4 the original design is the same one as in example 1: 0.08×0.16 m rectangular cross section; glulam GL24h wood; subjected to distributed loads of 0.825 and 2.475 kN/m (dead and live values, respectively). The maximum span possible for the unstrengthened design is 3 m.

Table 5 shows the results for increasing lengths in example 4. Again, CFRP became necessary only when SLS was active as a design constraint. Considering ULS only, the maximum span length compatible with FRP strengthening would be 3.6 m (a 20% higher than 3 m). With that range of lengths (and for the 0.08×0.16 m glulam GL24h cross section), shear

Table 3: Optimum solutions for Example 2: 2.5 m simply supported beam.

		CFRP		**GFRP**					
	Δq (%)	Type	L_C (mm)	Type	n. layers	H_G (mm)	$L_{G,A}$ (mm)	$L_{G,C}$ (mm)	**Cost** (€/€)
	10	1×512E	220	ULE630	1	90	0	790	0.65
SLS	20	1×512E	680	ULE630	2	90	0	990	1.80
+	30	1×512E	1010	ULE630	2	160	130	1190	3.22
ULS	40	1×512E	1400	ULE630	3	90	220	1450	4.45
	50	1×812E	1530	ULE630	3	140	290	1580	7.47
	10	-	-	BXE446	1	90	0	480	0.31
	20	-	-	ULE630	1	160	0	970	0.65
ULS	30	-	-	ULE630	2	90	130	1170	1.34
	40	-	-	ULE630	3	90	220	1320	2.48
	50	1×812E	390	ULE630	3	90	290	1430	3.59

Table 4: Optimum solutions for Example 3: 2 m simply supported beam.

	Δq (%)	CFRP Type	L_C (mm)	GFRP Type	n. layers	H_G (mm)	$L_{G,A}$ (mm)	$L_{G,C}$ (mm)	Cost (€/€)
SLS	5	-	-	BXE300	1	90	80	0	0.09
+	10	-	-	ULE630	1	90	130	360	0.36
ULS	15	-	-	BXE446	1	90	170	550	1.04
	20	-	-	ULE630	2	90	220	690	1.33
ULS	25	-	-	ULE630	2	90	250	790	1.52
only	30	-	-	ULE630	3	90	310	910	2.70

Table 5: Optimum solutions for Example 4: $q = 2.475$ kN/m; original span length of 3 m.

	ΔL (%)	CFRP Type	L_C (mm)	GFRP Type	n. layers	H_G (mm)	$L_{G,A}$ (mm)	$L_{G,C}$ (mm)	Cost (€/€)
SLS	5	1×512E	760	ULE630	3	160	0	810	2.03
+	10	1×512E	1680	BXE800	3	160	0	1550	4.42
ULS									
	5	-	-	BXE300	1	90	0	640	0.28
ULS	10	-	-	ULE630	1	140	0	1200	0.67
	15	-	-	ULE630	3	90	0	1610	1.33
	20	-	-	ULE630	3	90	0	1930	2.29

Table 6: Optimum solutions for Example 5: $q = 6.49$ kN/m; original span length of 2 m.

	ΔL (%)	CFRP Type	L_C (mm)	GFRP Type	n. layers	H_G (mm)	$L_{G,A}$ (mm)	$L_{G,C}$ (mm)	Cost (€/€)
SLS	5	-	-	BXE300	1	90	80	430	0.38
+	10	-	-	ULE630	1	130	150	810	0.89
ULS	15	1×512E	760	ULE630	2	90	220	1100	3.18
	20	1×512E	1280	ULE630	3	90	280	1340	5.43
	5	-	-	BXE300	1	90	80	430	0.38
ULS	10	-	-	ULE630	1	130	150	810	0.89
	15	-	-	ULE630	2	90	220	1100	1.90
	20	-	-	ULE630	3	90	280	1340	3.38

strengthening would not be needed. When considering GFRP sheets for flexural reinforcement (with SLS not active), the tendency of the genetic algorithm is to start increasing the height H_G of the 'U' sheets; however, for longer spans (with moments increasing proportionally to the square of the span), the genetic algorithm finds it more adequate to maintain a shorter height H_G and increase the number of layers.

Example 5 starts with the same cross section, with a short span of 2 m and fixed loads of 0.825 kN/m and 6.49 kN/m (dead and live, respectively). Optimum results for increasing span lengths are listed in Table 6. The most expensive solution corresponds to a design that needs reinforcement for both SLS (vertical deformations) and ULS (shear failure), with a span length increase of 20%.

6 CONCLUSIONS

A genetic algorithm for optimum design of bending and shear FRP reinforcement of low cost timber beams has been presented. This GA serves the purpose of minimizing the material cost associated with the reinforcement materials for ULS and SLS requirements according to the Spanish standards.

Adaptive schemes for both the penalty function and the mutation operator showed adequate performance when applying the GA to this particular design problem.

The analysis of the solutions found by the GA for a set of practical examples showed that for structures with flexural failure behaviour, the ultimate load can be increased up to 50-60% over the ultimate load for the unreinforced beam. However, for beams with shear failure controlled behaviour, the ultimate load should not be increased over 30% of the unreinforced original load, due to the failure on the supports area.

Finally, GFRP sheets should be considered not only for shear strengthening but also for flexural strengthening, since they allow reducing the amount of CFRP laminate needed and so, they allow to reduce the retrofitting costs. The GA proved effective on deciding whether this could be achieved by using higher 'U' sheets or by adding layers of FRP.

ACKNOWLDEGEMENTS

The authors express deep gratitude to GRESMES Research Group at the University of Alicante for their useful assistance. This work was partially financed by the University of Alicante by means of the GRE12-04 Research Project and Generalitat Valenciana, grant GV/2014/079.

REFERENCES

[1] Triantafillou, T.C., Shear reinforcement of wood using FRP materials. *Journal of Materials in Civil Engineering*, **9**(2), pp. 22–25, 1997. https://doi.org/10.1061/(asce)0899-1561(1997)9:2(65)

[2] Bru, D., Baeza, F.J., Varona, F.B., García-Barba, J. & Ivorra, S., Static and dynamic properties of retrofitted timber beams using glass fibre reinforced polymers. *Materials and Structures*, **49**(1), pp. 181–191, 2014. https://doi.org/10.1617/s11527-014-0487-0

[3] Holland, J.H., *Adaptation in natural and artificial systems*, MIT Press: Cambridge, Mass., 1975.

[4] Perera, R. & Varona, F.B., Flexural and shear design of FRP plated RC structures using a genetic algorithm. *ASCE Journal of Structural Engineering*, **135**(11), pp. 1418–1429, 2009. https://doi.org/10.1061/(asce)0733-9445(2009)135:11(1418)

[5] Bru, D., Baeza, F.J., Varona, F.B. & Ivorra, S., *Numerical and experimental evaluation of FRP reinforcement on the mechanical behaviour of timber beams*. Proceedings of the 16th European Conference on Composite Materials, pp. 1–8, 2014.

[6] Yang, Y., Liu, J. & Xiong, G., Flexural behaviour of wood beams strengthened with HFRP. *Construction and Building Materials*, **43**, pp. 118–124, 2013. https://doi.org/10.1016/j.conbuildmat.2013.01.029
[7] Gen, M. & Cheng, R., *A survey of penalty techniques in genetic algorithms*. Proceedings of the 1996 International Conference on Evolutionary Computation, pp. 804–809, 1996.
[8] Radcliffe, N.J., Equivalence class analysis of genetic algorithms. *Complex Systems*, **5**, pp. 183–205, 1991.
[9] Juvandes, L.F.P. & Barbosa, R.M.T., Bond analysis of timber structures strengthened with FRP systems. *Strain: An International Journal for Experimental Mechanics*, **48**, pp. 124–135, 2012.
https://doi.org/10.1111/j.1475-1305.2011.00804.x

SIMULATION OF THE VIBRATIONS OF A NON-UNIFORM BEAM LOADED WITH BOTH A TRANSVERSELY AND AXIALLY ECCENTRIC TIP MASS

DESMOND ADAIR[1] & MARTIN JAEGER[2]
[1]School of Engineering, Nazarbayev University, Republic of Kazakhstan.
[2]School of Engineering & ICT, University of Tasmania, Australia.

ABSTRACT

The main purpose of this work is to employ the Adomian modified decomposition method (AMDM) to calculate free transverse vibrations of non-uniform cantilever beams carrying a transversely and axially eccentric tip mass. The effects of the variable axial force are taken into account here, and Hamilton's principle and Timoshenko beam theory are used to obtain a single governing non-linear partial differential equation of the system as well as the appropriate boundary conditions. Two product non-linearities result from the analysis and the respective Cauchy products are computed using Adomian polynomials. The use of AMDM to make calculations for such a cantilever beam/tip mass arrangement has not, to the authors' knowledge, been used before. The obtained analytical results are compared with numerical calculations reported in the literature and good agreement is observed. The qualitative and quantitative knowledge gained from this research is expected to enable the study of the effects of an eccentric tip mass and beam non-uniformity on the vibration of beams for improved dynamic performance.
Keywords: Adomian decomposition method, cantilever beam, mechanical vibrations, tip mass

1 INTRODUCTION

Important components of engineering structures can be idealized as cantilever beams with a concentrated mass at its free end. Examples of this are antenna structures, large aspect aircraft wings fitted with external tanks or stores, flexible robot arms and atomic force microscopes. The presence of the tip mass plays an important role in the dynamic characteristics of the beam and exerts an inertial force, which is a function of the system motion [1]. The problem of a uniform cantilever beam with an eccentric tip mass has received much attention, for example, Prescott [2], Goel [3], and, Rama Bhat and Wagner [4]. Exact solutions have been found, for example, To [5] derived an exact expression for natural frequencies and mode shapes of a uniform beam with base excitation and an axially eccentric tip mass and Auciello [6] has suggested an exact analysis of free vibration of a linearly tapered cantilever beam with tip mass with rotary inertia and axial eccentricity. Work has been carried out for non-uniform beams which can achieve a better distribution of strength and weight. The non-uniformity, which usually arises from a variable beam cross section or inhomogeneous material properties, leads to a fourth-order partial differential equation with variable coefficients. In general these fourth-order equations cannot be solved analytically. However, for some specific non-uniform beams, exact solutions of the eigenvalue problem have been reported in terms of power series [7], orthogonal polynomials [8] and Bessel functions [9]. More recent work has been carried on beams with tip mass and with uniform [10] and non-uniform [1] cross sections. Matt [10] simulated transverse vibrations of a cantilever beam, with an eccentric tip mass, in the axial direction using integral transforms which included an implicit filter scheme. Malaeke and Moeenfard [1] modelled large amplitude flexural-extensional free vibrations of non-uniform beams carrying a transversely and axially eccentric tip mass with the effects of the variable axial force taken into account.

DOI: 10.2495/CMEM-V6-N4-679-690

The method of solution here is the Adomian modified decomposition method, which is a wide ranging method of solution of problems involving algebraic, differential [11], integro-differential [12] and partial differential equations [13]. Specific to this work, the Adomain decomposition and Adomian modified decomposition method have been used by several groups [14, 15] for uniform and non-uniform beams, starting with either the Euler-Bernoulli or Timoshenko formulations. Mao [14] applied the AMDM to rotating uniform beams and included a centrifugal stiffening term while Adair and Jaeger [15] applied the AMDM to rotating non-uniform beams which also included a centrifugal stiffening term. Yaman [16] has used the Adomian decomposition method to investigate the influence of the orientation effect on the natural frequency of a cantilever beam carrying a tip mass.

In this study we use the computational approach of AMDM to analyze free transverse vibrations of non-uniform cantilever beams carrying a transversely and axially eccentric tip mass. As the beam is assumed to be under the effect of gravity, the spatial variation of axial force is taken into account. The natural frequencies and mode shapes of the beam for three test cases were found and compared with results reported in the literature.

2 MATHEMATICAL FORMULATION

The schematic view of a non-uniform beam with an eccentric tip mass is shown in Fig. 1. The distances a and b are the transverse (z) and axial (x) eccentric values, respectively, CoG is the centre of gravity of the tip mass M and l is the length of the beam. The polar mass moment of inertia of the tip mass around CoG is J and, in this analysis, the axial stretching of the beam due to both gravitational and inertial axial forces is considered.

In order the find the governing equations of motion of the beam, Hamilton's principle is employed. Every system behaves in such a way to satisfy

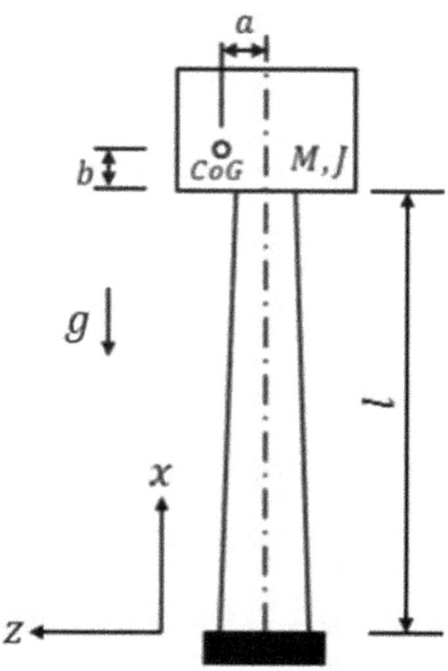

Figure 1: Schematic view of a non-uniform cantilever beam with an eccentric tip mass.

$$\delta\int_{t_1}^{t_2}\left(K-U+W_e\right)dt=0, \tag{1}$$

where δ is the variation operator, t_1 and t_2 are arbitrary times, K and U are the total kinetic and potential energies of the system respectively and W_e is the work done by external forces.

For this system, the Euler-Bernoulli assumption that the rotation of cross-sections of the beam is small in comparison to the bending deformation is made. Also it is assumed that the angular distortion due to shear is considered small compared to the bending deformation and can be ignored. Therefore, the beam in question is long and slender with the length much greater than the thickness. With the assumption that the material of the beam is linear elastic, then the overall strain energy can be found from

$$U=\frac{1}{2}\int_0^l EA(x)\left(\frac{\partial u}{\partial x}+\frac{1}{2}\left(\frac{\partial w}{\partial x}\right)^2\right)^2 dx+\frac{1}{2}\int_0^l EI(x)\left(\frac{\partial^2 w}{\partial x^2}\right)^2 dx, \tag{2}$$

where $A(x)$ and $I(x)$ are the area and the second moment of area of the beam's cross-section around the neutral axis and E is Young's modulus of elasticity of the beam's material.

The total kinetic energy of the system is given as $K=K_{column}+K_{mass}$, where

$$K_{column}=\frac{1}{2}\int_0^l \rho A(x)\left(\frac{\partial w}{\partial t}\right)^2 dx+\frac{1}{2}\int_0^l \rho A(x)\left(\frac{\partial u}{\partial t}\right)^2 dx, \tag{3}$$

$$\begin{aligned} K_{mass}&=\frac{1}{2}MV_{CoG}^2+\frac{1}{2}J_{CoG}\dot{\theta}^2=\frac{1}{2}M\left(\left(\frac{dw(l,t)}{dt}\right)^2+\left(a^2+b^2+r_G^2\right)\left(\frac{\partial^2 w}{\partial x\partial t}(l,t)\right)^2\right.\\ &\left.+2b\frac{dw(l,t)}{dt}\frac{\partial^2 w}{\partial x\partial t}(l,t)\right)+\frac{1}{2}M\left(\left(\frac{du(l,t)}{dt}\right)^2-2a\frac{du(l,t)}{dt}\frac{\partial^2 w}{\partial x\partial t}(l,t)\right). \end{aligned} \tag{4}$$

Here, ρ is the material density of the beam, r_G is the radius of gyration of the tip mass, and $\dot{\theta}$ and V_{CoG} are the rotational and translational velocities of the tip mass and its centre of gravity, respectively. Here V_{CoG} is obtained using the relative velocity between the tip mass's centre of gravity and the point of attachment to the beam.

The virtual work done on the system with the assumption that no viscous damping is present is

$$\delta W_e=\delta W_{g_1}+\delta W_{g_2}, \tag{5}$$

where δW_{g_1} and δW_{g_2} are the virtual gravitational work done on the tip mass and the beam, respectively. The virtual gravitational work done on the system can be written as

$$\delta W_{g_1}=-Mg\delta u(l,t)-Mga\delta\left(\frac{\partial w}{\partial x}(l,t)\right), \delta W_{g_2}=-\int_0^l g\rho A(x)\delta u(x,t)dx. \tag{6}$$

Hamilton's principle is now applied to obtain the governing equations of vibration of the system

$$\frac{\partial^2}{\partial x^2}\left(EI(x)\frac{\partial^2 w(x,t)}{\partial x^2}\right)-\frac{\partial}{\partial x}\left(N(x,t)\frac{\partial w(x,t)}{\partial x}\right)+\frac{\partial}{\partial t}\left(\rho A(x)\frac{\partial w(x,t)}{\partial t}\right)=0. \tag{7}$$

$$\frac{\partial}{\partial t}\left(\rho A(x)\frac{\partial u(x,t)}{\partial t}\right)-\frac{\partial N(x,t)}{\partial x}+g\rho A(x)=0, \tag{8}$$

The variable axial force of the beam, $N(x,t)$ is defined as

$$N(x,t)=EA(x)\left(\frac{\partial u(x,t)}{\partial x}+\frac{1}{2}\left(\frac{\partial w(x,t)}{\partial x}\right)^2\right). \tag{9}$$

For a long and slender beam, with a small slenderness ratio, the longitudinal inertia is small compared to the restoring force and so

$$\frac{\partial N(x,t)}{\partial x}\approx g\rho A(x). \tag{10}$$

Substitution for the terms $N(x,t)$ and $\partial N(x,t)/\partial x$, as defined in eqns (9) and (10) into the expanded form of eqn (7) gives

$$\begin{aligned}&\left[EI(x)\right]\frac{\partial^4 w(x,t)}{\partial x^4}+2\left[EI(x)\right]'\frac{\partial^3 w(x,t)}{\partial x^3}+\left[EI(x)\right]''\frac{\partial^2 w(x,t)}{\partial x^2}-g\rho A(x)\frac{\partial w(x,t)}{\partial x}\\&-EA(x)\left(\frac{\partial u(x,t)}{\partial x}\frac{\partial^2 w(x,t)}{\partial x^2}+\frac{1}{2}\left(\frac{\partial w(x,t)}{\partial x}\right)^2\frac{\partial^2 w(x,t)}{\partial x^2}\right)+\rho A(x)\frac{\partial^2 w(x,t)}{\partial t^2}=0,\end{aligned} \tag{11}$$

where the superscript $'$ mean differentiation w.r.t. x.

According to modal analysis for harmonic free vibration, $w(x,t)$ and $u(x,t)$, can be separable in space and so eqn (11) can be reduced to

$$\begin{aligned}&\left[EI(x)\right]\frac{\mathrm{d}^4 w(x)}{\mathrm{d}x^4}+2\left[EI(x)\right]'\frac{\mathrm{d}^3 w(x)}{\mathrm{d}x^3}+\left[EI(x)\right]''\frac{\mathrm{d}^2 w(x)}{\mathrm{d}x^2}-g\rho A(x)\frac{\mathrm{d}w(x)}{\mathrm{d}x}\\&-EA(x)\left(\frac{\mathrm{d}u(x)}{\mathrm{d}x}\frac{d^2 w(x)}{\mathrm{d}x^2}+\frac{1}{2}\left(\frac{\mathrm{d}w(x)}{\mathrm{d}x}\right)^2\frac{\mathrm{d}^2 w(x)}{\mathrm{d}x^2}\right)-\rho A(x)\omega^2 w(x)=0.\end{aligned} \tag{12}$$

The dimensions of the beam in the y and z direction may be given as

$$d(x)=d_0\left[1+(\alpha_d-1)\frac{x}{l}\right],\ d'(x)=d_0'\left[1+(\alpha_{d'}-1)\frac{x}{l}\right], \tag{13}$$

where $\alpha_d=d_1/d_0$, $a_{d'}=d_1'/d_0'$; d_0, d_0' are the cross-sectional widths at $x=0$ and d_1, d_1' are the cross-sectional widths at $x=l$. The area and moment of inertia of the beam vary according to

$$A(x)=A_0\left[1-\beta_{d'}\frac{x}{l}\right]\left[1-\beta_d\frac{x}{l}\right], I(x)=I_0\left[1-\beta_{d'}\frac{x}{l}\right]\left[1-\beta_d\frac{x}{l}\right]^3, \tag{14}$$

where $A_0=d_0d_0'$ and $I_0=d_0'd_0^3/12$ are the cross-sectional area and moment of inertia at $x=0$, respectively, and $\beta_d=1-\alpha_d$, $\beta_{d'}=1-\alpha_{d'}$.

Without loss of generality, the following dimensionless quantities are now introduced.

$\frac{w}{l}\to\phi, \frac{u}{l}\to\psi, \frac{x}{l}\to\eta, \frac{t}{T}\to\tau, \frac{a}{l}\to\bar{a}, \frac{b}{l}\to\bar{b}, \frac{r_G}{l}\to\bar{r}_G, \frac{gT^2}{l}\to\bar{g}, \frac{M}{l\rho A(0)}\to\bar{M}$, where

$T=l^2\sqrt{\frac{\rho A_0}{EI_0}}$.

$$\lambda_1=\frac{EI(\eta)}{EI_0}, \lambda_2=\frac{A(\eta)}{A_0}, \lambda_3=\frac{EA(\eta)l^2}{EI_0}, \lambda_4=\frac{\rho A_0\omega^2 l^4}{EI_0}, \tag{15}$$

$$\lambda_1'=\left[\frac{EI(\eta)}{EI_0}\right]', \lambda_1''=\left[\frac{EI(\eta)}{EI_0}\right]'' \tag{16}$$

Equation (12) can now be written in the form

$$\begin{aligned}&\frac{d^4\phi(\eta)}{d\eta^4}+2\frac{\lambda_1'}{\lambda_1}\frac{d^3\phi(\eta)}{d\eta^3}+\frac{\lambda_1''}{\lambda_1}\frac{d^2\phi(\eta)}{d\eta^2}-\bar{g}\frac{\lambda_2}{\lambda_1}\frac{d\phi(\eta)}{d\eta}\\&-\frac{\lambda_3}{\lambda_1}\left(\frac{d\psi(\eta)}{d\eta}\frac{d^2\phi(\eta)}{d\eta^2}+\frac{1}{2}\left(\frac{d\phi(\eta)}{d\eta}\right)^2\frac{d^2\phi(\eta)}{d\eta^2}\right)-\omega^2\frac{\lambda_4}{\lambda_1}\phi(\eta)=0.\end{aligned} \tag{17}$$

The governing equation, eqn (17), cannot be solved as a closed-form solution and so some approximate approach must be adopted. Here the Adomian modified decomposition method (AMDM) is chosen.

3 ADOMIAN MODIFIED DECOMPOSITION METHOD

The Adomian decomposition method (ADM) is an iterative method, which has proved successful in treating non-linear equations. It is based on the search for a solution in the form of a series in which the non-linear terms are calculated recursively using the Adomian polynomials. The ADM gives the solutions $\phi(\eta)$ of eqn (17) in a series form of the infinite sum

$$\phi(\eta)=\sum_{j=0}^{+\infty}\phi_j(\eta), \tag{18}$$

Substitution of eqn (18) into the linear and nonlinear terms of eqn (17) yields

$$L\sum_{j=0}^{+\infty}\phi_j(\eta)+R\sum_{j=0}^{+\infty}\phi_j(\eta)-\frac{\lambda_3}{\lambda_1}N_1[\psi,\phi]-\frac{1}{2}\frac{\lambda_3}{\lambda_1}N_2[\phi]=0, \tag{19}$$

where

$$L = \frac{d^4}{\partial \eta^4}, R = \frac{\lambda_1'}{\lambda_1}\frac{\partial^3}{\partial \eta^3} + \frac{\lambda_1''}{\lambda_1}\frac{\partial^2}{\partial \eta^2} - g\frac{\lambda_2}{\lambda_1}\frac{\partial}{\partial \eta} - \omega^2 \frac{\lambda_2}{\lambda_1}. \tag{20}$$

Here L is an invertible operator, which is taken as the highest-order derivative and R is the remainder of the linear operator.

$N_1[\psi,\phi]$ in eqn (19) is a product nonlinearity and the respective Cauchy product can be computed using Adomian polynomials (A_n) as

$$N_1[\psi,\phi] = \sum_{j=0}^{+\infty} A_j(\psi_0,\ldots,\psi_j;\phi_0,\ldots,\phi_j) =$$

$$\frac{\partial \psi}{\partial \eta}\frac{\partial^2 \phi}{\partial \eta^2} = \sum_{j=0}^{+\infty}\frac{\partial \psi_n}{\partial \eta}\sum_{m=0}^{+\infty}\frac{\partial^2 \phi_m}{\partial \eta^2} = \sum_{j=0}^{+\infty}\sum_{m=0}^{j}\frac{\partial \psi_m}{\partial \eta}\frac{\partial^2 \phi_{j-m}}{\partial \eta^2}.$$

$N_2[\phi]$ in eqn (19) is also a product non-linearity, and accordingly the respective Cauchy product can be computed using Adomian polynomials (B_n) as

$$N_2[\phi] = \sum_{j=0}^{+\infty} B_j(\phi_0,\ldots,\phi_j) = \left(\frac{\partial \phi}{\partial \eta}\right)^2 \frac{\partial^2 \phi}{\partial \eta^2} = \frac{\partial^2 \phi}{\partial \eta^2}\left(\frac{\partial \phi}{\partial \eta}\right)^2 =$$

$$\sum_{j=0}^{+\infty}\frac{\partial^2 \phi_n}{\partial \eta^2}\sum_{m=0}^{+\infty}\frac{\partial \phi_m}{\partial \eta}\sum_{l=0}^{+\infty}\frac{\partial \phi_l}{\partial \eta} = \sum_{j=0}^{+\infty}\frac{\partial^2 \phi_n}{\partial \eta^2}\sum_{m=0}^{+\infty}\sum_{l=0}^{+\infty}\frac{\partial \phi_l}{\partial \eta}\frac{\partial \phi_{m-l}}{\partial \eta} = \tag{22}$$

$$\sum_{j=0}^{+\infty}\sum_{m=0}^{j}\frac{\partial^2 \phi_m}{\partial \eta^2}\sum_{l=0}^{j-m}\frac{\partial \phi_l}{\partial \eta}\frac{\partial \phi_{j-m-l}}{\partial \eta}.$$

According to the Adomian modified decomposition method (AMDM) the terms $\phi_j(\eta)$ in eqn (18) can be expressed as $C_j\eta^j$ so giving the infinite series

$$\phi(\eta) = \sum_{j=0}^{+\infty} C_j\eta^j, \tag{23}$$

where the unknown coefficients C_j are determined concurrently. Using a similar approach $\psi(\eta)$ can also be equated to the series $\sum_{j=0}^{+\infty} C_j'\eta^j$.

Using the linear operator L of Eq. (20) then the inverse operator of L is a 4-fold integral operator defined as

$$L^{-1} = \int_0^\eta\int_0^\eta\int_0^\eta\int_0^\eta(\cdots)\,d\eta\, d\eta\, d\eta\, d\eta. \tag{24}$$

Equation (19) can be written as

$$\phi(\eta) = \Phi(\eta) - L^{-1}\left[R\sum_{j=0}^{+\infty} C_j\eta^j + \frac{\lambda_3}{\lambda_1}\sum_{j=0}^{+\infty} A_j\eta^j + \frac{1}{2}\frac{\lambda_3}{\lambda_1}\sum_{j=0}^{+\infty} B_j\eta^j\right], \tag{25}$$

where R is now

$$R=\frac{\lambda_1'}{\lambda_1}\sum_{j=0}^{+\infty}(j+1)(j+2)(j+3)C_{j+3}\eta^j+\frac{\lambda_1''}{\lambda_1}\sum_{j=0}^{\infty}(j+1)(j+2)C_{j+2}\eta^j$$
$$-\bar{g}\frac{\lambda_2}{\lambda_1}\sum_{j=0}^{\infty}(j+1)C_{j+1}\eta^j-\omega^2\frac{\lambda_4}{\lambda_1}\sum_{j=0}^{\infty}C_j\eta^j, \quad (26)$$

The Adomian's polynomials in eqn (25) are found using the standard method [17] for $B_j(\phi_0,\ldots,\phi_j)$ and the generalized method for several variables as reported by Adomian and Rach [18] is used for $A_j(\psi_0,\ldots,\psi_j;\phi_0,\ldots,\phi_j)$.

The term $\Phi(\eta)$ found in eqn (25) is the initial term and defined as

$$\Phi(\eta)=\sum_{j=0}^{3}C_j\eta^j=\phi(0)+\phi'(0)\eta+\phi''(0)\eta^2/2+\phi'''(0)\eta^3/6 \quad (27)$$

On substituting eqns (23), (26) and (27) into eqn (25), the following is obtained

$$\sum_{j=0}^{\infty}C_j\eta^j=\phi(0)+\phi'(0)\eta+\phi''(0)\eta^2/2+\phi'''(0)\eta^3/6$$
$$-\sum_{j=0}^{\infty}\left\{\frac{\eta^{j+4}}{(j+1)(j+2)(j+3)j+4)}\right.$$
$$\sum_{m=0}^{j}\left[(m+1)(m+2)(m+3)\frac{\lambda'_1}{\lambda_1}C_{m+3}+(m+1)(m+2)\frac{\lambda''_1}{\lambda_1}C_{m+2}\right. \quad (28)$$
$$\left.\left.-(m+1)\bar{g}\frac{\lambda_2}{\lambda_1}C_{m+1}-\omega^2\frac{\lambda_4}{\lambda_1}C_m\right]+\frac{\lambda_3}{\lambda_1}\left(A_j+\frac{1}{2}B_j\right)\right\}$$

The recurrence relation for the coefficients C_j can now be stated and the solution for $\phi(\eta)$ can be calculated from eqn (28). The series solution is $\phi(\eta)=\sum_{j=0}^{\infty}C_j\eta^j$, although all of the coefficients C_j cannot yet be determined, and thus, the solutions must be approximated by the truncated series $\sum_{j=0}^{n-1}C_j\eta^j$, with the successive approximations being $\phi^{[n]}(\eta)=\sum_{j=0}^{n-1}C_j\eta^j$ as n increases and the boundary conditions are met.

Thus $\phi^{[1]}(\eta)=C_0,\phi^{[2]}(\eta)=\phi^{[1]}(\eta)+C_1\eta,\phi^{[3]}=\phi^{[2]}(\eta)+C_2\eta^2$ serve as approximate solutions with increasing accuracy as $n\to\infty$. The three coefficients $C_j\ (j=0,1,2,3)$ depend on the boundary equations. In this case, the two coefficients C_0 and C_1 can be chosen as arbitrary constants, and the other two coefficients C_2 and C_3 are stated in terms of the problem parameters or as the functions of the other coefficients.

The initial term $\Phi(\eta)$ in eqn (25) is a function of C_0 and C_1 and from the recurrence relation obtained from eqn (28) the coefficients $C_j\ (j\geq 4)$ are functions of C_0 and C_1 and λ_4. By substituting $\phi^{[n]}(\eta)$ into the boundary conditions we then have

$$f_{r0}^{[n]}(\lambda_4)C_0 + f_{r1}^{[n]}(\lambda_4)C_1 = 0, r = 1,2. \tag{29}$$

For non-trivial solutions, C_0 and C_1, the frequency equation is given a

$$\begin{vmatrix} f_{10}^{[n]}(\lambda_4) & f_{11}^{[n]}(\lambda_4) \\ f_{20}^{[n]}(\lambda_4) & f_{21}^{[n]}(\lambda_4) \end{vmatrix} = 0 \tag{30}$$

The ith estimated eigenvalue $\lambda_{2(i)}^{[n]}$ corresponding to n is obtained from eqn (30), i.e. the ith estimated dimensionless natural frequency $\Omega_{n(i)}^{[n]} = \sqrt{\lambda_{2(i)}^{[n]}}$ is also obtained and n is determined by the following equation

$$\left| \Omega_{n(i)}^{[n]} - \Omega_{n(i)}^{[n-1]} \leq \varepsilon \right| \tag{31}$$

where $\Omega_{n(i)}^{[n-1]}$ is the ith estimated dimensionless natural frequency corresponding to $n-1$, and ε is a preset sufficiently small value. If eqn (31) is satisfied, then $\Omega_{n(i)}^{[n]}$ is the ith dimensionless natural frequency $\Omega_{n(i)}$. By substituting $\Omega_{n(i)}^{[n]}$ into eqn (29)

$$C_1 = -\frac{f_{r0}^{[n]}\left(\Omega_{n(i)}^{[n]}\right)}{f_{r1}^{[n]}\left(\Omega_{n(i)}^{[n]}\right)} C_0, r = 1\, or\, 2. \tag{32}$$

and all of the other coefficients C_j can be obtained from recurrence relations. Furthermore, the ith mode shape $\phi_i^{[n]}$ corresponding to the ith eigenvalue $\Omega_{n(i)}^{[n]}$ is obtained by

$$\phi_i^{[n]}(\eta) = \sum_{j=0}^{n-1} C_j^{[i]} \eta^j \tag{33}$$

where $C_j^{[i]}(\eta)$ is $C_j(\eta)$ in which λ_4 is substituted by $\lambda_{4(i)}$ and $\phi_i^{[n]}$ is the ith eigenfunction corresponding to the ith eigenvalue $\lambda_{4(i)}$. By normalizing eqn (33) the ith normalized eigenfunction is defined as

$$\bar{\phi}_i^{[n]}(\eta) = \frac{\phi_i^{[n]}(\eta)}{\sqrt{\int_0^1 \left[\phi_i^{[n]}(\eta)\right]^2 d\eta}} \tag{34}$$

where $\bar{\phi}_i^{[n]}(\eta)$ is the ith mode shape function of the beam corresponding to the ith natural frequency $\omega_i^{[n]} = \sqrt{\lambda_i^{[n]}}\sqrt{EI_0 / \rho A l^4} = {}^{[n]}_{n(i)}\sqrt{EI_0 / \rho A l^4}$.

4 RESULTS

Results are given for the calculations of natural frequencies and mode shapes for a non-uniform beam, a uniform beam with constant axial load, i.e. the non-linear terms of eqn (17) were made constant, and an eccentric mass, and finally, a non-uniform beam with an eccentric mass. The calculated results are compared with results found in the literature.

4.1 Case 1: non-uniform beam

The test case present here is a clamped-free constant width tapered beam (wedge) and the boundary conditions used were

$$\phi(0)=0,\frac{\mathrm{d}\phi(0)}{\mathrm{d}\eta}=0,\frac{\mathrm{d}^2\phi(1)}{\mathrm{d}\eta^2}=0,\frac{\mathrm{d}^3(1)}{\mathrm{d}\eta^3}=0. \tag{35}$$

For the wedge beam $\alpha_{d'}=1,\alpha_d=\alpha;\beta_{d'}=0,\beta_d=\beta$ the area and moment of the section can now be written as

$$A(\eta)=A_0\left[1+(\alpha-1)\eta\right]=A_0(1-\beta\eta),I(\eta)=I_0\left[1+(\alpha-1)\eta\right]^3=I_0(1-\beta\eta)^3,$$

where β is known as the taper ratio.

In this case the coefficients C_0 and C_1 were set to zero and the coefficients C_2 and C_3 are arbitrary constant. By substituting $\phi^{[n]}(\eta)=\sum_{j=0}^{n-1}C_j\eta^j$ into the last two boundary conditions of eqn (35) the following algebraic equations involving C_2 and C_3 are obtained

$$\begin{aligned}&\sum_{j=0}^{n-3}(j+1)(j+2)C_{j+2}=f_{12}^{[n]}(\lambda_4)C_2+f_{13}^{[n]}(\lambda_4)=0,\\&\sum_{j=0}^{n-4}(j+1)(j+2)(j+3)C_{j+3}=f_{22}^{[n]}(\lambda_4)C_2+f_{23}^{[n]}(\lambda_4)C_3=0.\end{aligned} \tag{36}$$

and for non-trivial solutions of C_2 and C_3 the frequency equation can be written as

$$f_{12}^{[n]}(\lambda_4)f_{23}^{[n]}(\lambda_4)-f_{22}^{[n]}(\lambda_4)f_{13}^{[n]}(\lambda_4)=0 \tag{37}$$

By solving cqn (37) for n and taking the real root for $\lambda_4=\Omega_n^2$, it was found that for $n-30$

$$\left|\Omega_{n(1)}^{[30]}-\Omega_{n(1)}^{[29]}\le\varepsilon=0.00001.\right| \tag{38}$$

with $\Omega_{n(1)}\approx\Omega_{n(1)}^{[30]}=3.517$.

The results for the wedge beam for different taper ratios are compared with those reported by Banerjee *et al.* [19], as shown in Table 1. The calculated results are in reasonable agreement with those reported by Banerjee *et al.* [19]. Values for $\beta=0$ are not reported by Banerjee *et al.* [19] as their method suffered numerical overflow when $\beta=0$.

Reasonable agreement was found between the two methods, although the current calculations generally gave slightly larger values. The maximum difference between the two methods was of the order of 1% and occurred for $\Omega_{n(3)}$.

4.2 Case 2: Uniform beam with eccentric tip mass experiencing constant axial force

Following [1] calculations were made for a uniform beam experiencing a constant axial loading with an eccentric tip mass. The material and geometric properties are listed in Table 2. The boundary conditions used for these calculations and those for Case 3 are

Table 1: Normalized natural frequencies of a wedge beam.

β	$\Omega_{n(1)}$		$\Omega_{n(2)}$		$\Omega_{n(3)}$	
	Current	Ref.[21]	Current	Ref.[21]	Current	Ref.[21]
0	3.517	–	22.100	–	61.003	–
0.1	3.562	3.559	21.346	21.338	59.001	58.980
0.2	3.612	3.608	20.639	20.621	56.293	56.192
0.3	3.671	3.667	19.905	19.881	53.592	53.322
0.5	3.830	3.824	18.355	18.317	47.482	47.265
0.7	4.090	4.082	16.667	16.625	41.005	40.588
0.9	4.641	4.631	14.987	14.931	33.231	32.833

Table 2: Properties of the uniform beam with tip mass.

Parameter	E	ρ	M	l	a	b	r_G	D
Value	210 GPa	7680 kg/m^3	5000 kg	30 m	1 m	1 m	0.5 m	3 m

$$\phi(0)=0,\frac{\mathrm{d}\phi(0)}{\mathrm{d}\eta}=0,\frac{\mathrm{d}^2\phi(1)}{\mathrm{d}\eta^2}=-\frac{\bar{M}\bar{g}\bar{a}}{\lambda_1},\lambda_1\frac{\mathrm{d}^3\phi(1)}{\mathrm{d}\eta^3}+\lambda_1'\frac{\mathrm{d}^2\phi(1)}{\mathrm{d}\eta^2}+\bar{M}\bar{g}\frac{\mathrm{d}\phi(1)}{\mathrm{d}\eta}=0 \quad (39)$$

An exact solution for this eigenvalue problem can be found [1] and is used here for comparison.

It can be seen form Table 2 that generally the current calculations give a closer result to the exact solution.

4.3 Case 3: Non-uniform beam with eccentric tip mass and rotary inertia

Here a linearly varying cross section cantilever beam with a both transversely and axially eccentric tip mass with rotary inertia is considered. The beam here is tubular and the physical properties are given in Table 4.

Using the present method, the normalized natural frequencies and mode shapes of the beam are obtained as on Fig. 2.

Table 3: Normalized natural frequencies of uniform beam.

N		$\Omega_{n(1)}$	$\Omega_{n(2)}$	$\Omega_{n(3)}$
0	Exact	0.7670	4.7763	13.2822
	Present	0.7678	4.7764	13.2955
	FEA Ref. [1]	0.7675	4.7707	13.2460
10^5	Exact	0.7657	4.7749	13.2811
	Present	0.7663	4.7763	13.2965
	FEA Ref. [1]	0.7662	4.7693	13.2451

Table 4: Physical and geometrical properties of non-uniform beam with eccentric tip mass.

Parameter	E	ρ	M	l	β	a	b	r_G
Value	210 GPA	7680 kg/m3	5000 kg	30 m	0.8	0.5 m	0.5 m	0.3 m

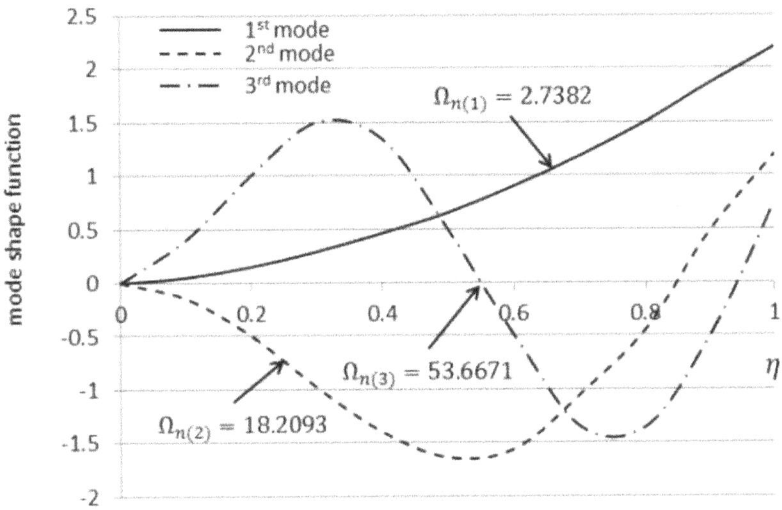

Figure 2: First three normalized natural frequencies and mode shapes for Case 3.

5 CONCLUSIONS

A method has been developed for calculating characteristics of a vibrating non-uniform cantilever beam with an eccentric tip mass. The developed approach is nontrivial and complicated due to the nature of the problem. No exact solution of this general problem is available. Several test cases were performed with quite reasonable agreement between the results of the current approach and those reported in the literature found.

REFERENCES

[1] Malaeke, H. & Moeenfard, H., Analytical modeling of large amplitude free vibration of non-uniform beams carrying a both transversely and axially eccentric tip mass. *Journal of Sound and Vibration*, **366**, pp. 211–229, 2016. https://doi.org/10.1016/j.jsv.2015.12.003

[2] Prescott, J., *Applied elasticity*, Dover Publications, New York, 1946.

[3] Goel, R., Vibrations of a beam carrying a concentrated mass. *Journal of Applied Mechanics*, **40**, pp. 821–822, 1973. https://doi.org/10.1115/1.3423102

[4] Rama Bhat, B., & Wagner, H., Natural frequencies of a uniform cantilever with a tip mass slender in the axial direction. *Journal of Sound and Vibration*, **45**(2), pp. 304–307, 1976. https://doi.org/10.1016/0022-460x(76)90606-4

[5] To, C.W.S., Vibration of a cantilever beam with a base excitation and tip mass. *Journal of Sound and Vibration*, **172**, pp. 289–304, 1994.

[6] Auciello, N.M., Transverse vibration of a linearly tapered cantilever beam with tip mass of rotary inertia and eccentricity. *Journal of Sound and Vibration*, **194**, pp. 25–34, 1996. https://doi.org/10.1006/jsvi.1996.0341
[7] Mable, H. & Rogers, C., Transverse vibrations of double-tapered cantilever beams with end support and with end mass. *The Journal of the Acoustical Society of America*, **55**, pp. 986–991, 1974.
https://doi.org/10.1121/1.1914673
[8] Caruntu, D., On bending vibrations of some kinds of variable cross-section using orthogonal polynomials. *Revue Roumaine des Sciences Techniques-Série Délelött Mécanique Appliquée*, **41**, pp. 265–272, 1996.
[9] Sanger, D., Transverse vibration of a class of non-uniform beams. *Journal of Mechanical Engineering Science*, **10**, pp. 111–120, 1968.
https://doi.org/10.1243/jmes_jour_1968_010_018_02
[10] Matt, C. F. T., Simulation of the transverse vibrations of a cantilever beam with an eccentric tip mass in the axial direction using integral transforms. *Applied Mathematical Modelling*, **37**, pp. 9338–9354, 2013.
https://doi.org/10.1016/j.apm.2013.04.038
[11] Adomian, G., *Nonlinear stochastic system theory and application to physics*. Kluwer Academic, Dordrecht, 1988.
[12] Adomian, G. & Rach, R., On composite nonlinearities and the decomposition method. *Journal of Mathematical Analysis and Applications,* **113**(2), pp. 504–509, 1986.
https://doi.org/10.1016/0022-247x(86)90321-5
[13] Wazwaz, A. M., Analytic treatment for variable coefficient fourth-order parabolic partial differential equations. *Applied Mathematics and Computation*, **123**, pp. 219–227, 2001. https://doi.org/10.1016/s0096-3003(00)00070-9
[14] Mao, Q., Application of Adomain modified decomposition method to free vibration analysis of rotating beams. *Mathematical Problems in Engineering*, **2013**, pp. 1–10, 2013.
https://doi.org/10.1155/2013/284720
[15] Adair, D. & Jaeger, M., Simulation of tapered rotating beams with centrifugal stiffening using the Adomian decomposition method. *Applied Mathematical Modelling*, **40**, pp. 3230–3241, 2016.
https://doi.org/10.1016/j.apm.2015.09.097
[16] Yaman, M., Adomian decomposition method for solving a cantilever beam of varying orientation with tip mass. *Journal of Computational and Nonlinear Dynamics*, **2**, pp. 52–57, 2007.
https://doi.org/10.1115/1.2389167
[17] Adomian, G., A new approach to nonlinear partial differential equations. *Journal of Mathematical Analysis and Applications*, **102**(2), pp. 420–434, 1984.
https://doi.org/10.1016/0022-247x(84)90182-3
[18] Adomian, G. & Rach, R., Generalization of Adomian polynomials to functions of several variables. *Computers & Mathematics with Applications*, **24**(5/6), pp. 11–24, 1992.
https://doi.org/10.1016/0898-1221(92)90037-i
[19] Banerjee, J. R., Su, H. & Jackson, D. R., Free vibration of rotating tapered beams using the dynamic stiffness method. *Journal of Sound and Vibration*, **298**, pp. 1034–1054, 2006. https://doi.org/10.1016/j.jsv.2006.06.040

RELATIONSHIP BETWEEN SHEAR PLANE OF THE FINAL PRESSING AND FATIGUE CRACK GROWTH BEHAVIOUR OF ROUND-BAR SPECIMENS OF CU PROCESSED BY ECAP

MASAHIRO GOTO[1], TAKAEI YAMAMOTO[1], SEUNG ZEON HAN[2], JEE HYUK AHN[2], JUNICHI KITAMURA[1], KUSNO KAMIL[3], TERUTOSHI YAKUSHIJI[4], TOSHIKI MASUDA[1], TAKASHI IWAMURA[1] & SANGSHIK KIM[5]
[1]Department of Mechanical Engineering, Oita University, Oita, Japan.
[2]Korea Institute of Materials Science, Changwon, Republic of Korea.
[3]Universitas Muslim Indonesia, Makassar, Indonesia.
[4]National Institute of Technology, Oita College, Oita, Japan.
[5]Gyeongsang National University, Chinju, Republic of Korea.

ABSTRACT
The formation mechanism of inclined fatigue cracks in ultrafine-grained Cu processed by equal channel angular pressing was studied by using a smooth specimen with a small blind hole. The crack growth direction depended on the location of drilling hole along the circumferential direction of the round bar specimen and on the applied stress amplitudes. Although the low-cycle fatigue crack growth paths inclined 45° and 90° to the loading-axis were observed in the different locations on the surface, crack faces in these cracks were extended along one set of maximum shear stress planes, corresponding to the shear plane of the final processing. To study the crack growth behaviour, surface damage around the crack paths formed by the two-step fatigue stress tests was observed. Profile of crack face was examined, showing the aspect ratios (b/a) of b/a = 0.38 and 1.10 for the cracks with 45° and 90° inclined path directions with respect to the loading axis, respectively. The role of the microstructure and deformation mode at the crack-tip areas on the formation of crack paths parallel to the shear plane of the final pressing was discussed in terms of the microstructural evolution caused by cyclic stressing and the mixed-mode stress intensity factor.
Keywords: copper, crack propagation, equal channel angular pressing, fatigue.

1 INTRODUCTION

Ultrafine grained (UFG) materials, having the grain sizes in the sub-micron and nanoscopic range, are of considerable interest. To date, many investigators have focused on optimizing processing conditions, on determining the underlying microstructural mechanisms, and on the static mechanical properties [1–5].

For the envisaged structural applications, extensive studies have been made on the fatigue crack growth behaviour of UFG metals and alloys processed by equal channel angular pressing (ECAP). For ECAP-prepared samples, the *yz*-, *zx*- and *xy*-planes are defined by three mutually orthogonal planes of sectioning that lie perpendicular to the longitudinal axis of the pressed sample, and parallel to either the sample side, or the top faces at the point of exit from the die, respectively. In low-cycle fatigue tests, many investigators have showed that the crack in the *zx*- and *xy*-plane created a 45°- [6–8] and a 90°- incline [9, 10] to the loading axis, respectively, and these crack path directions were parallel to the shear direction of the final ECAP processing. In addition, the growth rate of cracks in the *zx*-plane tended to be larger than that in the *xy*-plane [10]. Up to now, low-cycle fatigue (LCF) crack growth behaviour of UFG materials has been mainly discussed from the viewpoints of microstructure and morphological features of surface damage. On the other hand, the discussion from the mechanical viewpoints should be done for a better understanding of the fatigue damage of UFG materials. However, such studies are few and certain questions remain unanswered.

DOI: 10.2495/CMEM-V6-N4-691-702

In the present study, fatigue tests of copper (99.99% Cu) processed by ECAP through eight passages were carried out. The objective of this study is to investigate the physical background of the formation of crack growth paths and fast crack growth in the *zx*-plane in terms of the mixed-mode deformation at the crack tip.

2 EXPERIMENTAL PROCEDURES

The material used was pure oxygen-free (99.99 wt %) Cu. Before ECAP, the materials were annealed at 500°C for 1 h (grain size, 100 μm). The ECAP die had a 90° angle between intersecting channels. The angles at the inner and outer corners of the channel intersection were $\Phi = 90°$ and $\Psi = 45°$, respectively. Repetitive ECAP was accomplished according to the Bc route (after each pressing, the billet bar was rotated 90° around its longitudinal axis). Each rod was subjected to eight sequential passes of pressing at room temperature. MoS_2 was used as a lubricant for each pressing, and the pressing speed was 5 mm/sec. The tensile test specimens were cylindrical, 4 mm in diameter and 40 mm long. The tensile tests were conducted on all the specimens at room temperature on an Instron-4208 testing machine at a strain rate of $1.6 \times 10^{-3}\ s^{-1}$. The pre-ECAP mechanical properties were 232 MPa tensile strength, 65% elongation and Vickers hardness of 63 (load: 2.9 N). After eight ECAP passages, the properties changed to 435 MPa, 22.6%, and 136, respectively. The grain size was measured using a transmission electron microscope. The average grain/cell size of ECAP-prepared samples was measured as 320 nm, here for elongated grains; the thickness was measured as grain size and not as the diameter of equiaxed grains.

An analysis has shown that the equivalent strain, ε, after one pass is given by the following relationship [11]:

$$\varepsilon = \frac{1}{\sqrt{3}}\left\{2\cot\left(\frac{\Phi}{2}+\frac{\Psi}{2}\right)+\Psi\,\mathrm{cosec}\left(\frac{\Phi}{2}+\frac{\Psi}{2}\right)\right\}. \tag{1}$$

Total of eight extrusion passes resulted in an equivalent strain of about 7.8.

Fatigue specimens 5 mm in diameter were machined from their respective processed bars. Although the specimens had a shallow circumferential notch (20-mm notch radius and 0.25-mm notch depth), the fatigue strength reduction factor for this geometry was close to 1, meaning that they could be considered plain. The fatigue specimens were electrolytically polished (approximately ≈ 25 μm from the surface layer) prior to mechanical testing to remove any preparation-affected surface layer.

Figure 1 shows the location of monitoring areas and the definition of orientation of fatigue specimens relative to the final pressing plane described as a dark elliptical-area. The monitoring area was set on the middle surfaces where an intersection between the shear plane of the final pressing and the specimen surface makes an angle of 45° (the *zx*-plane) or 90° (the *xy*-plane) with respect to the loading axis. The fatigue damage and crack growth behaviour were monitored on both planes. Prior to testing, in some specimens, a small blind hole (both diameter and depth of 0.1 mm) was drilled on the monitoring area as a crack starter. The cracks/hole on the *zx*- and *xy*-planes are referred to hereafter as the '*zx*-plane crack/hole' and '*xy*-plane crack/hole', respectively.

All fatigue tests were performed at room temperature using a rotating-bending fatigue machine (constant bending-moment type) operating at 50 Hz. The fatigue damage on the specimen surface was observed using an optical microscope (OM) and a scanning electron microscope (SEM). The crack length, l, was measured along the circumferential direction of the surface using a plastic replication technique. The stress value referred to is that of the nominal stress amplitude, σ_a, at the minimum cross-section (5-mm diameter).

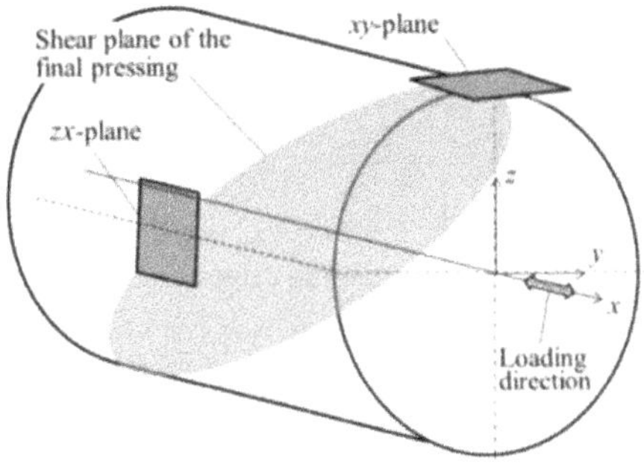

Figure 1: The location of monitoring areas and definition of orientation of fatigue specimens relative to the final pressing.

3 EXPERIMENTAL RESULTS

Figure 2 shows the crack growth paths in the *zx*- and *xy*-planes under high- and low-stress amplitudes (σ_a = 240 and 90 MPa). At σ_a = 240 MPa, the *zx*-plane crack that initiated from the hole created a 45° incline to the loading axis and this crack path direction was parallel to the shear direction of the final ECAP processing (Fig. 1). The 45° inclined crack growth direction has been commonly observed in the *zx*-plane of LCF UFG metals [6–8]. However, the *xy*-plane crack grew nearly perpendicular to the loading axis. At σ_a = 90 MPa, on the other hand, the macroscale growth direction was perpendicular to the loading axis regardless of the plane where the crack initiated, nevertheless, the crack propagated in a zigzag manner at the microscale. There was no difference in the crack growth behaviour and the morphological features of surface damage around crack paths between the UFG and conventional grain-sized copper. This was because that the grain coarsening as a result of dynamic recrystallization due to a long term of cyclic stressing. Grain coarsening during fatigue is a time-, amplitude-, purity- and temperature-dependent microstructural process. It has been suggested that a certain time and cumulative strain are required for the release of sufficient stored strain energy to allow dislocation motion, recrystallization and subsequent grain coarsening [12].

To study the unique crack growth behaviour at a high-stress amplitude, the behaviour of microcrack initiation was monitored by using plane specimens (without a hole). Figure 3a and b show the change in surface states around a fatal crack, which led to the final fracture of the specimen, in the *zx*- and *xy*-planes under a high-stress amplitude (σ_a = 240 MPa) corresponding to the LCF regime. In the *zx*-plane, at an early fatigue stage, a large number of shear bands (SBs) with a length of less than a few micrometres [13] were created a 45° incline to the loading axis and this SB orientation was parallel to the shear plane of the final ECAP processing. Fatigue cracks were initiated from SBs and propagated along the SBs with a coalescence of microcracks generated in nearby SBs. It is reasonable to assume that the SB formation and microcrack growth were strongly affected by the microstructural inhomogeneity retained in the matrix because of the incomplete dynamic recovery/recrystallization. It has been shown that unlike the microstructure that evolved at σ_a = 90 MPa, the defect structure produced during ECAP processing should be still moderately retained in the matrix fatigued at σ_a = 240 MPa [14, 15]. Indeed, among innumerable maximum shear stress planes, SBs and microcracks were initiated along one set of the maximum shear stress planes corresponding to the final ECAP shear plane. To successfully release the local high strain energy caused by the defects distributed along the final ECAP streamline, the maximum shear stress planes

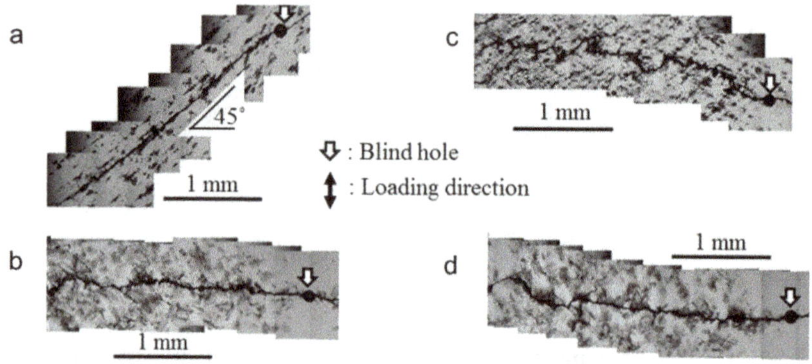

Figure 2: Effect of stress amplitudes on crack growth paths in the *zx*- and *xy*-planes: (a and b) the *zx*-plane crack at σ_a = 240 and 90 MPa; (c and d) the *xy*-plane crack at σ_a = 240 and 90 MPa.

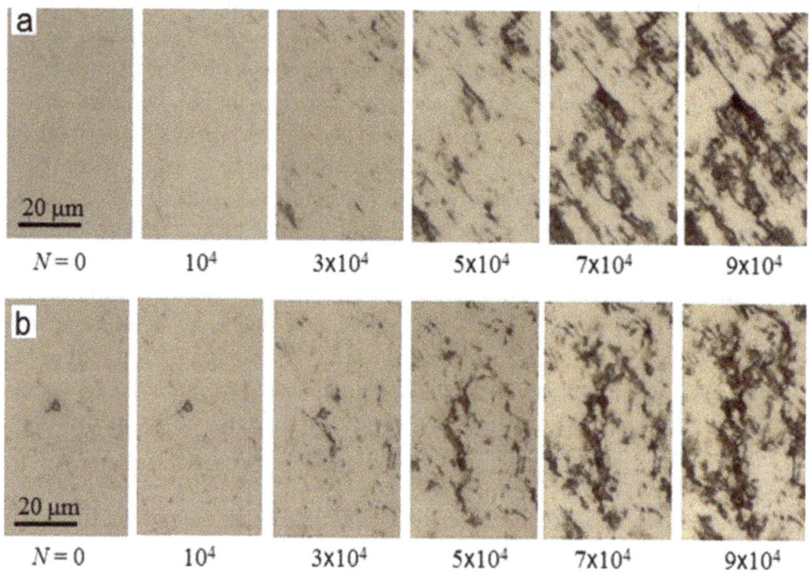

Figure 3: OM micrograph showing the change in surface states around a fatal crack under cyclic stressing at σ_a = 240 MPa: (a) the *zx*-plane; (b) the *xy*-plane.

corresponding to the final ECAP shear plane may be a reasonable selection. The fatigue tests of UFG Cu in the LCF regime showed the 45° inclined crack paths in the *zx*-plane [6–10]. Regarding the *xy*-plane, the crack initiation and growth behaviour was like that observed for the *zx*-plane; however, the orientation of SBs and the crack growth direction were nearly perpendicular to the loading axis.

Figure 4 shows the crack growth curve of drilled specimens at σ_a = 240 MPa. For both planes, there was no distinct difference in the number of cycles required to initiate fatigue crack from the hole regardless of the location of drilling hole. The slop of growth curve was steeper in the *zx*-plane crack than the *xy*-plane crack for crack length of less than 1 mm, however, the curves for crack length of longer than 1 mm had same slop for both planes. This is

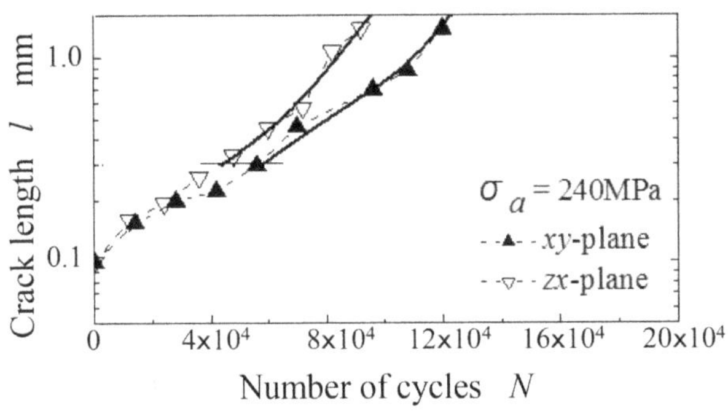

Figure 4: Crack growth curve of drilled specimens at σ_a = 240 MPa.

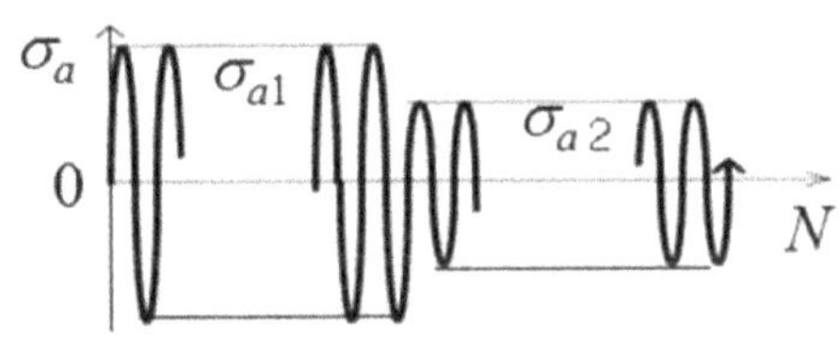

Figure 5: Loading pattern (high-to-low block stressing).

because that the assumption of the *zx*- and *xy*-plane cracks holds for a limited area along the circumference of the specimen. For the present specimen with 5 mm diameter, the assumption appears to be meaningless for a long crack in excess of l = 1 mm.

To clarify the difference in crack growth mechanism between the *zx*- and *xy*-planes, two-step fatigue stress tests (Fig. 5) were conducted, i.e. after the formation of a crack with an arbitrary length at the high-stress amplitude (σ_{a1}), the crack continued to grow at the low-stress amplitude (σ_{a2}). Figure 6a and b shows the crack growth paths under two-step stress tests. For the *zx*-plane (Fig. 6a), a crack with a growth path at a 45° incline to the loading axis was formed under 5.3×10^4 repetitions of σ_{a1} = 240 MPa, leaving a linear crack path. After a stress amplitude change to σ_{a2} = 90 MPa, the crack grew nearly perpendicular to the loading axis. For the *xy*-plane (Fig. 6b), the crack grew perpendicular to the loading axis under 6.0×10^4 repetitions of σ_{a1} = 240 MPa. Under the second stress, the crack continued to grow perpendicular to the loading axis.

Figure 7 shows SEM micrograph of surface states around crack paths just before and after the stress change. Figure 7a shows the surface damage in the *zx*-plane. The surface at the crack edges that were formed under high stress was comparatively flat, but the surface that formed under low stress had traces of localized plastic deformations, suggesting change in the crack growth mechanism between high and low stresses. Figure 7b shows the surface damage in the *xy*-plane. Before the stress change, the crack paths showed severe zigzag manner. The traces of heavy localized plastic deformations were left around the crack paths. After the stress change, the crack continued to grow with zigzag manner, but the degree of zigzag was moderate as compared with that before the stress change.

To discuss the physical background of crack growth behaviour, the specimens were sectioned in the longitudinal direction (parallel to the *x*-axis), at specific crack lengths, followed

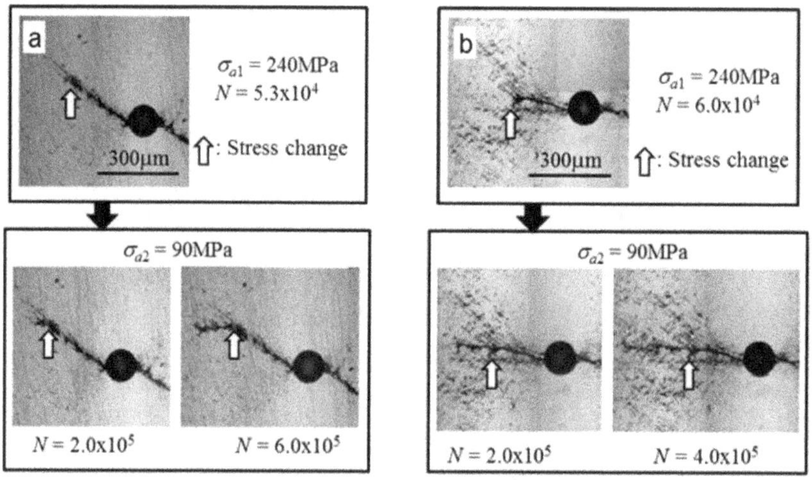

Figure 6: Crack growth paths under high-to-low block stressing: (a) the *zx*-plane crack; (b) the *xy*-plane crack.

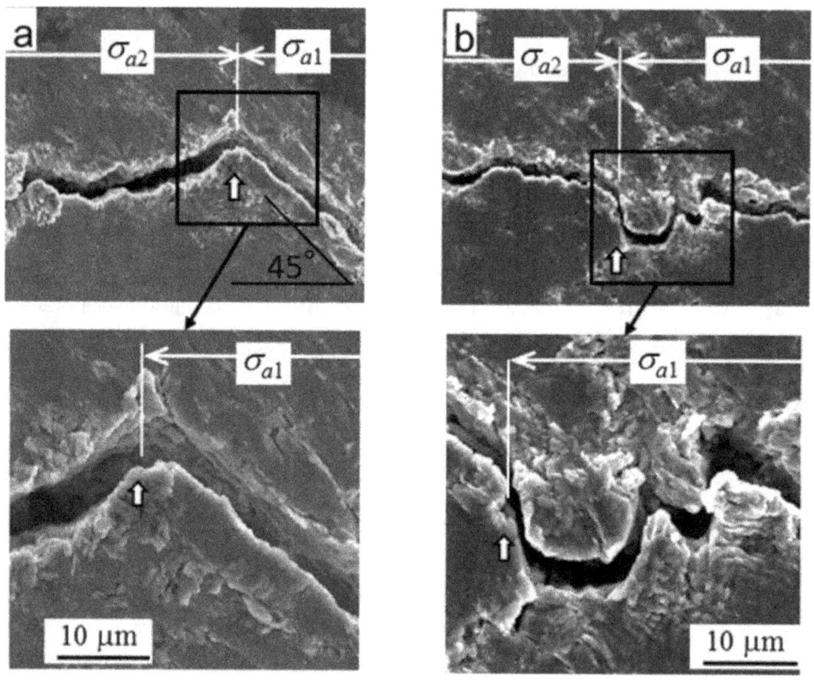

Figure 7: Surface state around the crack path just before and after the stress change (σ_{a1} = 240 MPa, σ_{a2} = 90 MPa (a) the *zx*-plane crack; (b) the *xy*-plane crack.

by etching of the sectioned planes. Figure 8 shows the crack growth paths formed at σ_a = 240 MPa in the surface and the sectioned plane; (a) a *zx*-plane crack with 0.975-mm length and (b) a *xy*-plane crack with 0.918-mm length. The inner growth direction of the *zx*-plane crack was perpendicular to the surface. For the *xy*-plane cracks, the inner growth direction was inclined 45° to the surface. Although the inclination direction of the crack path to the loading

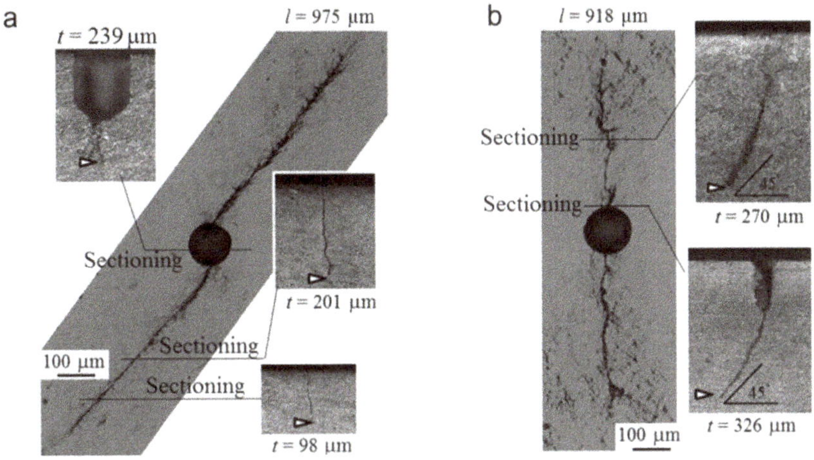

Figure 8: Sectioning position described in the crack paths on the surface and the inner growth paths under σ_a = 240 MPa: (a) a 0.975-mm-long crack in the *zx*-plane; (b) a 0.918-mm-long crack in the *xy*-plane.

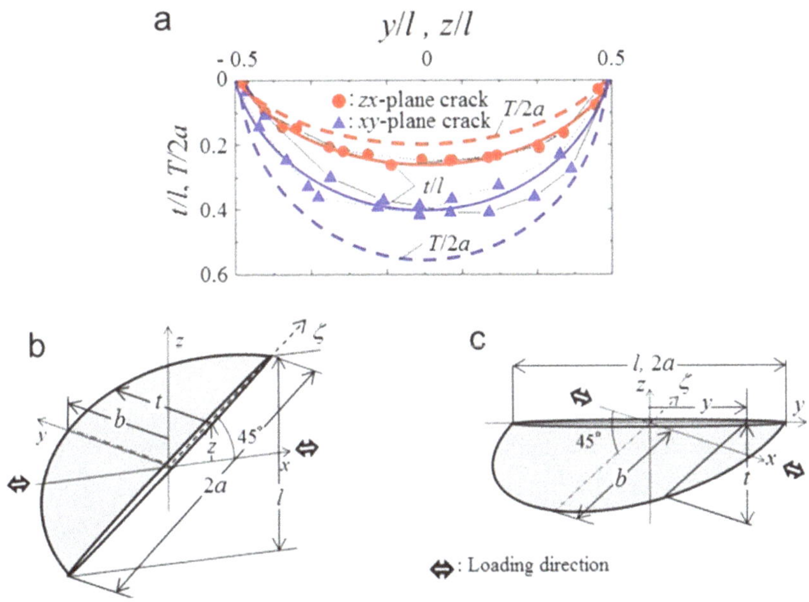

Figure 9: Schematics of crack growth behavior inside the specimen at σ_a = 240 MPa: (a) crack face profile in terms of dimensionless sizes for the *zx*- and *xy*-plane cracks; (b, c) definition of the crack length/depth for an inclined crack on the *zx*-plane and a deflected crack on the *xy*-plane.

axis was different between the *zx*- and *xy*-plane cracks, the paths for cracks in each plane were parallel to the shear plane of the final ECAP processing. Regarding the inner growth direction at low-stress amplitudes, like the crack paths commonly formed in the conventional grain-sized materials, the macroscale crack paths for the surface and sectioned planes were nearly perpendicular to the loading direction, regardless of the location of the hole [16].

Figure 9a shows the profile of the crack face, y/l or z/l versus t/l relationship, where t is the crack depth measured along the vertical direction to the specimen surface. The crack face shape was approximated by a semi-ellipse with the value of $t/l = 0.266$ at the deepest point for the zx-plane crack and $t/l = 0.389$ for the xy-plane crack. Considering the experimental results of Fig. 8, a declined surface-crack on the zx-plane and a deflected surface-crack on the xy-plane were defined as described in Figs 9b and c, where terms a, T and b are a half crack-length measured along the crack path direction, the crack depth measured along the crack face direction at a specific value of y and z, and the maximum depth of a semielliptical crack measured along the crack face, respectively. A red- and a blue-dashed lines in Fig. 9a show the crack face profile for the zx- and xy-plane cracks replotted by using a term $T/2a$ instead of t/l. The values of b/a, aspect ratio, were 0.378 for the zx-plane crack and 1.10 for the xy-plane crack. Incidentally, at a low-stress amplitude ($\sigma_a = 90$ MPa), it has been shown that no deference in the crack face profile between the zx- and xy-plane cracks and the aspect ratio was $b/a = 0.76$ which was nearly equivalent to the value of b/a measured in annealed conventional grain-sized Cu [16].

4 DISCUSSION

To study the mechanical background of the unique crack growth directions of UFG Cu from the viewpoints of the deformation mode at the crack tip, the stress intensity factor (SIF) values were evaluated, by assuming a semi-infinite body with inclined semi-elliptical surface cracks, subjected to tension stress in the x-direction at infinity. The applied tension stress is resolved into two stress components perpendicular and tangential to the crack face. Accordingly, a 45° inclined surface crack is subjected to both normal, σ (= 1), and shear stress, τ (= 1), as illustrated in Figs 10a and b. Noda *et al.* [17] analyzed the SIFs of a semi-elliptical surface-crack subjected to modes I, II and III loading. The values of dimensionless SIF (DL-SIF), F_I, F_{II}, F_{III}, for the current surface-crack were taken from Noda's solution calculated for Poisson's ratio: $\nu = 0.3$. The DL-SIF, $F_I(\alpha)$, $F_{II}(\alpha)$, $F_{III}(\alpha)$, were defined as;

$$F_i(\alpha) = \frac{K_i(\alpha)}{K_{iE}(\alpha)} \qquad i = \text{I, II, III}. \tag{2}$$

where $K_i(\alpha)$ is SIF along the crack front and $K_{iE}(\alpha)$ is SIF of an elliptical crack [17]. The location along a crack front was defined by the eccentric angle of the ellipse, α, defined as the

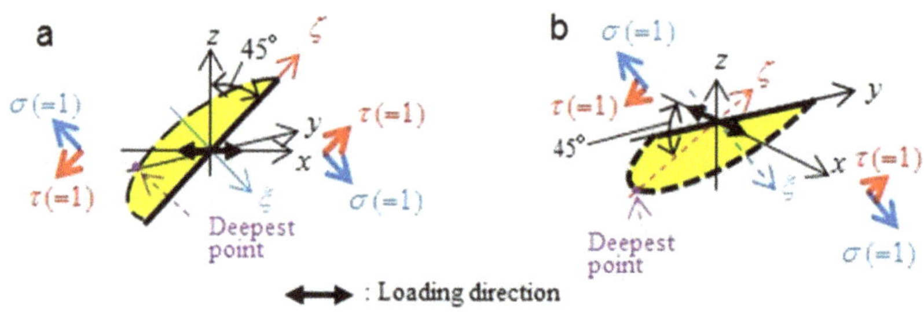

Figure 10: Inclined semi-elliptical surface cracks in a semi-infinite body subjected to tension load in the x-direction at infinity: (a) a model for a surface crack on the zx-plane; (b) a model for a surface crack on the xy-plane.

angle for mapping a semi-circle [17]. The crack front at the surface and the deepest point are shown by $\alpha = 0$ and 90°, respectively. However, the DL-SIF values at the surface were evaluated by the values at $\alpha = 1°$ instead of $\alpha = 0°$, because in three-dimensional surface cracks, the point where the crack front intersects a free surface is known as a corner point. The stress singularity at this point is different from that of an ordinary crack [18].

Table 1 shows the DL-SIF values of cracks with the aspect ratio, $b/a = 0.38$ for the *zx*-plane crack, and 1.1 for the *xy*-plane crack, which are given from Fig. 9. The value of F_I for the *zx*-plane crack is larger at the crack bottom ($\alpha = 90°$) than at the surface ($\alpha = 1°$), giving rise to an aspect ratio larger than 1. However, the actual aspect ratio was much smaller than 1. Regarding the F_{II}, the large F_{II} value (= 0.77) at the surface promotes in-plane shear mode crack growth. At the bottom, F_{III} takes a value of 0.83, but it seems reasonable to assume that antiplane shear mode deformation at the deepest point is negligible for the inward crack growth. Consequently, the *zx*-plane crack predominantly propagated at the surface due to in-plane shear mode deformation, which brings the generation of new SBs and extension of preexistent SBs around the crack tip areas, producing an aspect ratio much smaller than 1. The tensile-mode deformation reflected by large F_I values should play the role of crack growth. From the perspective of the formation of the shallow semielliptical cracks, the tensile-mode deformation might assist shear mode growth, through the debonding of SBs around the crack tip [15, 19].

For the *xy*-plane crack (Table 2), the F_I value at the surface was 0.74, i.e. larger than 0.60 at the bottom. This should lead to a superior extension of crack length, compared with the crack depth ($b/a < 1$). However, the actual crack extension was accelerated at the bottom, rather than at the surface. On the other hand, the F_{II} value was higher at the bottom (0.53) than at the surface (–0.08), suggesting the deep semielliptical shape of the *xy*-plane crack caused by the in-plane shear mode deformation at the bottom. At the surface, mode-I crack growth appeared to be predominant, because of negligible values of F_{II} and F_{III}. Therefore, mode-II crack growth plays an important role in determining the crack face shape.

The unique growth direction of microcracks at high stresses was attributed to inhomogeneous microstructures due to ECAP; microcracks initiated from SBs grew along the maximum shear stress plane corresponding to the shear plane of the final pressing, resulting in the inclined semielliptical surface cracks. As the crack grew, although the contribution of microstructural inhomogeneity on crack growth weakened, in-plane shear-mode deformation at the

Table 1: Dimensionless stress intensity factors for the *zx*-plane crack ($b/a = 0.38$).

Type of crack	α(deg)	F_I	F_{II}	F_{III}
zx-plane crack	1	0.68	0.77	0.31
$b/a = 0.38$	90	0.93	0	0.83

Table 2: Dimensionless stress intensity factors for the *xy*-plane crack ($b/a = 1.10$).

Type of crack	α(deg)	F_I	F_{II}	F_{III}
xy-plane crack	1	0.74	–0.08	0.09
$b/a = 1.10$	90	0.60	0.53	0

crack tip promoted the formation of a large inclined crack with a unique face profile. The fast crack growth rate in the *zx*-plane crack resulted from in-plane shear mode deformation in addition to the tensile mode deformation.

As indicated in Fig. 7b, the crack paths in the *xy*-plane under high stress showed a heavy zigzag manner. The zigzag growth paths were formed by the coalescence in preexistent SBs around the crack paths under the tensile mode deformation (The ND-SIF values were $F_I = 0.74$, F_{II} (= −0.08) and F_{III} (= 0.09)). The retarded crack growth on the *xy*-plane was therefore believed to be due to enhanced roughness-induced crack closure. For the *zx*-plane crack, the crack paths at high-stress amplitude ($\sigma_a = 240$ MPa) were linear (Fig. 7a) because of the coalescence of pre-existent collinear-SBs/shear-cracks ahead of the crack tip under in-plane shear mode deformation ($F_{II} = 0.77$).This should contribute to the acceleration of the *zx*-plane crack.

5 CONCLUSIONS

The main findings of this study on the formation mechanism of inclined surface cracks in UFG copper can be summarized as follows:

1. At high-stress amplitudes, the crack path/face in both *zx*- and *xy*-planes was extended along one set of the maximum shear stress planes, corresponding to the final ECAP shear plane, among innumerable maximum shear stress planes, showing a high microstructural sensitivity to crack-path formation. At low-stress amplitudes, the crack path was perpendicular to the loading direction regardless of the ECAP shear plane.
2. Two-step stress tests indicated that the surface at the *zx*-plane crack edges that were formed under high stress was comparatively flat, whereas the surface around the *xy*-plane crack paths showed traces of severe plastic deformation, meaning a difference in predominant crack growth mechanism: shear mode for the former and tensile mode for the later.
3. At high-stress amplitudes, the cracks in the *zx*- and *xy*-planes had a shallow and deep semielliptical face, respectively. The normal aspect ratio, b/a, was 0.38 for the *zx*- and 1.1 for the *xy*-plane cracks. The inclined crack path and crack face profile were attributed to the inhomogeneity in ECAP microstructure and the in-plane shear-mode deformation around the crack-tip areas.
4. The fast crack growth rate in the *zx*-plane crack resulted from in-plane shear mode deformation in addition to the tensile mode deformation.

ACKNOWLEDGEMENTS

The authors wish to express their gratitude to Professor Nao-aki. Noda of the Kyushu Institute of Technology, Japan, for fruitful discussion on mixed-mode stress intensity factors. This study was supported by a Grant-in-Aid for Scientific Research (C) (KAKENHI: No. 26420021) from the Japan Society for the Promotion of Science, as well as a National Research Foundation of Korea (NRF) grant funded by the Global Frontier R&D Program (2013M3A6B1078874) at the Global Frontier Hybrid Interface Materials R&D Centre funded by the Ministry of Science, ICT, and a Future Planning and National Research Foundation of Korea (NRF) grant funded by the Korea government (MSIP) (no. 2011-0030058). The authors are very grateful to the members of the Strength of Materials Laboratory of Oita

University for their excellent experimental assistance. Thanks are also extended to the members of the Korea Institute of Materials Science, for performing the ECAP processing of our copper rods.

REFERENCES

[1] Segal, V.M., Materials processing by simple shear. *Materials Science & Engineering: A*, **197**, pp. 157–164, 1995.
https://doi.org/10.1016/0921-5093(95)09705-8

[2] Valiev, R.Z., Structure and mechanical properties of ultrafine-grained metals. *Materials Science & Engineering: A*, **234–236**, pp. 59–66, 1997.
https://doi.org/10.1016/S0921-5093(97)00183-4

[3] Agnew, S.R. & Weertman, J.R., Cyclic softening of ultrafine grain copper. *Materials Science & Engineering: A*, **244**, pp. 145–153, 1998.
https://doi.org/10.1016/S0921-5093(97)00689-8

[4] Vinogradov, A. & Hashimoto, S., Multiscale phenomena in fatigue of ultra-fine grain materials-an overview. *Materials Transactions*, **42**, pp. 74–84, 2000.
https://doi.org/10.2320/matertrans.42.74

[5] Brebbia, C.A., Telles, J.C.F. & Wrobel, L.C., (eds), *Boundary Element Techniques*, Springer-Verlag: Berlin and New York, pp. 11–13, 1984.

[6] Höppel, H.W., Zhou, Z.M., Mughrabi, H. & Valiev, R.Z., Microstructural study of the parameters governing coarsening and cyclic softening in fatigued ultrafine-grained copper. *Philosophical Magazine -A*, **82**, pp. 1781–1794, 2002.
https://doi.org/10.1080/01418610208235689

[7] Wu, S.D., Wang, Z.G., Jiang, C.B., Li, G.Y., Alexandrov, I.V. & Valiev, R.Z., The formation of PSB-like shear bands in cyclically deformed ultrafine grained copper processed by ECAP. *Scripta Materiala*, **48**, pp. 1605–1609, 2003.
https://doi.org/10.1016/S1359-6462(03)00141-6

[8] Kunz, L., Lukáš, P. & Svpboda, M., Fatigue strength, microstructural stability and strain localization in ultrafine-grained copper. *Materials Science & Engineering: A*, **434**, pp. 97–104, 2006.
https://doi.org/10.1016/j.msea.2006.02.029

[9] Fang, D., Zhang, P., Duan, Q., Wu, S., Zhang, Z., Li, J. & Zhao, N., Fatigue behavior of Al-Cu alloy subjected to different number of ECAP passes. *Advanced Engineering Materilas*, **9**, pp. 860–866, 2007.
https://doi.org/10.1002/adem.200700110

[10] Zhang, Z.F., Wu, D.S., Li, Y.J., Liu, S.M. & Wang, Z.G., Cyclic deformation and fatigue properties of Al–0.7 wt. % Cu alloy produced by equal channel angular pressing. *Materials Science & Engineering, A*, **412**, pp. 279–286, 2005.
https://doi.org/10.1016/j.msea.2005.08.221

[11] Goto, M., Han, S.Z., Ahn, J.H., Yakushiji, T., Euh, K., Kim, S.S. & Lee, J., The role of mixed-mode deformation at the crack tip on shear banding and crack propagation of ultrafine-grained copper. *International Journal of Fatigue*, **66**, pp. 220–228, 2014.
https://doi.org/10.1016/j.ijfatigue.2014.04.007

[12] Iwahashi, Y., Wang, J., Horita, Z., Nemoto, M. & Langdon, T.G., Principle of equal-channel angular pressing for the processing of ultra-fine grained materials. *Scripta Materialia*, **35**, pp. 143–146, 1996.
https://doi.org/10.1016/1359-6462(96)00107-8

[13] Khatibi, G., Horky, J., Weiss, B. & Zehetbauer, M.J., High cycle fatigue behavior of copper deformed by high pressure torsion. *International Journal of Fatigue*, **32**, pp. 269–278, 2010.
https://doi.org/10.1016/j.ijfatigue.2009.06.017

[14] Goto, M., Han, S.Z., Yakushiji, T., Lim, C.Y. & Kim, S.S., Formation process of shear bands and protrusions in ultrafine grained copper under cyclic stresses. *Scripta Materialia*, **54**, pp. 2101–2106, 2006.
https://doi.org/10.1016/j.scriptamat.2006.03.001

[15] Goto, M., Han, S.Z., Euh, K., Kang, J-H., Kim, S.S. & Kawagoishi, N., Formation of a high-cycle fatigue fracture surface and a crack growth mechanism of ultrafine grained copper with different stages of microstructural evolution. *Acta Materiala*, **58**, pp. 6294–6305, 2010.
https://doi.org/10.1016/j.actamat.2010.07.051

[16] Goto, M., Kamil, K., Han, S.Z., Euh, K., Kim, S.S. & Lee, J., Fatigue-induced grain coarsening and crack growth behavior in ultrafine grained copper under different loading histories. *International Journal of Fatigue*, **51**, pp. 57–67, 2013.
https://doi.org/10.1016/j.ijfatigue.2013.02.008

[17] Goto, M., Han, S.Z., Yamamoto, T., Kitamura, J., Ahn, J.H., Yakushiji, T., Kim, S.S. & Lee, J., Formation mechanism of inclined fatigue-cracks in ultrafine-grained Cu processed by equal channel angular pressing. *International Journal of Fatigue*, **92**, pp. 577–587, 2016.
https://doi.org/10.1016/j.ijfatigue.2016.02.006

[18] Noda, N.A. & Kagita, M., Variations of stress intensity factors of a semi-elliptical surface crack subjected to mode I, II, III loading. *International Journal of Pressure Vessels Piping*, **81**, pp. 635–644, 2004.
https://doi.org/10.1016/j.ijpvp.2004.03.008

[19] Benthem, J.P., State of stress at the vertex of crack in a half-space. *International Journal of Solids and Structures*, **13**, pp. 479–492, 1977.
https://doi.org/10.1016/0020-7683(77)90042-7

[20] Zhang, J.Z., A shear band decohesion model for small fatigue crack growth in an ultrafine grain aluminium alloy. *Engineering Fracture Mechanics*, **65**, pp. 665–681, 2000.
https://doi.org/10.1016/S0013-7944(99)00148-4

PREDICTING MODULUS OF ELASTICITY OF RECYCLED AGGREGATE CONCRETE USING NONLINEAR MATHEMATICAL MODELS

JUNIOR A. REYES-SÁNCHEZ[1], ANTONIO J. TENZA-ABRIL[2], FERRAN VERDU[3] & JOSÉ A. REYES PERALES[3]
[1]Department of Civil Engineering, National Autonomous University of Honduras, Honduras.
[2]Department of Civil Engineering, University of Alicante, Spain.
[3]Department of Applied Mathematics, University of Alicante, Spain.

ABSTRACT

It is estimated that currently the consumption of natural aggregates used annually in the production of concrete in the world is around 10 billion tons. Moreover, more than 10 million tons of waste is generated annually from the construction industry. The incorporation of recycled aggregates in the production of concrete arises mainly due to an environmental factor, because it emphasizes the reduction in the consumption of raw materials, reduction of the emission of pollutants to the atmosphere derived from the processes of extraction of natural aggregates, between others. Several studies quantifies the decrease of mechanical properties according to the percentage of replacement of natural aggregate by recycled concrete aggregate. In the present study the authors provide several nonlinear models, which are able to predict the modulus of elasticity behaviour of the concrete manufactured with recycled aggregate. A database was composed of 147 different mixtures of recycled aggregate concrete collected from publications of scientific journals. The database has been used to introduce it to the software Polimodels. Polimodels is able to generate different models using different nonlinear regression algorithms. Six different models for the modulus of elasticity are proposed, dependents on certain physical and mechanical parameters of the recycled aggregate, as the following; the percentage of absorption, Los Angeles abrasion coefficient, and the percentage of substitution of natural aggregate by recycled aggregate. It is possible to appreciate the remarkable reduction in the modulus of elasticity due to the increase of recycled aggregates in the concrete. When the models have more independent variables a better adjustment is noticed that help us to improve the prediction of the modulus of elasticity of recycled aggregate concrete.
Keywords: elastic modulus, nonlinear models, prediction, recycled aggregates, recycled concrete

1 INTRODUCTION

The use of recycled aggregates (RA) in the production of concrete has many kinds of purposes: the reduction in the consumption of raw materials, the reduction of CO_2 emissions into the atmosphere due to natural aggregate (NA) extraction processes, as well as the valorisation of construction and demolition waste (C&W), among others.

It is estimated that the annual consumption in the world of NA used in the production of concrete is about 10 billion tons. In addition, more than 10 million tons of C&W is generated annually [1]. As a consequence, the environment has to absorb a very important amount of waste materials from the construction sector. Moreover, the construction sector consumes a high quantity of raw materials for the production of concrete.

The use of recycled aggregates concrete (RAC), provides an alternative of high environmental value and more rational from an economic point of view [2]. Extraction of NA, cement manufacturing and all production processes related to the manufacture of concrete, with the consequent energy consumption of these processes, are associated with the emission of tons of CO_2 into the atmosphere. For that reason, the use of RA in the manufacture of concrete constitutes a significant step towards a more sustainable society.

DOI: 10.2495/CMEM-V6-N4-703-715

The main difference between the recycled aggregates from concrete (RCA) and the NA is the mortar adhered to the first, which causes a higher water absorption (WA), lower density, higher porosity and a lower wear resistance. Those properties determine the greatest divergence between the two kinds of aggregates. Higher mortar content adhered to the RCA, results in a greater negative effect on the concrete made with it [3]. Usually, RCA has higher roughness, angular shape and porosity and lower density than NA [4].

The use of RCA in concrete production greatly affects the properties of concrete in a fresh state. Workability of concrete decreases with increase of RCA content. This rapid loss of workability is associated with the high absorption of RCA and the increase of fine content during the mixing process as the adhered mortar to the RCA is detached [5]. Currently, the causes of the low utilization of RCA are based on its lower performance against the use of NA [1, 6, 7]. The lower performance in the RCA is mainly due to the higher absorption properties (WA) and higher permeability than in NA [8]. Whit the increase of RCA content in the concrete mixture the modulus of elasticity and compressive strength of RAC decrease [9]. For these reasons, the NA replaced by RCA is limited to a maximum of 20% in structural concrete by several regulations such as EHE-08 [10].

There are a large number of experimental investigations on RAC that allow the use of the results obtained to describe the behaviour of RAC through mathematical modelling. However, in most of the proposed prediction models, the dependent variable is determined according to the values of the RAC dosage to be produced (w/c ratio, cement quantity, etc.) or mechanical values of the same concrete that is studied (compressive strength, indirect tensile strength or modulus of elasticity). The prediction of the mechanical properties, that can be found several regulations [10,11], implies a high number of experimental measures of the RAC to predict the behaviour.

The main interest of the authors is to predict the modulus of elasticity using several non-linear mathematical models, knowing some physical properties of the RCA, such as the percentage of water absorption at 24 hours (WA), Los Angeles abrasion coefficient (LA) and percentage of recycled aggregate (RCA). In addition, the modulus of elasticity of a control concrete (concrete with 0% of RCA in its dosage) will be used as independent variable. Moreover, we pretend to analyze the influence of the number of independent variables in the different proposed nonlinear models.

2 MATERIALS AND METHODOLOGY

2.1 Database

To achieve the previously mentioned objective, a database has been generated from 147 different mixtures extracted from different researches where the main theme was RCA. The main topic of the selected research is the study of the addition of RCA.

The characteristics of the mixtures used are shown in Table 1. In this table are listed the range of the main materials used in the concrete mixture. As can be seen, the wide range of the dosage will allow the models to be applicable to virtually any type of concrete with recycled aggregate that can be manufactured.

The database reflects the broad spectrum of dosages covered by the study and will greatly influence the mechanical behaviour of the RACs by incorporating RCA [12]. As described above, the study focuses on the treatment of data to generate nonlinear mathematical models, which are able to predict the modulus of elasticity (E_{RC}) of a 28-day-old RAC. First, knowing

Table 1: Dosage ranges of total analyzed mixtures.

	Average value (kg/m^3)	Range of mix proportions (kg/m^3)	
		Minimum	Maximum
Cement content (kg/m^3)	364	249	520
Addition content (kg/m^3)	26	18	35
Water/Cement ratio	0.52	0.35	0.82
Water/Binder ratio	0.52	0.35	0.82
Effective Water/Cement ratio	0.46	0.24	0.72
Effective Water/Binder ratio	0.48	0.29	0.65
Additive content (kg/m^3)	5	0	72
Sand (kg/m^3)	698	395	1000
Gravel (kg/m^3)	531	0	1353
RCA (kg/m^3)	543	0	1149
% RCA (in weight)	50	0	100

some physical properties of the aggregate (WA and LA), the percentage of RAC and the compressive strength of the control concrete without incorporating RAC in its dosage (0% RCA).

To obtain the E_{RC} models we have used a database of investigations [1–3, 7, 13–24]. The following phases have been considered; a) knowing the compressive strength value of a concrete with the same RCA degree ($f_{c,RC}$) but we still do not know the physical properties of the recycled aggregate (Model 1); after we know the WA value of the RCA (Model 2); and finally we also know the LA value of the RCA (Model 3). b) the E_{CC} value of the control concrete is known (Model 4); then, we also know the WA of the RCA (Model 5); and finally the value LA of the RCA (Model 6). It should be noted that WA of the RCA is one of the most important parameters to be taken into account when designing the dosage [3,10] because of WA is greater in RCA than the NA and it will influence the final physical and mechanical properties of the concrete [12,25]. As it will be analyzed later in the models and results, most are linked to this parameter.

The main properties of the RA used for the generation of each of the models is shown in Table 2. This Table shows the range of apparent densities of the WA at 24 hours and the LA of the RCA used in the investigations and used for the construction of the models. As can be observed, the wide range of the RAC properties covers the production of any kind of RAC.

The range of the mechanical properties of the RAC used for the different models is shown in Table 3. In this table are listed, for each type of model, the value range and the mean value

Table 2: Physical properties of the aggregates used to generate each of the models.

Properties of the RCA	Average value (kg/m^3)	Range of results (kg/m^3)	
		Minimum	Maximum
Particle density (kg/m3)	2191	1239	2750
Water absorption 24 hours (%)	5.39	0.24	16.34
Los Angeles abrasion coefficient (%)	41.26	21.56	65.45

Table 3: Mechanical properties of the aggregates used to generate each of the models.

Mechanical properties of the RCA	Average value (kg/m^3)	Range of results (kg/m^3)	
		Minimum	Maximum
f_c (MPa)	42.7	17.3	79.0
$f_{c,RC}$ (MPa)	38.3	13.4	79.0
$f_{ct,CC}$ (MPa)	5.0	2.1	15.5
$f_{ct,RC}$ (MPa)	4.3	1.4	16.5
E_{CC} (MPa)	32937	18900	49000
E_{RC} (MPa)	28540	13800	49000

of the compressive strength, tensile strength and modulus of elasticity of the control concretes (RCA = 0%) and of the RACs used in the present study. As can be seen the range of values of these properties is wide, so the resulting models will be applicable to a wide range of RAC.

Where $f_{c,\ RC}$ is compressive strength of the concrete with recycled aggregate substitution (MPa), f_c is the compressive strength of the reference concrete without substitution of recycled aggregates (for compressive strength models, MPa), $f_{ct,\ CC}$ is Indirect tensile strength of the reference concrete (MPa), $f_{ct,\ RC}$ is the indirect tensile strength of recycled aggregate concrete (MPa).

2.2 Nonlinear models

In general, it is really difficult to find a law that is capable of predicting the mechanical behaviour of concrete, whether or not incorporating recycled aggregate, so that, previously, a study of the dosage of the concrete is done. Subsequently, an experimental campaign is carried out in which the different additions of RA are contemplated and work is carried out with specific modifications of the dosage to achieve the mechanical and durability requirements proposed in the design phase. With these data, although it is true that a prediction law cannot be described, we can make models that can be adjusted and validated adequately from a statistical point of view, which is very useful.

With the problem presented in this way, it is necessary to formally obtain the relation between a variable y, dependent on a set of variables $x_1, x_2, \ldots, x_n$, which can be expressed in the relation $y = g(x_1, x_2, \ldots, x_n)$ where g can be linear or non-linear. The mathematical formulation of the relations obtained from the experimental data is based on the techniques of linear and nonlinear regression analysis. In the case of the former, the most generalized method for obtaining the values of the parameters involved in the equation is the least squares technique, which, by its linearity, leads to the solution of a linear system of the parameters. In contrast, nonlinear regression analysis leads to the use of numerical computation techniques to obtain a solution, the most common methods are Gauss-Newton gradient and Levenberg-Marquardt [25]. Each of these three algorithms poses a series of questions for its computational implementation and concentrate on obtaining the values of the partial derivatives. The procedure becomes cumbersome and also impractical, which makes it essential to use computer programs to facilitate the work, which is why in this study and for practical purposes has been made use of a powerful tool called Polimodels [26,27], which use the algorithm with the modules and techniques mentioned above to obtain and validate the parameters, which is considered a hybrid of different computational techniques, numerical calculus and statistical methods.

2.2.1 Polymodels software

Polimodels is software that has been implemented using Borland Delphi 6, and therefore is the Pascal object-oriented programming language. The graphical interface is based on menus and windows, and allows the definition of the $MANL_{\Phi}$, as well as various algorithm parameters and specification criteria previously. This executable generates a result in the form of files [25]. The algorithm that applies is appreciated in Fig. 1 in which basically the Gauss-Newton algorithm is used to obtain the parameters of the non-linear models.

As described briefly in Section 2.1, the dependent variable analyzed was modulus of elasticity (ERC) of RAC as a function of the independent variables that, depending on the phase, will be 2, 3 or 4.

The ACI 318 [11], EHE-08 [10] and other international regulations provide various formulas to be able to determine the tensile strength and the modulus of elasticity of the concrete, from the compressive strength. In this sense, and as can be seen in the different models, the study was done in the first three cases using as reference parameter the compressive strength of the reference concrete ($f_{c,RC}$), with recycled aggregate substitutions. In general terms, the variables that intervene directly in each model are:

$f_{c,RC}$: Compressive strength of concrete with recycled aggregate substitution (MPa).
E_{CC}: Modulus of elasticity of the reference concrete (MPa).
E_{RC}: Modulus of elasticity of the concrete with percentage of substitution of recycled aggregate calculated from the model (MPa).
RCA: Percentage of recycled aggregate by weight replacing conventional coarse aggregate.
WA: Percentage of water absorption of recycled aggregate at 24 hours.
LA: Los Angeles wear coefficient of the recycled aggregate.

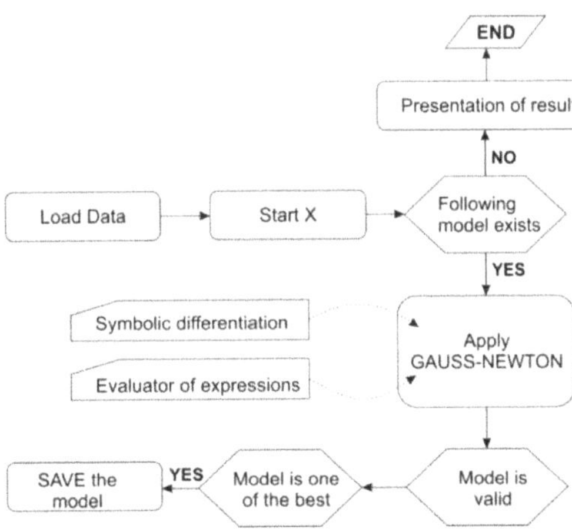

Figure 1: Summary of the process carried out in the research.

In addition for each of the independent variables we have analyzed the different models that are listed below:

a) $y = E_{RC}(f_{c,RC}, RCA)$
b) $y = E_{RC}(f_{c,RC}, RCA, WA)$
c) $y = E_{RC}(f_{c,RC}, RCA, WA, LA)$
d) $y = E_{RC}(E_{CC}, RCA)$
e) $y = E_{RC}(E_{CC}, RCA, WA)$
f) $y = E_{RC}(E_{CC}, RCA, WA, LA)$

In Table 4 all the variables analyzed in the study are grouped, in addition it is seen how each one of them has been combined to the different 6 generated models. As can be seen for the models there are two possible cases: the first is to know the value of the compressive strength of the concrete with its respective percentage of substitution and the second case is to know a reference parameter, that is, the modulus of elasticity but without substitution of NA by RCA.

2.2.2 Selection criteria of the generated models

Having the database completed and also the models that are intended to generate, a debugging is performed. The Polimodels gives us a series of possible models to use; a brief discussion of the different possible selection criteria will be made. We could consider the hypothesis-contrast design as to the advantages of one model over another, but the loss of normality derived from an absence of linearity in many of the resulting models, avoids the calculations of being rigorous and formal. For this reason, any attempt in this sense cannot be more than a mere approximation [26].

First, the value of the coefficient of determination R^2 is the most used estimator for regression in the field of research. However, the only consideration of the coefficient of determination as a measure of the goodness of the adjustment leads us to draw the wrong conclusions and the statement that additional measures are needed to complete the analysis [28].

Generally speaking, the model with the lowest average residue must be chosen as the 'best' model to explain the phenomenon [25]. In this sense, several 'distances' can be established

Table 4: Summary table of dependent and independent variables.

Model	Parameter studied	Dependent variable	Independent variables				
			$f_{c,RC}$	E_{CC}	*RCA*	*WA*	*LA*
1	*Modulus of Elasticity for RAC*	E_{RC}	x		X		
2			x		X	x	
3			x		X	x	x
4				x	X		
5				x	X	x	
6				x	X	x	x

* 28-day compressive strength, splitting strength and modulus of elasticity.

between the observed vector and the vector of estimated values; in other words, different rules for the residue vector can be considered, which are defined in eqxs. (1), (2) and (3).

$$\|e\| = \sqrt{e_1^2 + e_2^2 + e_3^2 + \ldots + e_k^2}. \quad (1)$$

$$\|e\|_1 = \max\{abs(e_1), abs(e_2), abs(e_3), \ldots, abs(e_k)\}. \quad (2)$$

$$\|e\|_2 = \frac{abs(e_1) + abs(e_2) + abs(e_3) + \ldots + abs(e_k)}{k}. \quad (3)$$

With $e_1, \ldots, e_k$ and k as the absolute value of the difference between the observed value and the estimated value; k=1..n. being the number of observations.

It is in this sense that new indexes are sought that incorporate a measure of interpretation more comfortable for the researcher, as is the case of the so-called Mean Absolute Percentage Error (MAPE) for regression models [29], which is defined by eqx. (4).

$$MAPE = \frac{abs\left(\frac{e_1}{y_1}\right) + abs\left(\frac{e_2}{y_2}\right) + abs\left(\frac{e_3}{y_3}\right) + \ldots abs\left(\frac{e_k}{y_k}\right)}{k}. \quad (4)$$

With $e_1, \ldots, e_k$ and k as the absolute value of the difference between the observed value and the estimated value; $y_1, \ldots, y_k$ are the observed value and k=1..n. is the number of observations.

As can be seen, $e_k / y_k \cdot 100$ is a measure of the percentage of error committed with respect to our observed variable, y_k, is in this sense that MAPE is chosen as the measure of discrimination and selection criteria.

3 RESULTS AND DISCUSSION

In the results presented below, we can see two parameters that have been crucial choosing the best model. It is provided the value of R^2 that although how is explained in detail, is not as representative, has helped us to make inferences as to which model to take but the parameter of choice has been the value of MAPE. Each model has been chosen in such a way that its MAPE is the minimum, as a direct measure that helps us to know how much is the percentage error that is committed between the observed variable and that of the model under study.

3.1 Models of modulus of elasticity taking as reference parameter $f_{c.RC}$

Usually, concrete incorporating recycled aggregates has a lower modulus of elasticity than concretes made of conventional aggregates. The modulus of elasticity is related to aggregates and the stiffness of the mortar. According to Bairagi *et al.* [30] the weakness of the new interfaces between the new mortar and the RAC with the old mortar can cause the development of cracks that affect the deformability of the concrete. This reduction is dependent of the compressive strength of the concrete giving a smaller reduction to higher strengths.

Figure 2 shows the relative modulus of elasticity of the concretes used for the study. The majority of RAC's have a modulus of elasticity lower than its reference (0% RAC) and only in specific cases exceed the value of the reference concrete.

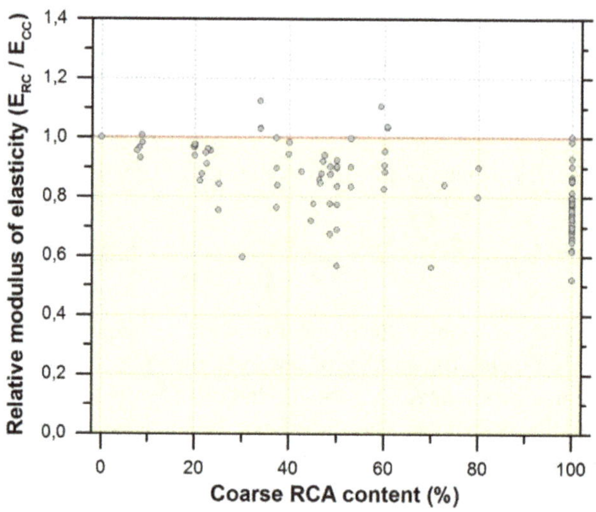

Figure 2: Relative modulus of elasticity of concrete vs coarse RCA content

Models 1, 2 and 3 for E_{RC} prediction based on the compressive strength of the RAC with the same percentage of substitution and the properties of the recycled aggregate are described in the following equations.

Model 1: $E_{RC} = \beta_1 \cdot RCA + \sqrt{\beta_2 \cdot f_{c,RC}} + \beta_3$

Model 2: $E_{RC} = \beta_1 \cdot WA + \beta_2 \cdot RCA + \sqrt{\beta_3 \cdot f_{c,RC}} + (\beta_4 \cdot WA \cdot RCA)$

Model 3: $E_{RC} = \sqrt{\beta_1 \cdot WA} + \beta_2 \cdot LA^4 + \beta_3 \cdot RCA^4 + \beta_4 \cdot f_{c,RC} + (\beta_5 \cdot f_{c,RC} \cdot RCA)$

Models 4, 5 and 6 for E_{RC} prediction as a function of the modulus of elasticity of the reference concrete E_{CC}, percentage of recycled aggregate replacement and its properties are described in the following equations.

Model 4: $E_{RC} = \beta_1 \cdot RCA^2 + \beta_2 \cdot E_{CC} + (\beta_3 \cdot RCA \cdot E_{CC})$

Model 5: $E_{RC} = \beta_1 \cdot WA + \beta_2 \cdot RCA^2 + \beta_3 \cdot E_{c,CC} + (\beta_4 \cdot E_{c,CC} \cdot RCA)$

Model 6: $E_{RC} = \beta_1 \cdot WA^3 + \beta_2 \cdot LA^4 + \beta_3 \cdot RCA + \beta_4 \cdot E_{CC} + (\beta_5 \cdot LA \cdot RCA)$

Table 5 shows the coefficients of the different models generated. The data obtained from the literature were compared with the results of the different models. Figure 3 shows the elastic modulus data observed and those calculated using the prediction models described above.

As can be appreciated, the models try to be as simple and clear as possible, so that their application is direct. In Fig. 3 the observed and calculated values are compared and it can be seen the 6 models with their respective MAPE and R^2, and their adjustment by adding each one of the variables. How was mentioned previously, the value of MAPE would be taken as the reference parameter because it is a more reliable measure of the correlation between observed and calculated values [28]. In addition, as a complementary measure, the value of R^2 is calculated.

Table 5: Prediction coefficients for models 1, 2, 3, 4, 5 y 6.

Models	β_1	β_2	β_3	β_4	β_5
Model 1	$-7.55{\cdot}10^1$	$2.04{\cdot}10^7$	–	–	–
Model 2	$8.05{\cdot}10^2$	$-3.78{\cdot}10^1$	$2.15{\cdot}10^7$	-8.92	–
Model 3	$6.71{\cdot}10^7$	$-7.45{\cdot}10^{-4}$	$3.75{\cdot}10^{-5}$	$4.88{\cdot}10^2$	-3.75
Model 4	$2.62{\cdot}10^{-1}$	$9.98{\cdot}10^{-1}$	$-0.34{\cdot}10^{-3}$	–	–
Model 5	$-4.68{\cdot}10^2$	$1.51{\cdot}10^{-1}$	1.06	$-2.74{\cdot}10^{-3}$	–
Model 6	2.03	$-3.81{\cdot}10^{-4}$	$1.25{\cdot}10^2$	1.00	-5.84

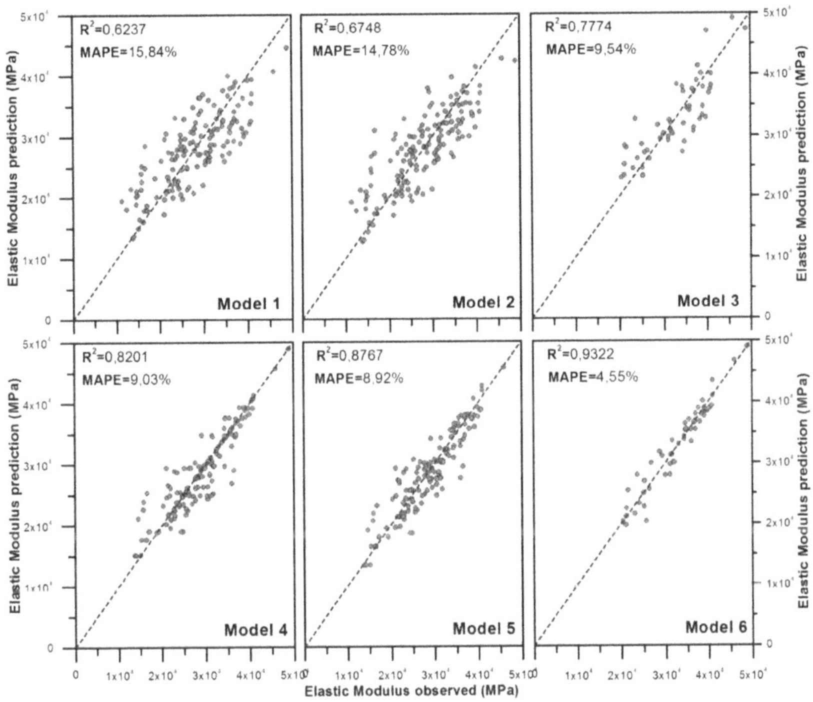

Figure 3: Prediction of calculated values and observed values of modulus of elasticity of the models 1, 2, 3, 4, 5 and 6.

Figure 4 shows the values of R^2 and MAPE for all the different models generated (1, 2, 3, 4, 5 and 6) in which, as variables are added to our prediction model. As can be observed there is a decrease in the MAPE values and at the same time an increase in the value of R^2. Mainly because R^2 and the number of data, usually vary inversely [28] in the same way number of variables also conditions this coefficient since R^2 is a non-decreasing function of the number of regressor variables present in the model, so that as the number of variables increases, R^2 increases [28]. According to this, R^2 measures the explanatory capacity of the different variables X, on the variable Y, this because introducing another regressive variable in the model, the explanatory level will be greater between the two, than with the first, or in any case, will not diminish because the first variable continues as explanatory. In

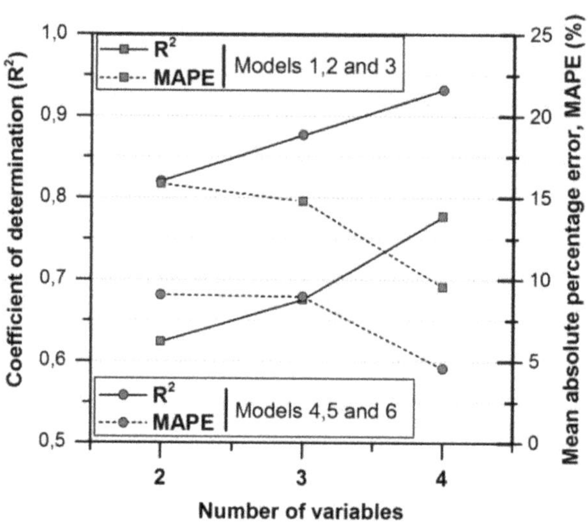

Figure 4: Coefficient of determination (R^2) and percentage of Mean Absolute Percentage Error (MAPE) as a function of the number of variables for the Modulus of Elasticity prediction models.

addition, it can be appreciated how MAPE decreases as the number of explanatory variables increases.

4 CONCLUSIONS

In this paper, the main objective has been to predict the modulus of elasticity of RAC. In order to do this, different studies that cover a wide range of results in the mechanical properties of RAC have been selected in order to obtain nonlinear models that can be adapted to the majority of RAC.

It has been taken into account, not only the percentage of incorporation of recycled aggregate and its physical properties, but also the modulus of elasticity of the reference concrete from which it is intended to replace part of the natural aggregate by recycled aggregate.

The nonlinear models obtained are able to predict the modulus of elasticity of RAC knowing the physical properties of the RCA, the RCA content and knowing the modulus of elasticity of the reference concrete (0% RCA). This is an economic saving for the producers since it would not be necessary to manufacture RAC to estimate the value of its modulus of elasticity.

The greater knowledge on the properties of the RCA, such as the WA and LA values, provides nonlinear models with a lower MAPE and a higher value of R^2, so we can state that the selection of these variables as explanatory of the modulus of elasticity is completely correct.

These models allow engineers to estimate the modulus of elasticity of recycled concrete as a function of the content of recycled aggregate and the knowledge of the properties of this one (WA and LA).

ACKNOWLEDGEMENTS

This research was financed by the University of Alicante through projects VIGROB-256 and GRE13-03. Thanks to the National Autonomous University of Honduras for subsidizing studies at the Masters level of one of the researchers at the University of Alicante.

REFERENCES

[1] Rahal, K., Mechanical properties of concrete with recycled coarse aggregate. *Building and Environment*, **42**, pp. 407–415, 2007.
https://doi.org/10.1016/j.buildenv.2005.07.033

[2] Corinaldesi, V., Mechanical and elastic behaviour of concretes made of recycled-concrete coarse aggregates. *Construction and Building Materials*, **24**, pp. 1616–1620, 2010.
https://doi.org/10.1016/j.conbuildmat.2010.02.031

[3] Etxeberria, M., Vazquez, E., Mari, A. & Barra, M., Influence of amount of recycled coarse aggregates and production process on properties of recycled aggregate concrete. *Cement and Concrete Research*, **37**, pp. 735–742, 2007.
https://doi.org/10.1016/j.cemconres.2007.02.002

[4] Sagoe-Crentsil, K.K., Brown, T. & Taylor, A.H., Performance of concrete made with commercially produced coarse recycled concrete aggregate. *Cement and Concrete Research*, **31**, pp. 707–712, 2001.
https://doi.org/10.1016/s0008-8846(00)00476-2

[5] Safiuddin, M., Alengaram, U.J., Rahman, M.M., Salam, M.A. & Jumaat, M.Z., Use of recycled concrete aggregate in concrete: a review. *Journal of Civil Engineering and Management*, **19**, pp. 796–810, 2013.
https://doi.org/10.3846/13923730.2013.799093

[6] Tabsh, S.W. & Abdelfatah, A.S., Influence of recycled concrete aggregates on strength properties of concrete. *Construction and Building Materials*, **23**, pp. 1163–1167, 2009.
https://doi.org/10.1016/j.conbuildmat.2008.06.007

[7] Rao, M.C., Bhattacharyya, S.K. & Barai, S.V., Influence of field recycled coarse aggregate on properties of concrete. *Materials and Structures*, **44**, pp. 205–220, 2011.
https://doi.org/10.1617/s11527-010-9620-x

[8] McNeil, K. & Kang, T.H.K., Recycled concrete aggregates: a review. *International Journal of Concrete Structures and Materials*, **7**, pp. 61–69, 2013.
https://doi.org/10.1007/s40069-013-0032-5

[9] Topcu, I.B. & Guncan, N.F., Using waste concrete as aggregate. *Cement and Concrete Research*, **25**, pp. 1385–1390, 1995.
https://doi.org/10.1016/0008-8846(95)00131-u

[10] Ministerio de Fomento, *EHE-08: Code on structural concrete: Articles and annexes*. Ministerio de Fomento, Madrid, 2011.

[11] American Concrete Institute, Building code requirements for structural concrete (ACI 318-08) and commentary, American Concrete Institute, Farmington Hills, Michigan, 2008.

[12] Padmini, A.K., Ramamurthy, K. & Mathews, M.S., Influence of parent concrete on the properties of recycled aggregate concrete. *Construction and Building Materials*, **23**, pp. 829–836, 2009.
https://doi.org/10.1016/j.conbuildmat.2008.03.006

[13] Barbudo, A., de Brito, J., Evangelista, L., Bravo, M. & Agrela, F., Influence of water-reducing admixtures on the mechanical performance of recycled concrete. *Journal of Cleaner Production*, **59**, pp. 93–98, 2013.
https://doi.org/10.1016/j.jclepro.2013.06.022

[14] Bravo, M., de Brito, J., Pontes, J. & Evangelista, L., Mechanical performance of concrete made with aggregates from construction and demolition waste recycling plants. *Journal of Cleaner Production*, **99**, pp. 59–74, 2015. https://doi.org/10.1016/j.jclepro.2015.03.012

[15] Cui, H.Z., Shi, X., Memon, S.A., Xing, F. & Tang, W., Experimental study on the influence of water absorption of recycled coarse aggregates on properties of the resulting concretes. *Journal of Materials in Civil Engineering*, **27**, pp. 1–9, 2015. https://doi.org/10.1061/(asce)mt.1943-5533.0001086

[16] Gomes, M., de Brito, J. & Bravo, M., Mechanical performance of structural concrete with the incorporation of coarse recycled concrete and ceramic aggregates. *Journal of Materials in Civil Engineering,* **26**, pp. 1–10, 2014. https://doi.org/10.1061/(asce)mt.1943-5533.0000973

[17] Ismail, S. & Ramli, M., Mechanical strength and drying shrinkage properties of concrete containing treated coarse recycled concrete aggregates. *Construction and Building Materials*, **68**, pp. 726–739, 2014. https://doi.org/10.1016/j.conbuildmat.2014.06.058

[18] Pepe, M., Toledo, R.D., Koenders, E.A.B. & Martinelli, E., Alternative processing procedures for recycled aggregates in structural concrete. *Construction and Building Materials*, **69**, pp. 124–132, 2014. https://doi.org/10.1016/j.conbuildmat.2014.06.084

[19] Purushothaman, R., Amirthavalli, R.R. & Karan, L., Influence of treatment methods on the strength and performance characteristics of recycled aggregate concrete. *Journal of Materials in Civil Engineering*, **27**, pp. 1–7 ,2015. https://doi.org/10.1061/(asce)mt.1943-5533.0001128

[20] Sheen, Y.N., Wang, H.Y., Juang, Y.P., Le, D.H., Assessment on the engineering properties of ready-mixed concrete using recycled aggregates, *Construction and Building Materials*, **45**, pp. 298–305, 2013. https://doi.org/10.1016/j.conbuildmat.2013.03.072

[21] Xiao, J.Z., Li, J.B. & Zhang, C., Mechanical properties of recycled aggregate concrete under uniaxial loading. *Cement and Concrete Research*, **35**, pp. 1187–1194, 2005. https://doi.org/10.1016/j.cemconres.2004.09.020

[22] Xuan, D.X., Zhan, B.J., Poon, C.S., Assessment of mechanical properties of concrete incorporating carbonated recycled concrete aggregates. *Cement & Concrete Composites*, **65,** pp. 67–74, 2016. https://doi.org/10.1016/j.cemconcomp.2015.10.018

[23] Zega, C.J. & Di Maio, A.A., Recycled concretes made with waste ready-mix concrete as coarse aggregate. *Journal of Materials in Civil Engineering*, **23**, pp. 281–286, 2011. https://doi.org/10.1061/(asce)mt.1943-5533.0000165

[24] Dilbas, H., Simsek, M. & Cakir, O., An investigation on mechanical and physical properties of recycled aggregate concrete (RAC) with and without silica fume. *Construction and Building Materials*, **61**, pp. 50–59, 2014. https://doi.org/10.1016/j.conbuildmat.2014.02.057

[25] Verdu, F., Un algoritmo para la construcción múltiple de modelos matemáticos no lineales y el estudio de su estabilidad, Thesis, Alicante, pp. 1–617, 2004.

[26] Verdu, F. & Villacampa, Y., A computational algorithm for the multiple generation of nonlinear mathematical models and stability study. *Advances in Engineering Software*, **39**, pp. 430–437, 2008.
https://doi.org/10.1016/j.advengsoft.2007.03.004

[27] Verdu, F. & Villacampa, Y., A computer program for a Monte Carlo analysis of sensitivity in equations of environmental modelling obtained from experimental data. *Advances in Engineering Software*, **33**, pp. 351–359, 2002.
https://doi.org/10.1016/s0965-9978(02)00023-6

[28] Martínez, E., Errores frecuentes en la interpretación del coeficiente de determinación lineal. *Anuario Jurídico y Económico Escurialense,* **38**, pp. 1133–3677, 2005.

[29] de Myttenaere, A., Golden, B., Le Grand, B. & Rossi, F., Mean absolute percentage error for regression models. *Neurocomputing,* **192**, pp. 38–48, 2016.
https://doi.org/10.1016/j.neucom.2015.12.114

[30] Bairagi, N.K., Ravande, K. & Pareek, V.K., Behavior of concrete with different proportions of natural and recycled aggregates. *Resources Conservation and Recycling*, **9**, pp. 109–126, 1993.
https://doi.org/10.1016/0921-3449(93)90036-f

FREE VIBRATIONS OF STEPPED NANO-BEAMS

JAAN LELLEP & ARTUR LENBAUM
Institute of Mathematics and Statistics, University of Tartu, Estonia.

ABSTRACT
Free vibrations of beams and rods made of nano-materials are investigated. It is assumed that the dimensions of cross sections of nano-beams are piecewise constant and that the beams are weakened with cracks. It is expected that the vibrational behaviour of the nano-material can be described within the non-local theory of elasticity and that the crack induces additional local compliance. The latter is coupled with the stress intensity coefficient at the crack tip.
Keywords: beam, crack, non-local elasticity, nano-material, vibration.

1 INTRODUCTION

In the recent decade there has been considerable progress in the use of nano-plates and nano-beams due to the need of micro- and nano-electromechanical systems. It is known that the behaviour of nano-structures can be modelled with non-local theories of elasticity (see Eringen [1, 2], Reddy [3]). Lim [4, 5] has developed non-local bending theories and applied these for quasistatically loaded nano-beams. Buckling of nano-beams was studied with the help of non-linear non-local models by Reddy [3], Emam [6], Challamel *et al.* [7]. Analytical solutions for the transverse vibration of simply supported and clamped at both ends nano-beams with axial force are obtained by Li *et al.* [8], Lu *et al.* [9]. In Ref. [10] the Ritz method is accommodated for buckling and vibration of non-local beams. The vibrations of nano-beams with cracks are studied in Ref. [11] making use of the non-local theory of elasticity. In the present paper the free vibrations of nano-beams clamped at both ends are investigated. It is assumed that the nano-beams have stepped cross sections and that the beams are weakened with cracks located at the corners of steps.

2 PROBLEM FORMULATION

Let us consider natural vibrations of a nano-beam of length l. The edges of the beam at $x = 0$ and $x = l$ are fully clamped. The coordinate axis Ox coincides with the axis of corresponding straight beam; the onset of coordinates is located at the centre of the left-hand end of the beam. It is assumed that the nano-beam has rectangular cross sections with the width b and the height

$$h = \begin{cases} h_0, x \in (0,a), \\ h_1, x \in (a,l). \end{cases} \tag{1}$$

In eqn. (1) the quantities h_0, h_1 and a are considered as given numbers. The nano-beam is weakened with a crack of length c at $x = a$. The crack length is expected to be constant. Thus the crack area is

$$S_c = cb. \tag{2}$$

The aim of the study is to determine the eigenfrequencies of natural vibrations of the nano-beam and to clarify the sensitivity of eigenfrequencies on the geometrical parameters of the beam and on the physical parameters of the material. The material of the nano-beam is

DOI: 10.2495/CMEM-V6-N4-716-725

assumed to be an elastic material obeying the constitutional equations of a non-local theory of elasticity (see Eringen [2], Lellep & Lenbaum [12]).

3 EQUATION OF MOTION

It was shown in the previous study by the authors [12] that the constitutional equations of non-local elasticity lead to the equation

$$M - \eta \frac{\partial^2 M}{\partial x^2} = M_C, \tag{3}$$

where M stands for the bending moment and M_c is the moment calculated by the rules of the classical bending theory of thin-walled beams. Evidently (see Ref. [12]) the bending moment M_c and the deflection w are coupled as

$$M_C = -EI \frac{\partial^2 w}{\partial x^2}, \tag{4}$$

where E is the Young's modulus and I stands for the moment of inertia of the cross section of the beam. In eqn (3) η is a material constant (it is connected with the dimensions of the lattice of the nano-material). Combining eqns (3) and (4) one can write

$$M = \eta \frac{\partial^2 M}{\partial x^2} - EI \frac{\partial^2 w}{\partial x^2}. \tag{5}$$

On the other hand, it follows from the equilibrium equations (see Ref. [8]), that

$$\frac{\partial^2 M}{\partial x^2} = \mu \frac{\partial^2 w}{\partial t^2} - N \frac{\partial^2 w}{\partial x^2}. \tag{6}$$

In eqn (6) $\mu = \rho bh$ stands for the mass per unit length of the nano-beam and ρ being the density of the material. Here t denotes time and N is the axial tension applied at the edges of the beam. Substituting eqn (6) into eqn (5) results in

$$M = \eta\left(-Nw'' + \mu\ddot{w}\right) - EIw'', \tag{7}$$

where the notation

$$w' = \frac{\partial w}{\partial x},$$

$$\dot{w} = \frac{\partial w}{\partial t} \tag{8}$$

is used. Combining eqns (6) and (7) one easily obtains

$$\eta\left(-Nw^{IV} + \mu\ddot{w}''\right) - EIw^{IV} = \mu\ddot{w} - Nw''. \tag{9}$$

The latter can be presented in the form

$$\left(\eta N + EI\right)w^{IV} + \mu\left(\ddot{w} - \eta\ddot{w}''\right) - Nw'' = 0. \tag{10}$$

This is the equation of motion for nano-beams. Taking $\eta = 0$ in eqn (10), one obtains

$$EIw^{IV} - Nw'' + \mu\ddot{w} = 0. \tag{11}$$

The latter is the equation of free vibrations of beams in the classical beam theory. The eqn (10) can be solved with the method of separation of variables making use of appropriate boundary conditions. In the case of beams clamped at both ends the boundary conditions are

$$w(0,t) = 0, w'(0,t) = 0 \tag{12}$$

and

$$w(l,t) = 0, w'(l,t) = 0. \tag{13}$$

4 SOLUTION OF THE GOVERNING EQUATIONS

For solution of the linear fourth order differential equation with partial derivatives let us assume that

$$w(x,t) = W_j(x) \cdot T(t), \tag{14}$$

where $j = 0$ for $x \in (0,a)$ and $j = 1$ if $x \in (a,l)$. In eqn (14) the function $W_j(x)$ depends only on x and the function $T(t)$ is the function of time t. Differentiating eqn (14) with respect to x and t one can easily recheck that

$$\frac{\partial w}{\partial x} = W_j'(x) \cdot T(t),$$

$$\frac{\partial^2 w}{\partial x^2} = W_j''(x) \cdot T(t),$$

$$\frac{\partial^3 w}{\partial x^3} = W_j'''(x) \cdot T(t),$$

$$\frac{\partial^4 w}{\partial x^4} = W_j^{IV}(x) \cdot T(t),$$

$$\frac{\partial w}{\partial t} = W_j \cdot \dot{T}(t),$$

$$\frac{\partial^2 w}{\partial t^2} = W_j \cdot \ddot{T}(t). \tag{15}$$

In eqns (15) on has to take $j = 0$, if $x \in (0,a)$ and $j = 1$, if $x \in (a,l)$. Substituting the derivatives in eqns (15) into eqn (10), one obtains

$$(\eta N + EI_j)W_j^{IV} \cdot T + \mu_j(W_j \cdot \ddot{T} - \eta W_j'' \cdot \ddot{T}) - NW_j'' \cdot T = 0. \tag{16}$$

The separation of variables in eqn (16) leads to the equation

$$\frac{\left(\eta N + EI_j\right)\frac{W_j^{IV}}{W_j} - N\frac{W_j''}{W_j}}{\left(\eta\frac{W_j''}{W_j} - 1\right)\mu_j} = \frac{\ddot{T}}{T} = -\omega^2, \tag{17}$$

where ω stands for the frequency of natural vibrations. It immediately follows from eqn (17) that

$$\ddot{T} + \omega^2 T = 0 \tag{18}$$

and

$$\left(\eta N + EI_j\right)W_j^{IV} - W_j''\left(N - \mu_j\eta\omega^2\right) - \mu_j\omega^2 W_j = 0. \tag{19}$$

It is reasonable to assume that at the initial time instant $w\left(x,0\right) = 0$;$\dot{w}\left(x_*,0\right) = v_0$. Taking eqn (14) into account, one can assume that

$$T\left(0\right) = 0, \dot{T}\left(0\right) = 1. \tag{20}$$

Evidently, the particular solution of eqn (18) satisfying the initial conditions in eqn (20) can be presented as

$$T\left(t\right) = \frac{1}{\omega}\sin\left(\omega t\right). \tag{21}$$

The eqn (19) is a linear fourth order differential equation with constant coefficients. The characteristic equation corresponding to eqn (19) has the form

$$\left(\eta N + EI_j\right)\lambda^4 - \lambda^2\left(N - \mu_j\eta\omega^2\right) - \mu_j\omega^2 = 0. \tag{22}$$

The roots of the characteristic equation (22) are

$$\lambda_{1,2} = \pm\vartheta_j\ ; \lambda_{3,4} = \pm i\beta_j, \tag{23}$$

where i is the imaginary unit,

$$\vartheta_j = \sqrt{\frac{1}{2\left(\eta N + EI_j\right)}\left[N - \mu_j\eta\omega^2 + A_j\right]} \tag{24}$$

and

$$\beta_j = \sqrt{\frac{-N + \mu_j\eta\omega^2 + A_j}{2\left(\eta N + EI_j\right)}}. \tag{25}$$

In eqns (24) and (25) for the briefness sake the notation

$$A_j = \sqrt{\left(N - \mu_j\eta\omega^2\right)^2 + 4\mu_j\omega^2\left(\eta N + EI_j\right)} \tag{26}$$

is used. Taking eqns (22)–(26) into account, one can present the general solution of eqn (19) as

$$W(x) = C_{1j}\cosh \vartheta_j x + C_{2j}\sinh \vartheta_j x + C_{3j}\cos \beta_j x + C_{4j}\sin \beta_j x \tag{27}$$

for $x \in S_j$, provided $S_0 = [0,a]$, $S_1 = [a,l]$. In order to define particular solutions for regions S_0 and S_1, one has to satisfy the boundary conditions in eqns (12) and (13), also the boundary conditions for $M(x,t)$ and the appropriate continuity requirements for the deflection and bending moment.

5 INTERMEDIATE CONDITIONS

For modelling the influence of the crack on the eigenfrequencies the method of distributed line springs suggested by Dimarogonas and his co-workers [13] will be used. According to this method, one has to introduce the additional compliance K, so that

$$w'(a+0,t) - w'(a-0,t) = -KM(a,t), \tag{28}$$

where

$$w'(a \pm 0,t) = \lim_{x \to a \pm 0} w'(x,t). \tag{29}$$

Making use of eqns (7) and (14), one can introduce functions $m_j(x)$, so that

$$M(x,t) = m_j(x)T(t), \tag{30}$$

where

$$m_j(x) = -\eta\mu_j\omega^2 W - (\eta N + EI_j)W'' \tag{31}$$

for $x \in S_j$; $j = 0,1$. The relations in eqns (14), (30) and (31) admit to present the requirement in eqn (28) as

$$W'(a+) - W'(a-) = K\left(\eta\mu_0\omega^2 W(a-) + (\eta N + EI_0)W''(a-)\right). \tag{32}$$

Since the bending moment M is considered at $x = a$, at any time instant, one gets $m_1(a+) = m_0(a-)$, or according to eqn (31)

$$\left[\eta\mu\omega^2 W(a)\right] + \left[(\eta N + EI)W''(a)\right] = 0, \tag{33}$$

here square brackets denote the finite jump of corresponding quantity. This means that

$$\left[y(a)\right] = y(a+0) + y(a-0) \tag{34}$$

for any variable $y = y(x)$. Due to their physical background, it is clear that the deflection $W(x)$ and the shear force $Q = M'$ are continuous as well. Thus one has

$$\left[W(a)\right] = 0 \tag{35}$$

and

$$\left[\eta\mu\omega^2 W'(a)\right] + \left[(\eta N + EI)W'''(a)\right] = 0. \tag{36}$$

Following Dimarogonas [13], also Lellep and Kraav [14], Lellep and Liyvapuu [15], the additional compliance caused by the crack is evaluated as

$$K = \frac{6\pi h\left(1-\nu^2\right)}{EI} f\left(s\right), \tag{37}$$

where $I = \min\left(I_0, I_1\right)$, $h = \min\left(h_0, h_1\right)$ and

$$f\left(s\right) = \int_0^s y F^2\left(y\right) dy. \tag{38}$$

The function $F = F\left(s\right)$ is obtained with the help of experimental data. In the current paper, as in [12]

$$F\left(s\right) = 1.93 - 3.07s + 14.53s^2 - 25.11s^3 + 25.8s^4. \tag{39}$$

6 NATURAL FREQUENCIES OF THE NANO-BEAM

In order to define the eigenfrequencies of the vibration, one has to specify the constants $C_{ij}\left(i = 1,\ldots,4\ ; j = 0,1\right)$ in eqn (27). Making use of the boundary conditions in eqns (12) and (13) and taking into account eqns (14) and (27), one easily obtains

$$C_{30} = -C_{10},$$

$$C_{40} = -\frac{\vartheta_0}{\beta_0} C_{20} \tag{40}$$

and

$$C_{11}\cosh\vartheta_1 l + C_{21}\sinh\vartheta_1 l + C_{31}\cos\beta_1 l + C_{41}\sin\beta_1 l = 0,$$

$$C_{11}\vartheta_1\sinh\vartheta_1 l + C_{21}\vartheta_1\cosh\vartheta_1 l - C_{31}\beta_1\sin\beta_1 l + C_{41}\beta_1\cos\beta_1 l = 0. \tag{41}$$

Due to the continuity of the displacement W and the shear force at $x = a$, one has to take into account the requirements

$$\left[W\left(a\right)\right] = 0,$$

$$\left[M'\left(a\right)\right] = 0, \tag{42}$$

where according to eqn (30) $M' = m_j'\left(x\right)T\left(t\right)$. It is worthwhile to mention, that the total set of boundary and intermediate requirements, which consists of eqns (32), (33) and (40) – (42), is a linear system of algebraic equations with respect to unknown constants $C_{10},\ldots,C_{40}\ ;C_{11},\ldots,C_{41}$. Since it is a linear homogenous system, the non-trivial solution exists under the condition, that the determinant Δ of this vanishes. Equalizing $\Delta = 0$ leads to the equation for determination of natural frequencies.

7 DISCUSSION

The obtained set of algebraic equations is solved numerically with the aid of the computer program Mathcad. Numerical results are obtained for the nano-beam with dimensions $l = 500nm, b = 10nm$ and the material constant $\sqrt{\eta} = 2nm$. The results of calculations are presented in Figs 1–5, where the lowest eigenfrequency ω is depicted. Here the stepped nano-beam with thicknesses $h_0 = 50nm$, $h_1 = \gamma \cdot h_0$ is investigated, where $s = \frac{c}{h_0}$ and a is the point at which the crack occurs. In Figs 1– 4 the density of the material is $\rho = 7850\frac{kg}{m^3}$ and Young's modulus $E = 200GPa$. Figures 1 and 3 correspond to the case where $\gamma = 1.2$. Different curves in Figs 1 and 3 correspond to the crack extension $s = 0.2$;$s = 0.3$ and $s = 0.4$. It can be seen from Figs 1 and 3 that the shorter is the crack the higher is the natural frequency of vibration. The dependence of the eigenfrequency on the axial force N and the crack location a can be seen in Figs 1 and 3, respectively. In Figs 2 and 4 different curves correspond to

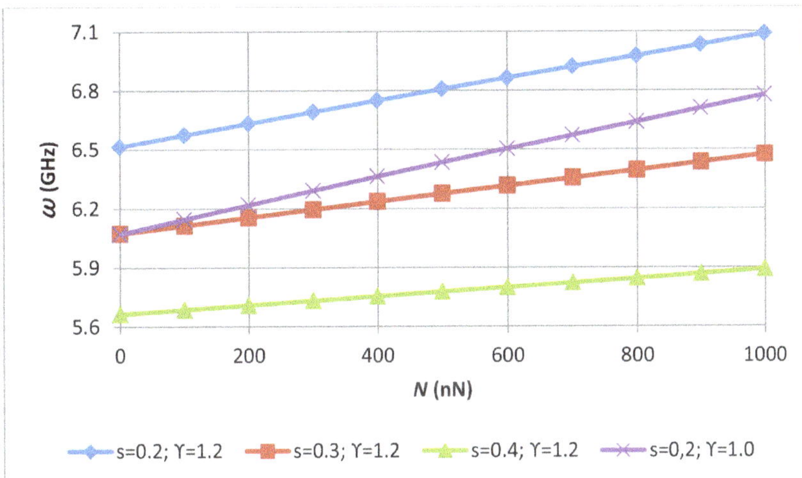

Figure 1: Influence of the axial force on the eigenfrequency for different crack extensions.

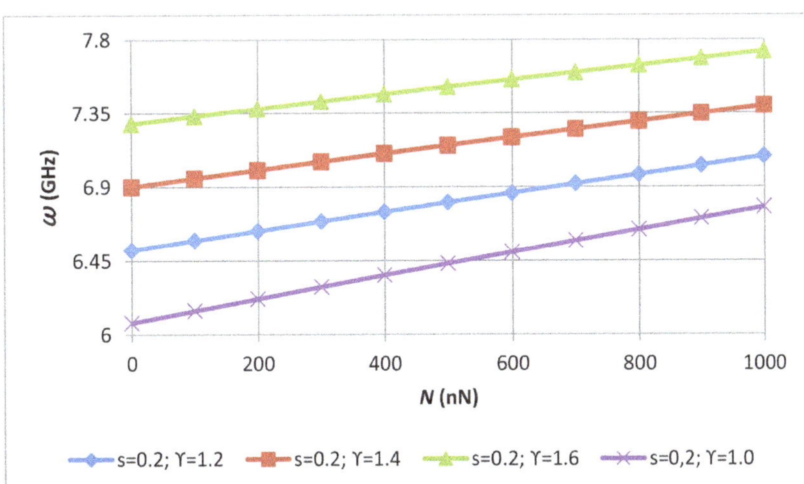

Figure 2: Influence of the axial force on the eigenfrequency for different ratios of thicknesses.

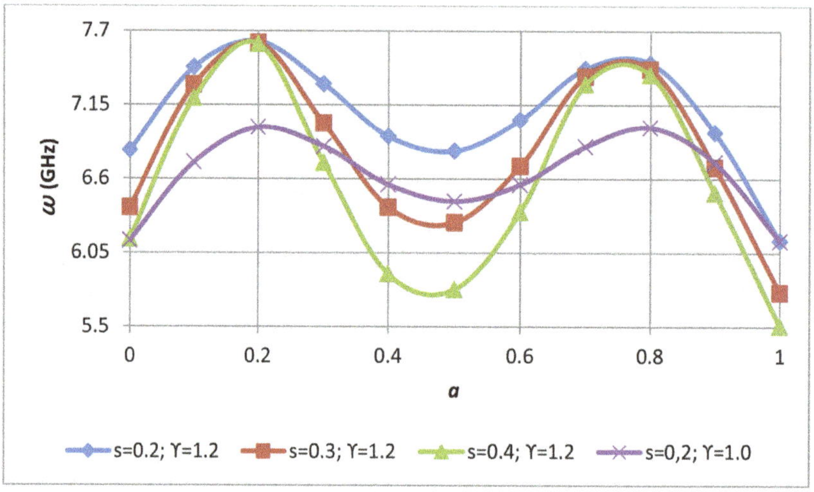

Figure 3: Influence of the crack location on the eigenfrequency for different crack extensions.

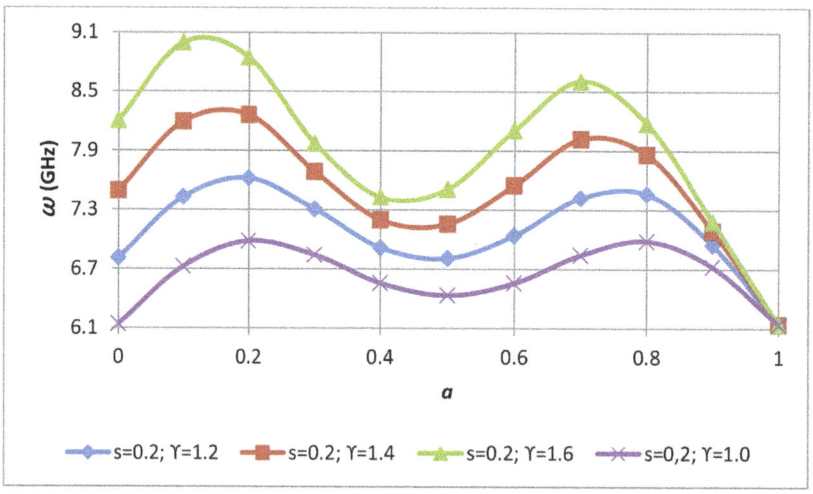

Figure 4: Influence of the crack location on the eigenfrequency for different ratios of thicknesses.

the different ratios of thicknesses and a fixed crack extension $s = 0.2$. It shows from Figs 2 and 4 that a higher ratio of thickness corresponds to a higher frequency. The eigenfrequency versus the axial force N (see Fig. 2) and the step location a (see Fig. 4) can be observed here as well. Also an example for the case of a nano-beam with constant thickness is presented in each of Figs 1–4. In Fig. 5 the dependence of the eigenfrequency on the material of the nano-beam is depicted for a case where $\gamma = 1.2$, $s = 0.2$ and the axial force $N = 500nN$. Different curves in Fig. 5 correspond to a nano-beam made of iron, copper, zinc or nickel, respectively.

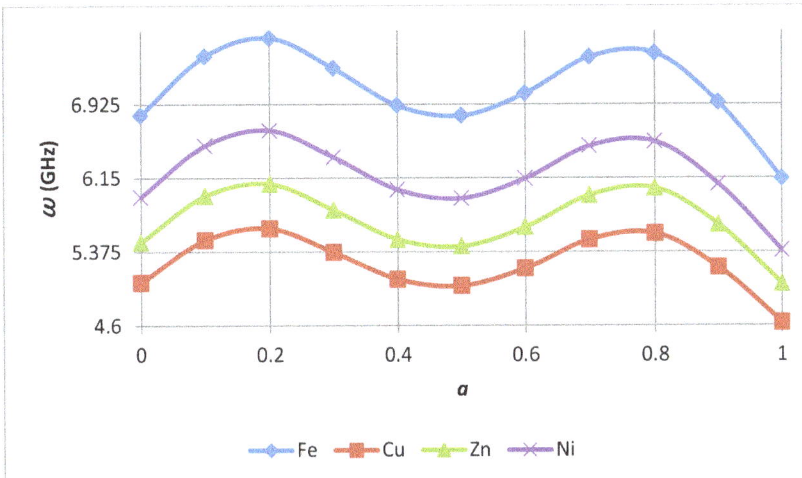

Figure 5: Influence of the crack location on the eigenfrequency for nano-beams made of different materials.

8 CONCLUDING REMARKS

A vibration analysis of nano-beams clamped at both ends was undertaken in the case of stepped beams with stable cracks or crack-like defects. It was shown that the defects have essential influence on the eigenfrequencies of nano-beams. Calculations carried out showed, that the lowest eigenfrequencies correspond to nano-beams with more severe defects. It was revealed that the higher is the tension applied to the nano-beam the higher is the frequency of natural vibrations.

ACKNOWLEDGEMENT

The partial support from the Institutional Research Funding IUT 20-57 of Estonian Ministry of Education and Research is gratefully acknowledged.

REFERENCES

[1] Eringen, A.C., *Nonlocal Continuum Field Theories*. Springer, New York, 2002.

[2] Eringen, A.C., On differential equations of nonlocal elasticity and solutions of screw dislocation and surface waves. *Journal Of Applied Physics*, **54**(9), pp. 4703–4710, 1983. https://doi.org/10.1063/1.332803

[3] Reddy, J.N., Nonlocal theories for bending, buckling and vibration of beams. *International Journal of Engineering Science*, **45**(2–8), pp. 288–307, 2007. https://doi.org/10.1016/j.ijengsci.2007.04.004

[4] Lim, C.W., Equilibrium and static deflection for bending of a nonlocal nanobeam. *Advances in Vibration Engineering*, **8**, pp. 277–300, 2009.

[5] Lim, C.W., On the truth of nanoscale for nanobeams based on nonlocal elastic stress field theory: equilibrium, governing equation and static deflection. *Applied Mathematics and Mechanics,* **31**(1), pp. 37–54, 2010. https://doi.org/10.1007/s10483-010-0105-7

[6] Emam, S.A., A general nonlocal nonlinear model for buckling of nanobeams. *Applied Mathematics Modelling*, **37**(10–11), pp. 6929–6939, 2013. https://doi.org/10.1016/j.apm.2013.01.043
[7] Challamel, N., Meftah, S.A. & Bernard,F., Buckling of elastic beams on non-local foundation: A revisiting of Reissner model. *Mechanics Research Communications*, **37**(5), pp. 472–475, 2010. https://doi.org/10.1016/j.mechrescom.2010.05.007
[8] Li, C., Lim, C.W., Yu, J.L. & Zheng, O.C., Analytical solutions for vibration of simply supported nonlocal nanobeams with an axial force. *International Journal Structural Stability and Dynamics*, **11**(2), pp. 257–271, 2011. https://doi.org/10.1142/s0219455411004087
[9] Lu, P., Lee, H.P. & Lu, C., Dynamic properties of flexural beams using a nonlocal elasticity model. *Journal Applied Physics*, **99**(7), p. 073510, 2006. https://doi.org/10.1063/1.2189213
[10] Ghannadpour, S.A.M. & Fazilati, B.M., Bending, buckling and vibration problems of nonlocal Euler beams using Ritz method. *Composite Structures*, **96**, pp. 584–589, 2013. https://doi.org/10.1016/j.compstruct.2012.08.024
[11] Roostai, H. & Haghpanahi, M., Vibration of nanobeams of different boundary conditions with multiple cracks based on nonlocal elasticity theory. *Applied Mathematical Modelling*, **38**(3), pp. 1159–1169, 2014. https://doi.org/10.1016/j.apm.2013.08.011
[12] Lellep, J. & Lenbaum, A., Natural vibrations of a nano-beam with cracks. *International Journal of Theoretical and Applied Mechanics*, **1**(1), pp. 247–252, 2016.
[13] Dimarogonas, A.D., Vibration of cracked structures: A state of the art review. *Engineering Fracture Mechanics*, **55**(5), pp. 831–857, 1996. https://doi.org/10.1016/0013-7944(94)00175-8
[14] Lellep, J. & Kraav, T., Buckling of beams and columns with defects. *International Journal of Structural Stability and Dynamics*, **16**(8), 2016. https://doi.org/10.1142/s0219455415500480
[15] Lellep, J. & Liyvapuu, A., Natural vibrations of stepped arches with cracks. *Agronomy Research*, **14**(3), pp. 821–830, 2016.

NUMERICAL MODEL FOR DESCRIBING THE SEGREGATION PHENOMENON IN LIGHTWEIGHT CONCRETE USING DENSITY SECTIONS

A. J. TENZA-ABRIL[1], Y. VILLACAMPA[2], F. BAEZA-BROTONS[1],
J. F. NAVARRO-GONZÁLEZ[2] & A. M. SOLAK[1]
[1]Department of Civil Engineering, University of Alicante, Spain.
[2]Department of Applied Mathematics, University of Alicante, Spain.

ABSTRACT

In this work, numerical models were obtained for describing the segregation phenomenon in lightweight aggregate concrete. To that end, a numerical methodology based on the generation of geometric models of finite elements has been applied, selecting those that describe better this phenomenon. The use of lightweight aggregate concretes (LWC) allows greater design flexibility and substantial cost savings. It is also well known that it contributes to a positive impact on the energy consumption of a building due to the high-thermal resistance values. However, lightweight concretes are susceptible to present aggregate segregation due to density differences between its components during concrete vibration. Segregation in concrete may strongly affect the concrete global properties. This fact justifies the needs for the identification and quantification of this phenomenon, in order to estimate the concrete segregation experimentally, a LWC was mixed in laboratory conditions. Controlled segregation was caused applying different times of internal vibration in a cylinder specimen. The specimens were horizontally sectioned in order to obtain the density in each section because the segregation index can be estimated obtaining a relation by comparing the densities of the upper and lower parts. Firstly, ANOVA test was performed to determine the statistical significance ($p<0.05$) of the differences in the density of the different sections, differences in the aggregate type and differences in the time of concrete vibration. Results show that there is a significant difference of each section and there is no significant difference of each lightweight aggregate used to mix the concrete in spite of their different density. In order to model the segregation in the LWC, at first, linear models were considered and rejected because for not explaining the phenomenon. However, the application of numerical models shows good results to describe the phenomenon of segregation in LWC.

Keywords: ANOVA, compactation, lightweight concrete, prediction models, segregation, vibration.

1 INTRODUCTION

The history of lightweight concrete dates back to over 3000 years ago [1, 2]. Structural LWC has been widely used following advances in production technology for lightweight aggregates (LWA). Several advantages of LWC can be achieved, to reduce dead load for structures [3], the reduction of the density produce an increase in the thermal resistance of the concrete and increase the energy efficiency of buildings with this kind of material. But this kind of concrete tends to segregate due to the low density of the aggregates of the mixture.

After concrete placement, concrete can contain up to 20% of entrapped air. This percentage varies according to the type of the concrete. The vibration of the concrete can improve the compressive strength about 3–5% for each percent of entrapped air removed. Vibration produce a settled concrete and entrapped air is forced to the surface and allow the concrete to move into formwork and eliminate the bigger voids.

A homogeneous and randomly oriented aggregate distribution can improve the mechanical properties, durability, stability and impermeability of concrete [4]. Eurolightcon [5] emphasizes the importance of homogeneity among the constituents of lightweight concrete. During the mixing of LWC, due to the low density of the aggregates used and the longer mixing times

DOI: 10.2495/CMEM-V6-N4-726-736

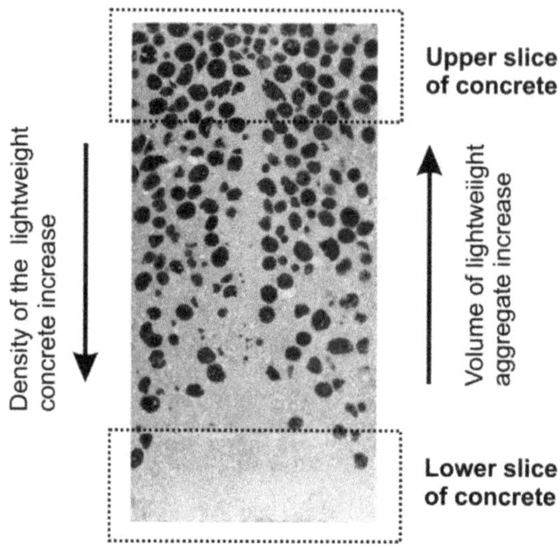

Figure 1: Segregation of concrete. Upper and lower slice in concrete specimen.

[6], the LWC is susceptible to the segregation of the aggregates as a result of the differences between the densities of their components. In fact, during the vibration of the concrete, lightweight aggregates tend to float.

According to Panesar and Shindman [7], segregated concretes are susceptible to an increased risk of cracking by cause of the separation of the aggregates from the rest of the mixture. To quantify the phenomenon of segregation, Ke and Beaucour propose several methods to determine segregation index [6]. In one of the methods to determine segregation index, the densities obtained from the upper and lower slices of a cylinder can be used. This segregation index can be calculated by Equation (1), as can be seen in Fig. 1.

$$I_s = \frac{\rho_{top}}{\rho_{bottom}} \tag{1}$$

where I_s is the segregation index, ρ_{top} is the density of the upper slice of the cylinder of concrete and ρ_{bottom} is the density of the lower slice of the cylinder of concrete.

The main objective of this study was to evaluate the possibility of using linear models or numerical models, based on experimental data, to determine the density of different sections of a segregated lightweight concrete in order to evaluate the segregation index knowing the vibration time of the concrete and the type of aggregate used in the mixture.

2 MATERIALS

Four different types of lightweight aggregates (LWA) of expanded clay having different densities as a coarse aggregate (M, LTM, HS and LTHS) and natural fine limestone aggregate were used to make several LWCs. Their main properties are listed in Table 1. Bulk density was determined by UNE EN 1097-3 [8]; dry particle density and 24 h absorption were determined by UNE EN 1097-6 [9] (pre-dried particles immersed in still water); absolute density was determined with Helio pycnometer; and the fraction of the aggregates was determined by UNE EN 933-1 [10].

Table 1: Geometrical and physical characteristics of aggregates used.

Aggregate type	Bulk density (kg/m³)	Absolute density (kg/m³)	Particle density (kg/ m³)	24 h water absorption (%)	Granulometric fraction (d_i/D_i)
M	269	2573	482	36.60	6/10
LTM	276	2656	613	29.55	4/12
HS	610	2674	1019	12.20	4/10
LTHS	676	2667	1118	11.05	4/10
Limestone	1610	2661	2708	0.12	0/4

Table 2: Mix proportions of the different LWCs.

Concrete type	Cement (kg/m³)	Water (kg/m³)	Fine aggregate (kg/m³)	LWA (kg/m³)
LW-M	350	210	991.1	148.9
LW-LTM	350	210	938.6	201.4
LW-HS	350	210	723.9	416.2
LW-LTHS	350	210	662.0	473.0

The experimental campaign involved the characterisation of concrete made with coarse LWA. Four types of concretes (with each LWA) with the same w/c ratio, cement content (CEM I 52.5 SR with an absolute density of 3176 kg/m^3) and fine aggregate were produced (Table 2) using the Fanjul method [11] in order to produce LWC with a target density of 1700 kg/m^3.

The w/c ratio relates to the effective water available for cement hydration. Due to the high absorption of the expanded clay, the LWA pre-soaked for 7 days in order to control the workability and effective w/c ratio of concrete [12]. For better control of the water content in the LWA, a day prior to mix the concrete, LWA were extended in a mesh for 20 min to reduce the superficial water content and placed into hermetic plastic bags to prevent water loss. Afterwards, a LWA sample with superficial and internal water is weighed and placed in a sieve covered by a paper filter sheet and vibrated for 15 s in order to remove the superficial water content. The LWA sample was then weighed without superficial water and immediately after that the sample was dried in an oven until a constant mass in order to obtain the internal water content.

3 METHODOLOGY

3.1 Experimental procedure

The procedure is summarised in Fig. 2. The concretes were mixed in a vertical shaft mixer (see Fig. 2, step 1). The mixing methodology for the concretes production consisted of: 1-min mixing cement and fine aggregate, addition of the total calculated water followed by 2-min mixing, addition of the coarse aggregate followed by 2-min mixing.

The concrete was vibrated using internal vibratory needle. Six 150x300 mm cylinders were cast for each mixture and were vibrated with six different times in one layer. Mixture samples were vibrated for 5, 10, 20, 40, 80 and 160 s (see Fig. 2, step 2).

After demoulding at 24 h, the specimens were kept in water until testing. After 28 days, each concrete specimen was saw-cut in horizontal slices in order to determine the density of all of them (see Fig. 2, step 4).

As mentioned above, the specimens manufactured were sectioned into seven slices of 40 mm thickness, as shown in Fig. 2, step 4. Subsequently, the density of each of these sections was determined by Equation (2).

$$\rho_i = \frac{m_{di}}{(m_{ssd} - m_{sub})} \tag{2}$$

where ρ_i is the section density, m_{di} is the dry mass of the concrete section, m_{ssd} is the saturated mass surface-dry of the concrete section and m_{sub} is the submerged mass of the concrete section.

3.2 Statistical analysis and numerical model approach

On the one hand, with the data obtained from the dry density of each section, type of aggregate, time of vibration and number of section, differences in the density of the sections of LWC in relation to the subpopulations defined in each of the other variable were analysed considering them as factors. On the other hand, mathematical models have been generated to obtain relationships between density of the LWC and factors, which will allow a better

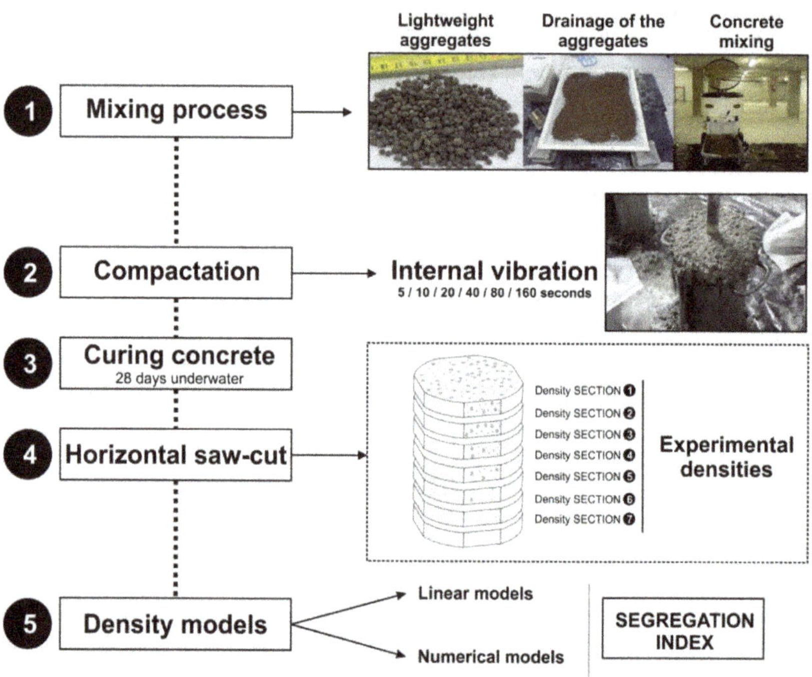

Figure 2: Summary of the process carried out in the research.

understanding of the phenomenon and make predictions. Generally, in the study and modelling of systems, it is necessary to analyse and determine the relationship between the variables defined by the Equation (3).

$$z = (x_1, x_2, \ldots, x_n) \tag{3}$$

and of which only the experimental data are known as Equation (4) shows.

$$\left\{z_i, x_i^1, x_i^2, x_1^3, \ldots, x_i^n\right\}_{i=1,2,\ldots,p} \tag{4}$$

There are different methodologies to obtain models from the experimental data and they can be analytically (mathematical equations) or numerically defined. The linear models did not fit well to the experimental data, so it has been necessary to generated numerical models in which the function is defined by its value at a finite set of points which allows the calculation of the function value at any other point. The numerical models have been generated applying the methodology presented by Villacampa *et al.* [13] and by Navarro-González and Villacampa [14].

3.2.1 Variance analysis

An analysis of the variance (one-way ANOVA) has been performed to determine if there are significant differences in the densities regarding the following factors: type of aggregate, time of vibration and section. SPPS software was used for the analysis.

For the study of the analysis of the variance, it is necessary that the dependent variable be normally distributed in each group; that the homogeneity of the variances be fulfilled, i.e., the variances must be equal in each group; and that the observations be independent. The test of Levene uses the level of significance set *a priori* for the ANOVA ($\alpha = .05$) to test the assumption of homogeneity of variance for any factor. The results for the factor '*section*' can be seen in Table 3.

Since the assumption of equality of variances does not have statistically significant difference, the statistics of Welch and Brown-Forsythe have been determined for each of the factors. They indicate that there are only significant differences for the '*section*' factor, as observed in Table 4.

In order to obtain information on the sections that present significant differences between them, the equality of variances between the subpopulations (or groups) is not verified.

Table 3: Levene's test for equality of variances.

Dry density LWC (kg/m^3)			Section
Levene	df1	df2	Sig.
4409	6	161	,000

Table 4: Robustness testing of the means equality (section factor).

Dry density LWC (kg/m^3) / section	Statistics	df1	df2	Sig.
Welch	11,417	6	70,868	,000
Brown-Forsythe	14,158	6	134,426	,000

The statistic of Games-Howell has been determined, and it has been used an *a priori* alpha level of significance of 0.05. This test allows to state that the difference of means is significant in the 0.05 level between the following sections:

- The density of the section 1 has significant differences with the densities of the sections 4, 5, 6 and 7.
- The density of the section 2 has significant differences with the densities of the sections 5, 6 and 7.
- The density of the section 3 has significant differences with the densities of the sections 5, 6 and 7.
- The density of the section 4 has significant differences with the density of the section 7.

This difference makes the segregation in concrete visible. Taking into account that the first section is the upper section and the seventh section is the lower section in the LWC cylinder specimen, significant differences can be obtained between the upper and the lower slices.

3.2.2 Linear models

Linear regression models are mathematical models used to approximate a relationship between a dependent variable, a set of independent or explanatory variables, and a random error or perturbation term, as can be seen in Equation (5).

$$Y = \beta_0 + \beta_1 x_1 + \beta_2 x_2 + \ldots\ldots + \beta_n x_n + \varepsilon \qquad (5)$$

In a linear regression model, the following assumptions must be fulfilled on the errors:

i) the mean of the error must be zero
ii) the variance of the error must be constant (homoscedasticity)
iii) the errors follow a normal distribution of zero mean and deviation σ
iv) the errors associated with the values Y are independent.

An adjustment measure is the coefficient of determination R^2 equal to the square of the correlation coefficient.

3.2.3 Numerical models

The numerical methodology developed by Villacampa *et al.* [14] generates n-dimensional representation models. The methodology is based on the definition and generation of a geometric model of finite elements described by González-Navarro and Villacampa [13]. The representation model is determined in the nodes, allowing the calculation at any point by using interpolation functions. The number of finite elements and nodes is determined by a number c called complexity of the model (Comp). For each value of complexity, a derived numerical model exists. So the methodology starts with the generation of a geometric model of finite elements defined in a hyper-cube. The meshing process generates a set of points called nodes, where the model is determined. To do this, each interval is divided into c sub-intervals, where c is called the *complexity of the model.*

The number of elements and nodes is determined by the value of c. Each model allows the estimation of relationship values at the experimental points and the correspondent sum of squared errors and determination coefficients can be calculated.

With the above-mentioned experimental data and the methodology, numerical models have been determined for different complexities. These models are models of representation of the relationship showed in Equation (6).

$$Density = f(x_1, x_2, x_3) \tag{6}$$

where x_1 x_2 y x_3 is the LWC section, time of vibration and aggregate type, respectively.

The following parameters have been compared in order to select the better models: the coefficient of determination that measures the goodness of the fit is defined by the Equation (7) and the absolute percentage error (MAPE) according to the Equation (8).

Numerical models have been generated for the complexities c = 10, 15, 20, 25, 30, 35, 40, 45 and 50.

$$R^2 = \frac{\sum_{i=1}^{n}\left(\tilde{y}_i - \bar{y}\right)^2}{\sum_{i=1}^{n}\left(y_i - \bar{y}\right)^2} \tag{7}$$

$$MAPE = \frac{\sum_{i=1}^{n}\left|\frac{y_i - \tilde{y}_i}{y_i}\right|}{n} \tag{8}$$

4 RESULTS AND DISCUSSION

4.1 One-way ANOVA

All the linear regression models between the density and the different variables were determined (section, time of vibration and type of aggregate) by applying the regression method backward that determines a model in which all the variables participate and after removing one by one all the variables. The backward method was used to generate three linear models that only explain a maximum of 34.6% of the data and in which the error is not normally distributed (it does not follow a normal distribution). The linear models are presented in Equations (9–11) and also can be seen in Fig. 3.

$$Density = 1428{,}586 + 56{,}923x_1 + 0.216x_2 - 12{,}244x_3 \tag{9}$$

$$Density = 1439{,}926 + 56{,}923x_1 - 12.244x_3 \tag{10}$$

$$Density = 1409{,}316 + 56{,}923x_1 \tag{11}$$

where x_1 is the section number, x_2 is the vibration time and x_3 is the aggregate type.

As can be observed, linear models to predict the density of the LWC in the different sections studied and each type of aggregate were not useful due to the low coefficient of determination (from 0.338 to 0.346) and do not describe the segregation phenomenon.

4.2 Numerical models

As mentioned above, the different complexity models up to 50 were studied in order to find the better explanation of the density of the LWC in the different sections. As can be seen in

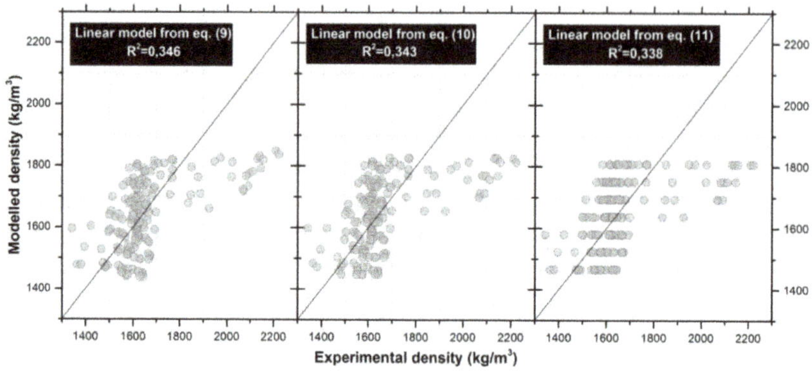

Figure 3: Linear models from Equations (9–11).

Fig. 4, the higher the complexity model the higher the coefficient of determination (R^2) and the lower the absolute percentage error (MAPE). For models higher than 30 slightly growth was observed, and probably, more noise is reproduced.

All models have a good fit because their coefficient of determination is between 0.88 and 0.99. Figure 5 shows the different models and the experimental density data.

The main goal in this research was to find a model capable of predicting the density knowing the vibration time and the type of the aggregate. It can predict the density of the LWC from the top to the bottom of the concrete specimen (seven different sections) so that the segregation index can be determined according to Equation (1), of the whole specimen (knowing the time of vibration and type of the aggregate). This makes it easier for engineers to know the state of the LWC placed without drill core.

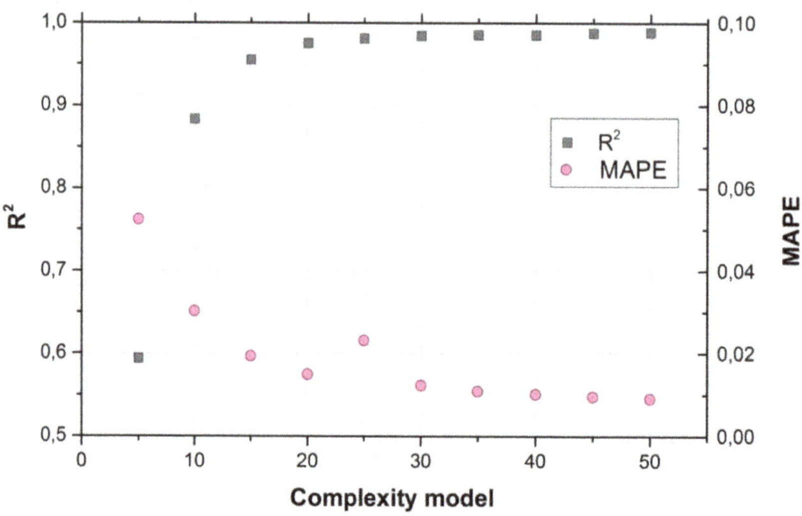

Figure 4: Evolution of MAPE and R^2 when the complexity model is increased.

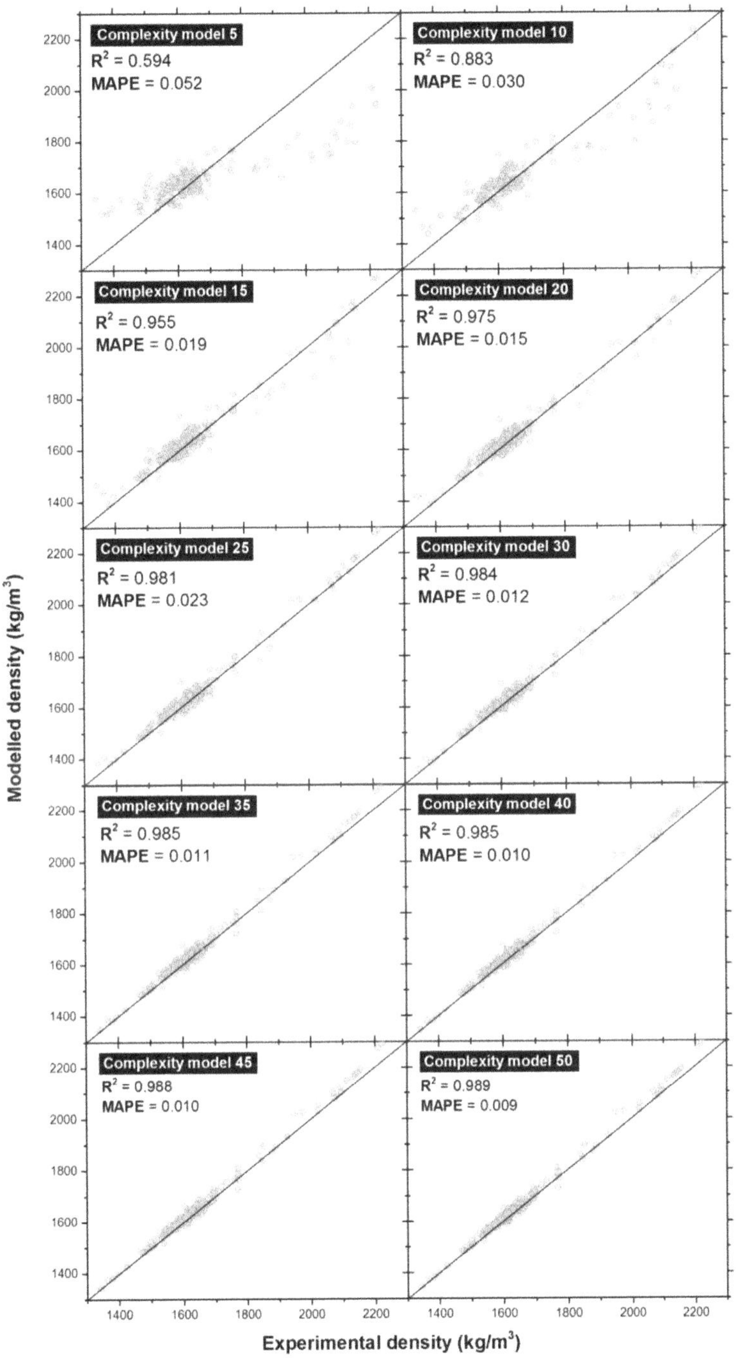

Figure 5: Experimental density vs. Modelled density with different complexity models.

5 CONCLUSIONS

In this research, linear models and numerical models were studied in order to predict the density of segregated lightweight concretes. Four types of lightweight aggregates, six different vibration time in the LWC and seven sections were studied in order to predict the density of the LWC in the different sections (from upper to lower). The statistical analysis showed that significant difference can be found in upper sections from lower sections. However, linear models were not suitable to describe the segregation phenomenon because explained up to the 34.6% of the experimental data. The described work provides an example on how numerical models, with different complexity, can achieve up to 98.9% of explanation of the experimental results. This research allows confirming that segregation phenomenon and segregation index can be described using numerical models.

ACKNOWLEDGEMENTS

This research was funded by the University of Alicante (GRE13-03) and (VIGROB-256). The authors wish to express their gratitude to Laterlite for supplying some samples used in the tests that were carried out.

REFERENCES

[1] Chandra, S. & Berntsson, L., *Lightweight aggregate concrete science, technology and applications*, Standard Publishers Distributors: Delhi, 2003.

[2] Yu, Q.L., Spiesz, P. & Brouwers, H.J.H., Ultra-lightweight concrete: conceptual design and performance evaluation. *Cement & Concrete Composites,* **61**, pp. 18–28, 2015. https://doi.org/10.1016/j.cemconcomp.2015.04.012

[3] Yoon, J.Y., Kim, J.H., Hwang, Y.Y. & Shin, D.K., Lightweight concrete produced using a two-stage casting process. *Materials*, **8**, pp. 1384–1397, 2015. https://doi.org/10.3390/ma8041384

[4] Han, J., Wang, K., Wang, X. & Monteiro, J.M.P., 2D image analysis method for evaluating coarse aggregate characteristic and distribution in concrete. *Construction and Building Materials,* **127**, pp. 30–42, 2016. https://doi.org/10.1016/j.conbuildmat.2016.09.120

[5] Eurolightcon. *Economic design and construction with lightweight aggregate concrete*, 1998.

[6] Barbosa, F.S., Farage, M.C.R., Beaucour, A.-L. & Ortola, S., Evaluation of aggregate gradation in lightweight concrete via image processing. *Construction and Building Materials,* **29**, pp. 7–11, 2012. https://doi.org/10.1016/j.conbuildmat.2011.08.081

[7] Panesar, D.K. & Shindman, B., The effect of segregation on transport and durability properties of self consolidating concrete. *Cement and Concrete Research,* **42**(2), pp. 252–264, 2012. https://doi.org/10.1016/j.cemconres.2011.09.011

[8] Asociación Española de Normalización y Certificación. *UNE EN 1097-3 Ensayos para determinar las propiedades mecánicas y físicas de los áridos. Parte 3: Determinación de la densidad aparente y la porosidad.* Madrid: AENOR, 1999.

[9] Asociación Española de Normalización y Certificación. *UNE EN 1097-6 Ensayos para determinar las propiedades mecánicas y físicas de los áridos. Parte 6: Determinación de la densidad de las partículas y la absorción de agua.* Madrid: AENOR, 2014.

[10] Asociación Española de Normalización y Certificación. *UNE EN 933-1 Ensayos para determinar las propiedades geométricas de los áridos. Parte 1: Determinación de la granulometría de las partículas. Métodos del tamizado*. Madrid: AENOR, 2006.

[11] Fernández-Fanjul, A. & Tenza-Abril, A.J., Méthode FANJUL: Dosage pondéral des bétons légers et lourds. *Annales du Bâtiment et des Travaux Publics*, **5**, pp. 32–50, 2012.

[12] Bogas, J.A. & Gomes, A., A simple mix design method for structural lightweight aggregate concrete. *Materials and Structures*, **46**(11), pp. 1919–1932, 2013. https://doi.org/10.1617/s11527-013-0029-1

[13] Villacampa, Y., Navarro-González, F.J. & Llorens, J., A geometric model for the generation of models defined in complex systems. *Ecosystems and Sustainable Development VII*, **112**, pp. 71–82, 2009. https://doi.org/10.2495/eco090081

[14] Navarro-González, F.J. & Villacampa, Y., A new methodology for complex systems using n-dimensional finite elements. *Advances in Engineering Software*, **48**(1), pp. 52–57, 2012. https://doi.org/10.1016/j.advengsoft.2012.02.001

EFFECT OF FUEL INJECTOR HOLE DIAMETER AND INJECTION TIMING ON THE MIXTURE FORMATION IN A GDI ENGINE - A CFD STUDY

PRIYANKA D. JADHAV & J. M. MALLIKARJUNA
Internal Combustion Engine Laboratory, Department of Mechanical Engineering, Indian Institute of Technology Madras, Chennai.

ABSTRACT
Performance and emission characteristics of a gasoline direct injection (GDI) engine are mainly influenced by the in-cylinder mixture preparation. However, in these engines, mixture formation depends upon many factors viz., fuel injection strategy and parameters, mode of operation, engine geometry, etc. Therefore, understanding the mixture formation, under various engine operating conditions and fuel system configurations, is very much essential. In this study, an attempt has been made to understand the effect of fuel injector-hole diameter and fuel injection timing on the mixture formation in a four-stroke, wall-guided GDI engine using computational fluid dynamics (CFD) analysis. The CFD simulations are carried out from inlet valve opening (IVO) to exhaust valve opening (EVO) period using the CONVERGE. The CFD models used are validated with the available data from the literature. The engine considered has a compression ratio (CR) of 11.5. All the CFD simulations are carried out at the engine speed of 2000 rev/min. Three fuel injector-hole diameters viz., 0.1, 0.14 and 0.18 mm and three fuel injection timings viz., 605, 620 and 635 crank angle degree (CAD) are considered for the analysis. The mixture formation is analyzed in the vicinity of the spark plug and at other parts of the combustion chamber. From the results, it is found that higher nozzle-hole diameter yielded very rich mixture zones near spark plug. Also, lower nozzle-hole diameter and retarded fuel injection timing showed higher indicated mean effective pressure (IMEP).
Keywords: combustion, fuel injection strategies, GDI engine, mixture stratification.

1 INTRODUCTION

Currently, gasoline direct injection (GDI) engines are preferred over the conventional port fuel injection (PFI) engines, because of better performance and emission characteristics. GDI engines are operated at two modes: homogeneous and stratified. In the stratification mode, engine performance and emissions are mainly affected by the in-cylinder mixture formation at the time of ignition [1]. But, the mixture stratification depends upon many factors viz., engine geometry, fuel injection strategy, engine operating conditions, etc., [2, 3]. Hence, it is important to study the mixture stratification inside the combustion chamber to optimize engine and operating parameters for better performance and emissions characteristics.

Previously, Krishna and Mallikarjuna [4] studied the effect of engine parameters on the mixture stratification in a wall-guided GDI engine using CFD analysis. They developed a parameter called 'stratification index' to quantify the mixture stratification. The simulations were carried out at the engine speeds varying from 2000 to 4000 rev/min., with CRs of 10.5, 11.5 and 12.5 and inlet air pressures of 1, 1.2 and 1.4 bar. Their results showed that the mixture stratification in a wall-guided engine varied significantly with the engine speed and inlet air pressure, whereas it was not much affected by CR. Costa *et al.* [5] analyzed mixture formation and early flame development in a GDI engine through numerical simulations and ultraviolet digital imaging under various fuel injection and ignition timings. They conducted experiments, in a four-valve, single-cylinder wall-guided GDI engine, at full-throttle operation with the CR of 10.5, the engine speed of 2000 rev/min., the fuel injection pressure of 100 bar and the

DOI: 10.2495/CMEM-V6-N4-737-748

equivalence ratio of 1.1±1%. They reported that the early fuel injection resulted in the formation of homogeneous mixture which led to stable combustion. Also, the stratified mixture increased cycle-by-cycle variations. Li *et al.* [6] investigated experimentally the mixture formation under split fuel injection in a GDI engine. The laser absorption and scattering (LAS) technique was used to measure the concentration of liquid and vapor phase sprays. The laser-induced fluorescence-particle image velocimetry (LIF-PIV) technique was used for analyzing the fuel spray and the flow of ambient air. The tests were carried out at various ambient pressures and temperatures, fuel injection pressures and total fuel quantities. They reported that an optimum combustion in a GDI engine could be achieved by using two fuel injection pulses. Costa *et al.* [7] investigated experimentally the mixture preparation and combustion, in an optically accessed turbocharged GDI engine, under various fuel injection pressures, and stoichiometric and lean mixture conditions. Their results showed that the stoichiometric mixture resulted in higher heat release and NO emissions, compared to that of lean mixture. Also, increasing fuel injection pressure and advancing the ignition timing improved power output and cyclic variations and reduced CO and soot emissions. Costa *et al.* [8] analyzed experimentally and numerically the effect of split fuel injection on engine power output and emissions in a four-valve single-cylinder turbocharged optical GDI engine. The tests were carried out at the engine speed of 2000 rev/min., fuel injection pressure of 100 bar, full-throttle conditions and the equivalence ratio of 1.1. Fuel injection was done in two stages during suction and compression strokes with equal mass of fuel. They reported that the heat release rate could be enhanced by the proper choice of spark and fuel injection timings.

From the above discussion, it is understood that, the mixture formation, in a GDI engine, is influenced significantly by many engine conditions. Also, in a wall-guided GDI engine, it mainly depends upon piston shape, engine speed, fuel injection strategy, etc. However, the evaluation of simultaneous effect of fuel injection strategy and nozzle geometry on mixture formation is rarely reported in the literature. Therefore, the present CFD analysis investigates the effect of fuel injection timings and nozzle-hole diameter on the mixture formation in a wall guided GDI engine.

2 CFD METHODOLOGY

2.1 Engine specifications

The present study considers a four-valve, four-stroke, pentroof head, single-cylinder wall guided GDI engine. Table 1 shows the engine specifications used in this study [5].

Table 1: Engine specifications (Costa *et al.*, [5]).

Stroke (mm)	79
Bore (mm)	81.3
Connecting rod length (mm)	143
Compression ratio	10.6:1
Number of valves	4
Inlet valve opening (IVO)	3 CAD before TDC
Inlet valve closing (IVC)	36 CAD after BDC
Exhaust valve opening (EVO)	27 CAD before TDC
Exhaust valve closing (EVC)	0 CAD after TDC

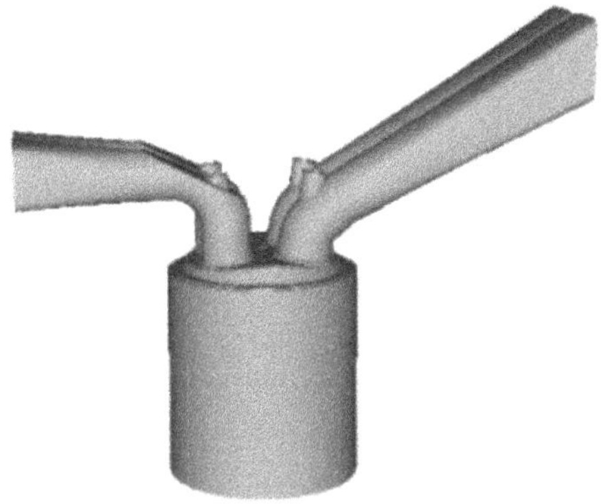

Figure 1: Engine computational domain.

2.2 Geometric Modelling and Meshing

The engine computational domain is modelled using the PTC-CREO as shown in Fig. 1. The full cycle CFD simulations are carried out using the CONVERGE. In the CONVERGE, the complex geometrical intersections are represented using a modified cut-cell Cartesian grid generation method. Two million cells are generated with the base mesh size of 4 mm. The velocity and temperature gradients during valve motion, fuel spray and combustion processes are captured using adaptive mesh refinement using maximum and minimum mesh sizes of 1 and 0.125 mm, respectively.

2.3 Boundary conditions

The intake pressure of 1 bar and intake temperature of 310 K are used. Exhaust gas is assumed to be delivered at the atmospheric pressure and at the temperature of 366 K. The engine speed of 2000 rev/min., compression ratio of 11.5 and equivalence ratio of 0.75 ± 0.05 are kept constant throughout the analysis.

2.4 CFD models considered

In-cylinder incompressible flow modelling considers the following governing equation [9].

$$\frac{\partial(\rho\varphi)}{\partial t} + div(\rho\varphi u) = div(\Gamma grad\varphi) + S_{\varphi} \tag{1}$$

The RNG k-ε turbulence model is used to capture the in-cylinder turbulent flow characteristics [10, 11]. The macroscopic fuel spray characteristics are predicted using the KH-RT spray breakup model [12]. Fuel spray wall impingement is predicted by the model developed by O'Rourke and Amsden [13, 14]. Table 2 summarises the other CFD models used in the present study.

2.5 Validation of the CFD models

The CFD models used in the present study are validated by comparing the CFD predicted in-cylinder pressures with that of the experimental results of Costa *et al.* [5]. Figure 2 shows

Table 2: CFD models used.

Spray break-up model	Kelvin-Helmholtz and Rayleigh-Taylor (KH-RT)
Spray evaporation model	Frossling model
Spray collision model	O'Rourke
Spray wall impingement model	O'Rourke and Amsden
Turbulence model	RNG k-ε
Combustion model	Detailed chemical kinetics SAGE

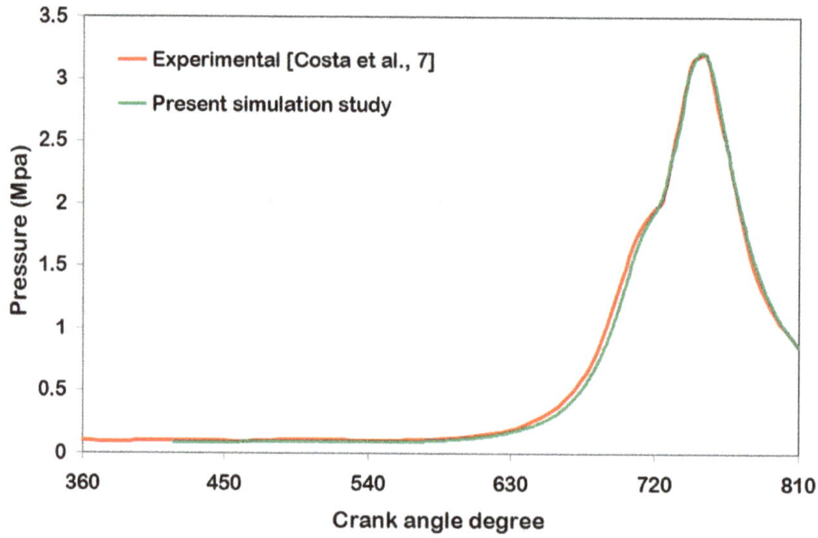

Figure 2: Comparison of in-cylinder pressures.

the comparison of the in-cylinder pressure traces obtained by Costa *et al.* [5] and that of the present study at the engine speed of 2000 rev/min., and the CR of 10.6. Both the results are in reasonably good agreement.

In addition, for the similar engine and the CFD models, Krishna and Mallikarjuna [4] predicted the spatial distribution of ER, at the time of ignition (710 CAD) using CFD analysis and compared it with that of the Costa *et al.* [5], as shown in Fig. 3. From Fig. 3, it is seen that, the ER distribution plots from the previous studies [4, 5] are in reasonably good agreement. Therefore, the CFD models can be used confidently for further CFD simulations.

2.6 Operating conditions

The effect of nozzle-hole diameter and start of injection (SOI) on in-cylinder pressure, in-cylinder mean temperature, heat release rate, etc., is investigated at 2000 rev/min., and at equivalence ratio of 0.75±0.05. The compression ratio of 11.5, the fuel injection pressure of 90 bar and the spark timing at 710 CAD are kept constant throughout the study. The three nozzle-hole diameters of 0.1, 0.14 and 0.18 mm are used along with the SOIs of 605, 620 and

(a) From the CFD analysis of Krishna and Mallikarjuna [4]

(b) From the CFD analysis of Costa et al. [5]

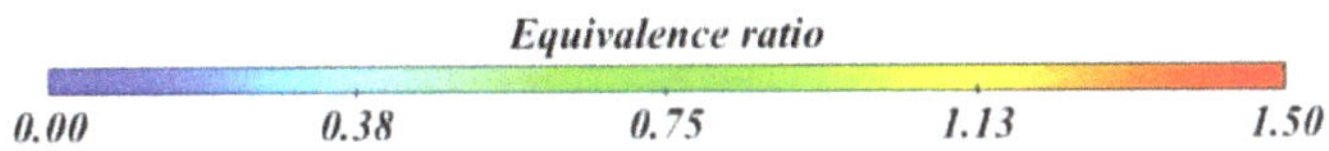

Figure 3: Comparison of ER distribution on the vertical central plane at the time of spark.

635 CAD. For comparison of results, the nozzle-hole diameter of 0.1 mm and the SOI of 605 CAD are considered as the basis.

3 RESULTS AND DISCUSSION

3.1 Effect of nozzle-hole diameter and fuel injection timing on in-cylinder pressure

Figure 4(a) shows the comparison of in-cylinder pressures at various fuel injector nozzle-hole diameters, at the SOI of 605 CAD. From Fig. 4(a), it is seen that, as the nozzle-hole diameter increases, the rate of pressure rise (slope of the curve) and the peak in-cylinder pressure reduces. This is because, when the nozzle-hole diameter increases, the fuel droplet size increases and thereby fuel atomization reduces, resulting in poor air-fuel mixing and combustion. Therefore, the nozzle-hole diameter of 0.1 mm is considered as the optimum one and used for the further analysis. From Fig. 4(a), it is found that, at the nozzle-hole diameters of 0.14 and 0.18 mm, in-cylinder peak pressures decrease by about 14 and 22.3% respectively, compared to that of the nozzle-hole diameter of 0.1 mm.

Figure 4(b) shows the comparison of in-cylinder pressures for various SOIs and the nozzle-hole diameter of 0.1 mm. From Fig. 4(b), it is seen that, for the nozzle-hole diameter of 0.1 mm, the peak in-cylinder pressure and rate of pressure rise are higher at a particular SOI (can be considered as the optimum). When the SOI is retarded or advanced from the optimum SOI, both the peak in-cylinder pressure and rate of pressure rise reduce. From Fig. 4(b), it is found

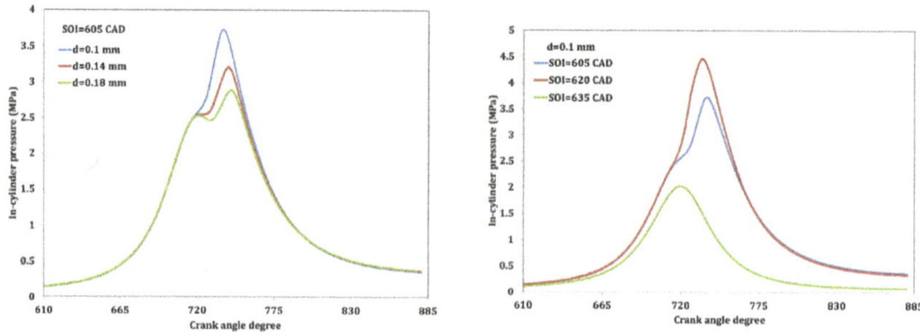

(a) Various nozzle-hole diameters (b) Various fuel injection timings

Figure 4: Comparison of in-cylinder pressures.

that, at the optimum SOI of 620 CAD, the in-cylinder peak pressure increases by about 19.6% and at the SOI of 635 CAD it decreases by about 45.8%, compared to that of the SOI of 605 CAD.

3.2 Effect of nozzle-hole diameter and fuel injection timing on mean in-cylinder temperature

Figure 5(a) shows the comparison of mean in-cylinder temperatures for various fuel injector nozzle-hole diameters, at the SOI of 605 CAD. From Fig. 5(a), it is seen that, at the SOI of 605 CAD, with the increase in nozzle-hole diameter, the in-cylinder mean temperature reduces. This is again because of poor fuel atomization at higher nozzle-hole diameters, which results in poor fuel evaporation, mixing and combustion. From Fig. 5(a), it is found that, the peak in-cylinder temperature decreases by about 3.3 and 6.5%, for the nozzle-hole diameters of 0.14 and 0.18 mm, respectively, compared to that of nozzle-hole diameter of 0.1 mm.

Figure 5(b) shows the comparison of mean in-cylinder temperatures for various SOIs and the nozzle-hole diameter of 0.1 mm. From Fig. 5(b), it is seen that the in-cylinder temperature is more at the optimum SOI (620 CAD), similar to the trend of in-cylinder pressure, compared to

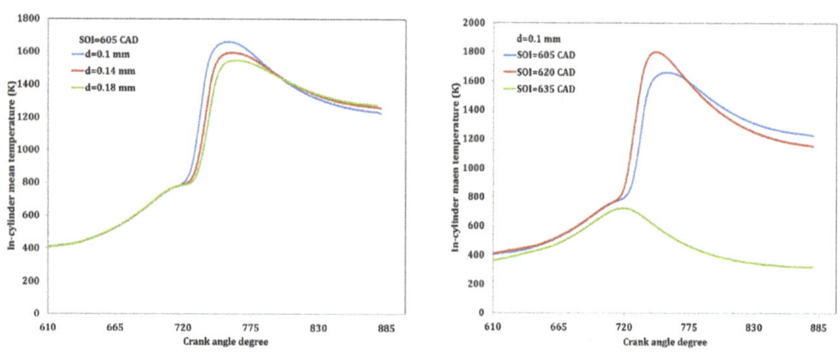

(a) Various nozzle-hole diameters (b) Various fuel injection timings

Figure 5: Comparison of mean in-cylinder temperatures.

that of the SOI of 605 CAD. However, at the SOI of 635 CAD, there is no combustion. This is because, when the SOI is more retarded, the mixture at vicinity of the spark plug may not be in the combustible range, which due to less time available for the mixture formation. From Fig. 5(b), it is found that at the SOI of 620 CAD, peak in-cylinder temperature increases by about 9%, whereas at the SOI of 635 CAD it decreases by about 56.3%, compared to that of the SOI of 605 CAD.

3.3 Effect of nozzle-hole diameter and fuel injection timing on heat release rate

Figure 6(a) shows the comparison of heat release rates for various fuel injector nozzle-hole diameters, at the SOI of 605 CAD. From Fig. 6(a), it is seen that at the SOI of 605 CAD, with the increase in the nozzle-hole diameter, the heat release rate reduces. This is because, as the nozzle-hole diameter increases, the fuel droplet size increases resulting in poor atomization and combustion, which leads to reduced heat release rate. From Fig. 6(a), it is found that the peak heat release rate decreases by about 6.6 and 13.1%, for the nozzle-hole diameters of 0.14 and 0.18 mm, respectively, compared to that of nozzle-hole diameter of 0.1 mm.

Figure 6(b) shows the comparison of heat release rates for various SOIs, at the nozzle-hole diameter of 0.1 mm. From Fig. 6(b), it is seen that the heat release rate is more at the optimum SOI (620 CAD). At the SOI of 635 CAD, as mentioned earlier, there may not be combustible mixture near the spark plug, leading to no combustion. From Fig. 6(b), it is found that at the SOI of 620 CAD, heat release rate increases by about 7.1%, compared to that of the SOI of 605 CAD.

3.4 Effect of nozzle-hole diameter and fuel injection timing on stratification index

Here, the mixture distribution in the engine combustion chamber is expressed quantitatively, using a parameter called stratification index (SI) [4]. Here, the SI equal to 1 refers to an ideally stratified mixture (representing equivalence ratio of the mixture near the spark plug is in the combustible range and decreases away from the spark plug). The SI less than 1 represents improperly stratified mixture (i.e., a lean mixture in the vicinity of the spark plug), whereas the SI greater than 1 represents a very rich mixture in the vicinity of the spark plug. Hence, an optimum combustion needs the mixture distribution that has the SI close to 1.

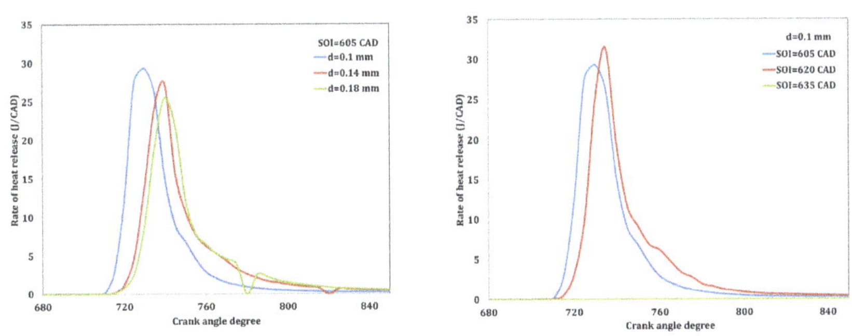

(a) Various nozzle-hole diameters (b) Various fuel injection timings

Figure 6: Comparison of heat release rates.

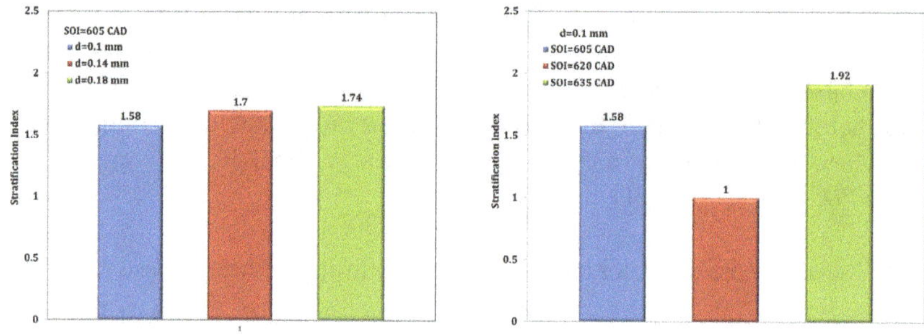

(a) Various nozzle-hole diameters (b) Various fuel injection timings

Figure 7: Comparison of stratification indices.

Figure 7(a) shows the comparison SI at various fuel injector nozzle-hole diameters and the SOI of 605 CAD. From Fig. 7(a), it is seen that as the nozzle-hole diameter increases, the SI increases. It means that the degree of richness of the mixture near the spark plug increases with the increase in nozzle-hole diameter. From Fig. 7(a), it is seen that for the nozzle-hole diameters of 0.14 and 0.18 mm, SI is about 1.7 and 1.74, respectively. Therefore, large nozzle-hole diameter is not advisable.

Figure 7(b) shows the comparison of SIs for various SOIs. From Fig. 7(b), it is seen that for the nozzle-hole diameter of 0.1 mm at the SOI of 605, the SI is about 1.58 which represents that the mixture is rich near the spark plug. At the SOI of 620 CAD, the SI is equal to 1 which represents that the mixture is properly stratified. At the SOI of 635, the SI is about 1.92 which represents the too rich mixture at the spark plug location and therefore no ignition and combustion. The SOI of 620 which results in the formation of a properly stratified mixture exhibits a high peak in-cylinder pressure and temperature compared to all the other conditions considered. From Fig. 7(b), it is found that at the SOI of 620 CAD, the SI decreases by about 0.58 and at the SOI of 635, it increases by about 0.34 compared to that of the SOI of 605 CAD.

3.5 Effect of nozzle-hole diameter and fuel injection timing on IMEP

Figure 8(a) shows the comparison IMEP at various fuel injector nozzle-hole diameters and the SOI of 605 CAD. From Fig. 8(a), it is seen that, as the nozzle-hole diameter increases, the

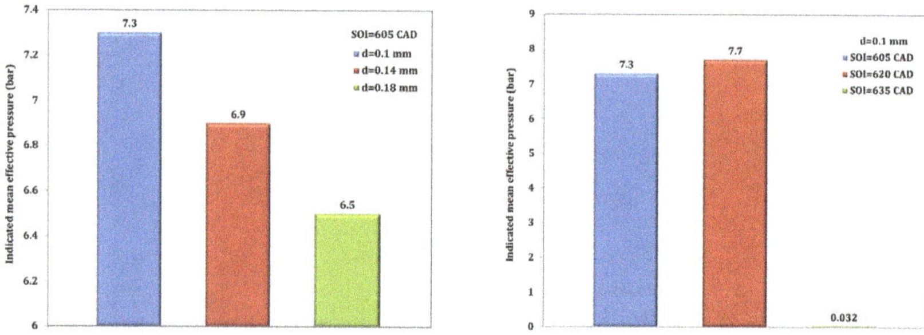

(a) Nozzle-hole diameter (b) Fuel injection timing

Figure 8: Comparison of IMEPs.

IMEP reduces. This is because, as discussed earlier, when the nozzle-hole diameter increases, fuel droplet size increases leading to poor atomization, mixing and combustion, and thereby the in-cylinder pressure decreases during the entire cycle and consequently reducing the engine IMEP. From Fig. 8(a), it is found that, at the nozzle-hole diameters of 0.14 and 0.18 mm, the IMEP decreased by about 5.5 and 10.95%, respectively, compared to that of the nozzle-hole diameter of 0.1 mm.

Figure 8(b) shows the comparison of the IMEP for various SOIs and the nozzle-hole diameter of 0.1 mm. From Fig. 8(b), it is seen that for the nozzle-hole diameter of 0.1 mm, at the optimum SOI of 620, the IMEP is higher, compared to that of the SOI of 605 CAD. Further retardation of the SOI to 635 CAD, resulted in no combustion. From Fig. 8(b), it is found that at the optimum SOI of 620 CAD, the IMEP increases by about 5.5% and it is insignificant at the SOI of 635 CAD.

3.6 Effect of nozzle-hole diameter and fuel injection timing on NOx emissions

Figure 9(a) shows the comparison of NOx emissions, at various fuel injector nozzle-hole diameters and the SOI of 605 CAD. From Fig. 9(a), it is seen that as the nozzle-hole diameter increases, the NOx emissions reduce. This is because of the reduction in in-cylinder temperature at higher nozzle-hole diameters (Fig. 5(a)). From Fig. 9(a), it is found that at the nozzle-hole diameters of 0.14 and 0.18 mm, the NOx emissions decrease by about 21 and 33%, respectively, compared to that of the nozzle-hole diameter of 0.1 mm.

Figure 9(b) shows the comparison NOx emissions for various SOIs and the nozzle-hole diameter of 0.1 mm. From Fig. 9(b), it is seen that at the nozzle-hole diameter of 0.1 mm, the NOx emissions are higher at the SOI of 620 CAD, where the in-cylinder temperatures are higher. At other two SOIs, the NOx emissions are lower because of lower in-cylinder temperatures (Fig. 5(b)). From Fig. 9(b), it is found that at the SOI of 620 CAD, the NOx emissions increase by about 56.25%, compared to that of the SOI of 605 CAD. Off course at the SOI of 635 CAD, there exist no NOx emissions because there is no combustion.

3.7 Effect of nozzle-hole diameter and fuel injection timing on CO emissions

Figure 10(a) shows the comparison CO emissions, at various fuel injector nozzle-hole diameters and the SOI of 605 CAD. From Fig. 10(a), it is seen that, as the nozzle-hole diameter increases,

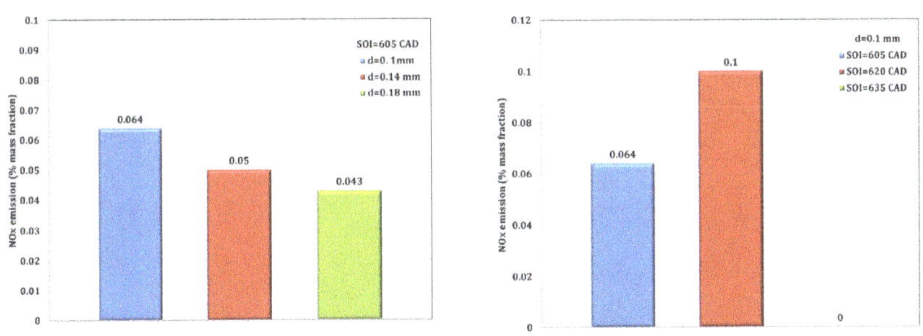

(a) Various nozzle-hole diameters (b) Various fuel injection timings

Figure 9: Comparison of NOx emissions.

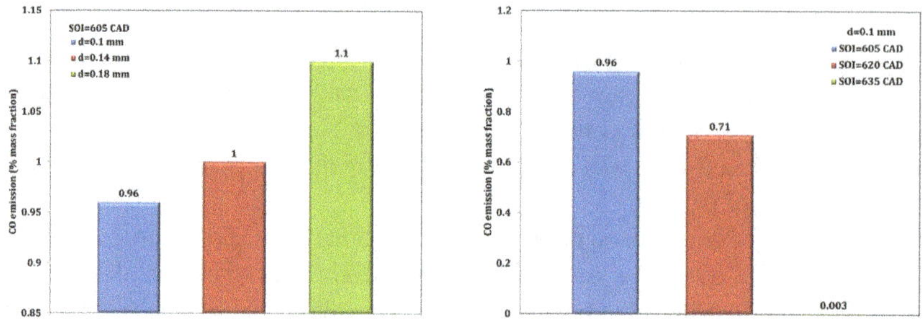

(a) Nozzle-hole diameter (b) Fuel injection timing

Figure 10: Comparison of CO emissions.

the CO emissions increase. This is because of local rich zones occurring due to poor mixing as a result of lower atomization with the increase in nozzle-hole diameter. From Fig. 10(a), it is found that at the nozzle-hole diameters of 0.14 and 0.18 mm, CO emissions increase by about 4 and 14.6%, respectively, compared to that of the nozzle-hole diameter of 0.1 mm.

Figure 10(b) shows the comparison of CO emissions, for various SOIs and at the nozzle-hole diameter of 0.1 mm. From Fig. 10(b), it is seen that for the nozzle-hole diameter of 0.1 mm, the CO emissions decrease at the SOI of 620 CAD, whereas no CO emissions occur, at the SOI of 635 CAD due to no combustion. From Fig. 10(b), it is found that at the SOI of 620 CAD, the CO emissions decrease by about 26%, compared to that of the SOI of 605 CAD.

3.8 Effect of nozzle-hole diameter and fuel injection timing on HC emissions

Figure 11(a) shows the comparison of HC emissions at various fuel injector nozzle-hole diameters and the SOI of 605 CAD. From Fig. 11(a), it is seen that as the nozzle-hole diameter increases, the HC emissions increase. This is because of improper fuel evaporation and mixing with air at large nozzle-hole diameter due to poor atomization as mentioned earlier. From Fig. 11(a), it is found that at the nozzle-hole diameters of 0.14 and 0.18 mm, the HC

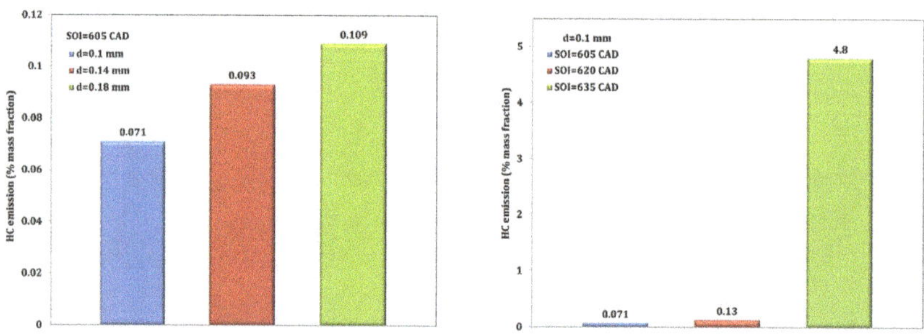

(a) Various nozzle-hole diameters (b) Various fuel injection timings

Figure 11: Comparison of HC emissions.

emissions increase by about 31 and 53.5%, respectively, compared to that of the nozzle-hole diameter of 0.1 mm.

Figure 11(b) shows the comparison of HC emissions for various SOIs. From Fig. 11(b), it is seen that for the nozzle-hole diameter of 0.1 mm, the HC emissions increase with the retarded of SOI. This again because of less time available for fuel evaporation and mixing with air, at the retarded SOI. From Fig. 11(b), it is found that at the SOI of 620 CAD, the HC emissions increase by about 83%, compared to that of the SOI of 605 CAD. Also, high HC emissions at the SOI of 635 CAD show the emission of unburnt fuel due to no combustion.

4 CONCLUSIONS

From the analysis of results, the following conclusions are drawn:

- At the SOI of 620 CAD, peak in-cylinder pressure increased by about 19.6%, and at the SOI of 635 CAD, it decreased by about 45.8% compared to that of the SOI of 605 CAD. At the nozzle-hole diameter of 0.14 and 0.18 mm, peak in-cylinder pressure decreased by about 14 and 22.3%, respectively, compared to that of the nozzle-hole diameter of 0.1 mm.
- At the SOI of 620 CAD, the peak heat release rate increased by about 7.1% as compared to that of the SOI of 605 CAD. At the nozzle-hole diameter of 0.14 and 0.18 mm, it decreased by 9.7 and 14.5%, respectively, compared to that of the nozzle-hole diameter of 0.1 mm.
- At the SOI of 620 CAD, the IMEP increased by about 5.5%, compared to that of the SOI of 605 CAD. At the nozzle-hole diameter of 0.14 and 0.18 mm, the IMEP decreased by about 5.5 and 11%, respectively, compared to that of the nozzle-hole diameter of 0.1 mm.
- At the SOI of 620 CAD, the NOx emissions increased by about 56.3%, compared to that of the SOI of 605 CAD. At the nozzle-hole diameter of 0.14 and 0.18 mm, the NOx emissions decreased by about 22 and 33%, respectively, compared to that of the nozzle hole diameter of 0.1 mm.
- At the SOI of 620 CAD, the CO emissions decreased by about 26%, compared to that of the SOI of 605 CAD. At the nozzle-hole diameter of 0.14 and 0.18 mm, the CO emissions increased by about 4.2 and 14.6%, respectively, compared to that of the nozzle-hole diameter of 0.1 mm.
- At the SOI of 620 CAD, the HC emissions increased by about 83%, compared to that of the SOI of 605 CAD. At the nozzle-hole diameter of 0.14 and 0.18 mm, the HC emissions increased by about 31 and 53.5%, respectively, compared to that of at nozzle-hole diameter of 0.1 mm.
- Finally, it is concluded that, the nozzle-hole diameter of 0.1 mm and the SOI of 620 CAD is the optimum configuration, to have a properly stratified mixture and to improve the indicated mean effective pressure, in the wall guided single-cylinder GDI engine.

REFERENCES

[1] Zhao, F.Q., Lai, M.C. & Harrington, D.L., Automotive spark-ignited direct-injection gasoline engines. *Progress in Energy and Combustion Science*, **25**, pp. 437–562, 1999. https://doi.org/10.1016/s0360-1285(99)00004-0

[2] Ganesan, V., *Internal Combustion Engines*. Tata McGraw-Hill Publishing Company Limited, New Delhi, 2003.

[3] Heywood, J.B., *Internal Combustion Engine Fundamentals*. McGraw-Hill Higher Education, New Delhi, 1988.

[4] Addepalli, S.K. & Mallikarjuna, J.M., Effect of engine performance on mixture stratification in a wall guided GDI engine: a quantitative CFD analysis. *SAE International Journal of Commercial Vehicles,* **10**, pp. 562–571, 2017. https://doi.org/10.4271/2017-01-0570

[5] Costa, M., Marchitto, L., Merola, S.S. & Sorge, U., Study of mixture formation and early flame development in a research GDI engine through numerical simulation and UV-digital imaging. *Energy,* **77**, pp. 88–96, 2014. https://doi.org/10.1016/j.energy.2014.04.114

[6] Li, T., Nishida, K., Zhang, Y. & Hiroyasu, H., Effect of split injection on stratified charge formation of direct injection spark ignition engines. *International Journal of Engine Research*, **8**, pp. 205–219, 2007. https://doi.org/10.1243/14680874jer02106

[7] Costa, M., Catapano, F., Sementaa, P., Sorge, U. & Vaglieco, B.M., Mixture preparation and combustion in a GDI engine under stoichiometric or lean charge: an experimental and numerical study on an optically accessible engine. *Applied Energy,* **180**, pp. 86–103, 2016. https://doi.org/10.1016/j.apenergy.2016.07.089

[8] Costa, M., Sorge, U., Merola, S., Irimescu, A., La Villetta, M. & Rocco, V., Split injection in a homogeneous stratified gasoline direct injection engine for high combustion efficiency and low pollutants emission. *Energy*, **117**, pp. 405–415, 2016. https://doi.org/10.1016/j.energy.2016.03.065

[9] Krishna, A., Mallikarjuna, J. M., Davinder, K. & Ramachandra Babu, Y., *In-cylinder flow analysis in a two-stroke engine - A comparison of different turbulence models using CFD* (SAE Technical Paper, 2013-01-1085), 2013. https://doi.org/10.4271/2013-01-1085

[10] Addepalli, S.K., Mallikarjuna, J.M. & Davinder, K., Effect of engine parameters on in-cylinder flows in a two-stroke gasoline direct injection engine. *Applied Energy,* **176**, pp. 282–294, 2000. https://doi.org/10.1016/j.apenergy.2016.05.067

[11] Yakhot, V. & Orszag, S.A., Renormalization group analysis of turbulence. I. basic theory. *Journal of Scientific Computing*, **1**, pp. 3–51, 1986. https://doi.org/10.1007/bf01061452

[12] Reitz, R.D., Modeling atomization processes in high-pressure vaporizing sprays. *Atomisation and Spray Technology*, **3**, pp. 309–337, 1987.

[13] O'Rourke, P.J. & Amsden, A.A., *A particle numerical model for wall film dynamics in port-injected engines* (SAE Technical Paper, 961961), 1996.

[14] O'Rourke, P. J. & Amsden, A., *A spray/wall interaction submodel for the KIVA-3 wall film model* (SAE Technical Paper, 2000-01-0271), 2000.

FINITE ELEMENT SIMULATION OF SPHERICAL INDENTATION EXPERIMENTS

S. SYNGELLAKIS[1], H. HABBAB[2] & B. G. MELLOR[2]
[1]Wessex Institute, UK.
[2]University of Southampton, UK.

ABSTRACT

The problem of indentation of ductile materials by ball indenters is, in this paper, addressed by numerical modelling. A finite element model is built using general purpose software. The axisymmetry of the problem is taken into account thus reducing its dimensionality. Particular attention is given to contact modelling as well as mesh design for optimal performance. The model is validated by comparing its predictions to the exact elastic solution as well as experimental measurements from elasto-plastic indentation tests. In the latter case, indenter imperfection is accounted for and material input are stress-strain curves originating from tensile tests. The sensitivity of numerical results to indenter elasticity is investigated. The effect of friction and specimen creep during indentation on load-displacement predictions is also assessed.
Keywords: creep, elasto-plastic deformation, finite element modelling, friction, spherical indentation.

1 INTRODUCTION

The response of linear elastic materials during indentation can be predicted by exact elasticity solutions as long as the indenter is described as a solid of revolution [1]. However, modelling the indentation process in the elasto-plastic region is a more complex problem. Since the constitutive equations are non-linear and a number of plastic properties must be included to describe material behaviour, an analytical solution is not easily obtained. As a result, understanding of elasto-plastic indentation behaviour can be better achieved through finite element (FE) simulation.

Computer simulations of indentation tests are often associated with efforts to develop material characterisation techniques based on indentation data [2–9]. In such context, FE modelling of conical, pyramidal and spherical indentations has been extensively used to simulate experiments and provide output suitable for the development of characterisation schemes applicable to materials of various constitutions and compositions. However, critical issues relating to the accuracy and reliability of the modelling have not always been discussed in great depth.

The subject matter of this paper is spherical indentations on ductile materials. The amount of published numerical work in this area is considerable. For this reason, brief reference is here made to a rather limited number of relevant publications. The aim is to highlight adopted tools and assumptions regarding materials and their properties.

General purpose software has been routinely used and the axisymmetry of the problem taken into account to reduce it into a two-dimensional one [2–3, 5–10]. The ball indenter has been modelled as rigid [2–3, 7–8, 10, 12] or elastic [2, 5–6, 9, 11]; the indented specimen elasto-plastic input was either obtained from tensile tests [8, 13] or conventionally built from known material parameters [3, 5–7, 9]; friction between indenter and specimen was sometimes ignored [3, 7, 9, 12] but more often accounted for [2, 5, 6, 8, 10, 13]. Various mesh schemes were applied to test and optimise the performance of the modelling [2, 9, 10].

The development of an FE analysis of the problem is here described in some detail. Particular attention is given to contact modelling as well as the shape, size and distribution of elements. The model was validated by comparing its predictions to exact elastic solutions as

DOI: 10.2495/CMEM-V6-N4-749-763

well as experimental results from stainless steel and carbon steel samples, whose stress-strain curves had been obtained by tensile testing. These curves were used as FE input to simulate the process of indenting these materials by a sphere of 150 μm nominal radius. Friction was found to have significant influence on pile-up. It was also shown that the agreement between numerical and experimental results could be enhanced if the creep occurring during indentation is taken into account.

2 MATERIAL SELECTION

Three types of steel were tested and analysed. They are two grades of mild steel, 43 A and 50 A, and a stainless steel, grade 316 L, supplied in plates of 8 mm thickness. Different grades of steel were considered in order to cover a range of hardness. Steel chemical composition is detailed in Table 1.

Sections taken from the supplied plates were metallographically prepared and examined by optical microscopy. The rolling direction was determined by observing the orientation of inclusions. The grain size was found equal to 10 to 15 μm. This was suitable as several grains were thus included within an indent, which might typically be 70–120 μm in diameter. Thus, the indentation data reflected the average behaviour of the bulk metal as determined by a tensile test, rather than the behaviour of a single grain.

3 TENSILE TESTING

Tensile tests provide reliable and realistic information on elasto-plastic characteristics, which is essential input to the planned FE analysis. Tensile test samples with a circular cross-section were machined from the as-received plates according to the ASTM specifications. They were cut in their rolling direction with a gauge length of 25 mm. Tensile testing was carried out using an Instron machine with the crosshead speed set at 1 mm/minute. The elongation was measured using an extensometer that can cover up to 40% strain. Elongation beyond this value was measured from the crosshead motion.

For a more accurate measurement of load and elongation, a digital data logger was connected to the Instron tensile tester for calibration and the test itself. The data produced by the Instron, corresponding to force and elongation values, were converted into engineering stress–strain points. Then, the true stress-logarithmic strain ($\sigma - \varepsilon$) relation, which is the final information sought from tensile testing, was obtained using the equations

$$\varepsilon = \ln(e+1) \quad (1)$$

$$\sigma = \sigma_0(e+1) \quad (2)$$

where σ_0 and e are the engineering stress and strain values, respectively.

Table 1: Details of chemical composition of the steel used.

Plate No	Grade	C	Si	Mn	P&S	Nb	V	Cr	Mo	Ni
3,4	43A	0.25	0.50	1.60	0.05	-	-	-	-	-
5,6,7	50A	0.23	0.50	1.60	0.05	$\frac{0.003}{0.10}$	$\frac{0.003}{0.10}$	-	-	-
Stainless steel	316L	0.03	1.0	2.0	-	-	-	$\frac{16.5}{18.5}$	$\frac{2.0}{2.5}$	$\frac{11.0}{14.0}$

The material was assumed to have the same elasto-plastic behaviour in compression as in tension. Thus the $\sigma - \varepsilon$ curve obtained from the tensile testing could be used as the input into the FE simulation of indentation, which causes a predominantly compressive stress field. Since the strains developing in an indentation test are expected to be higher than the fracture strain in tensile test, the $\sigma - \varepsilon$ curve was extrapolated over a range of higher strain values.

The maximum calculated logarithmic strain is at the engineering strain corresponding to maximum load where necking commences. The strain up to this point is considered uniform, but beyond that the strain becomes localised around the neck region hence its measured value depends on the considered gauge length, in which case eqns (1) and (2) are not valid. To determine the $\sigma - \varepsilon$ curve at higher strain values, the power law function

$$\sigma = K\varepsilon^n \tag{3}$$

was fitted to the $\sigma - \varepsilon$ curve; in eqn (3), K is the strength coefficient and n the strain hardening exponent. It was not always found possible to fit a single power law function to the whole post-yield region of the curve. In such a case, this part of the curve was split into two and a different function was fitted to each of them. Typical engineering and the true stress-logarithmic strain curves with their extrapolated parts are plotted in Figs 1 and 2.

The moduli of elasticity were taken equal to 210 and 185.1 GPa for mild steel and stainless steel, respectively. The values of material parameters for all specimens are summarised in Table 2. Although a single power law function is well fitted to the stress-strain curve for plates S6 and S7, S5 showed different plastic behaviour despite being of the same grade. Hence, obtaining the stress-strain curve for each plate is necessary for a precise description of material behaviour in FE simulations.

4 FINITE ELEMENT MODELLING

The indentation process was simulated by developing an elastic-plastic FE model using a version of ANSYS [14], a general purpose type of analysis software. Isotropic hardening plasticity and the von Mises yield criterion were assumed. Large strain and other non-linear features of ANSYS were adopted. The material data input for each specimen was the corresponding true stress-logarithmic strain curve. The diamond spherical indenter tip was modelled as both rigid

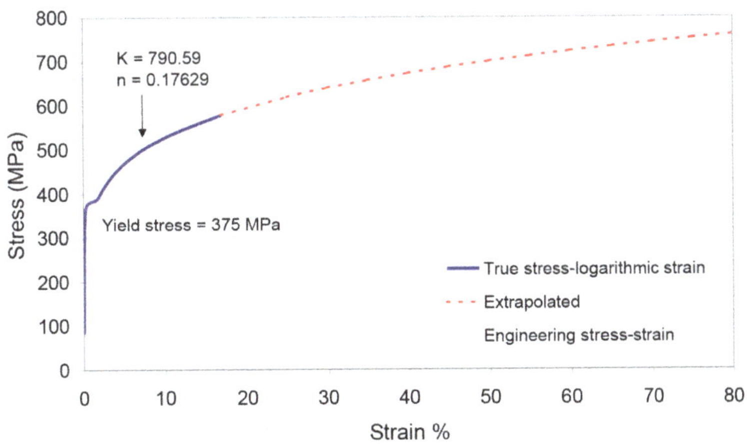

Figure 1: Stress–strain curves for the steel plate number 6, Grade 50 A. A single power function is fitted to the post-yield curve.

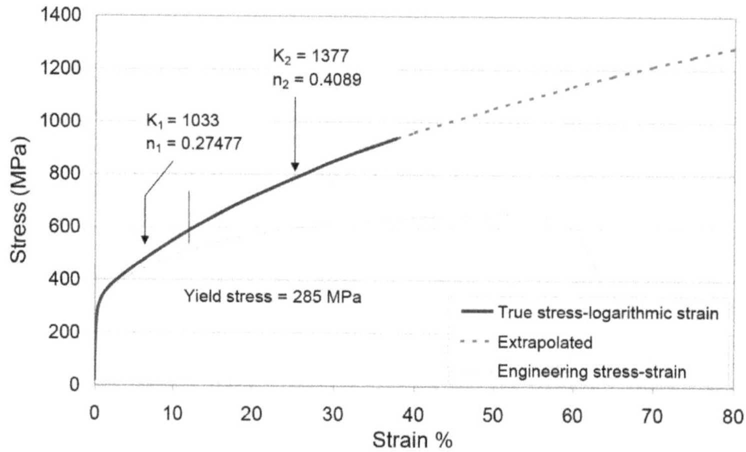

Figure 2: Stress–strain curves for the stainless steel plate, Grade 316 L. Two power functions fitted to the post-yield curve.

Table 2: Summary of the materials stress–strain curve obtained by tensile testing.

Plate number	Steel grade	Yield stress (MPa)	UTS (MPa)	Strain at UTS (%)	Failure strain (%)	Strength coefficient (MPa) K_1	K_2	Strain hardening exponent n_1	n_2
S3	43A	317	460	21	40	755.89	739.64	0.1883	0.1738
S4	43A	300	460	20.7	40	836.65	706.95	0.2156	0.1433
S5	50A	390	500	20.8	48	789.07	794.04	0.1693	0.1676
S6	50A	375	487	18.5	30	790.59		0.1763	
S7	50A	395	503	15.3	35	794.23		0.1607	
Stainless steel	316L	285	655	51.8	63.2	1033	1377	0.2748	0.401

and linear elastic; in the latter case, its modulus of elasticity and Poisson's ratio were taken equal to 1141 GPa and 0.07, respectively. These values were provided by the manufacturers of the instrumented indentation tester.

The simulation of indentation process is a 3D axisymmetric problem. It assumes a spherical indenter tip pressed into a cylinder of material that should be representative of a homogenous infinite half space. Loading the indenter is prescribed as applied displacement on its top surface. Running a 3D problem is expensive and requires large memory and disk space. Advantage can be taken of the axisymmetry of the problem, so that it can be simplified to a 2D problem instead, and only half of the indenter and the material cylinder section by a plane through the axis of symmetry are modelled. This reduces the number of elements and the number of degrees of freedom per node; hence, the running time and disk space required.

4.1 Element characteristics

A four-node quadrilateral element was used to mesh the deformable indenter and the sample. Each node of this element has two degrees of freedom.

The free mesh option in ANSYS was initially used to mesh both the indenter and the sample. Over the region where the two bodies are expected to be in contact, a regular mesh was mapped on both bodies for generating elements of the same size. This mesh regularity was expected to improve numerical performance and thus enhance accuracy over this region of interest. However, the indentation process involves high geometrical non-linearity. With increasing penetration depth, the induced strains increase and element distortion increases excessively and reaches a stage when the solution fails to converge even at shallow indentation due to violating element geometry limits.

The automatic free and mapped mesh options in ANSYS create a large number of elements with little control on their shape and interior angles that deviate from 90°. The steeper the transition from a fine mesh, in the contact region, to a coarse one, at the far edges, the greater the irregularity in element shape.

Decreasing the element geometric irregularity by introducing a smooth variation of mesh density produces a mesh with larger number of elements, hence longer running time and convergence problems. This necessitates the imposition of more control on all elements of the mesh, which can only be achieved by generating the mesh manually. Whilst this task is very complicated and tedious to do using ANSYS, another type of modelling software, PATRAN [15], offers a powerful facility of manual mesh generation with full control on all elements size and shape. In addition, it has the capability of doubling the element size during the transition from the fine mesh zone to the coarse mesh one without violating any element geometry limits. This is done by using trapezoidal elements as interlinks between the two-size zones, as illustrated in Fig. 3. As a result, fewer elements are created for the whole mesh compared with that of ANSYS.

Although this mesh decreased the running time for every load step, convergence difficulties persisted at deep penetration. This is due to the high strains imposed on the elements within the indentation zone of the material, especially those closer to the contact centre, which undergo excessive deformation and almost get squashed as a result. To counteract this excessive change in the aspect ratio, the initial element geometry of the specimen was modelled as

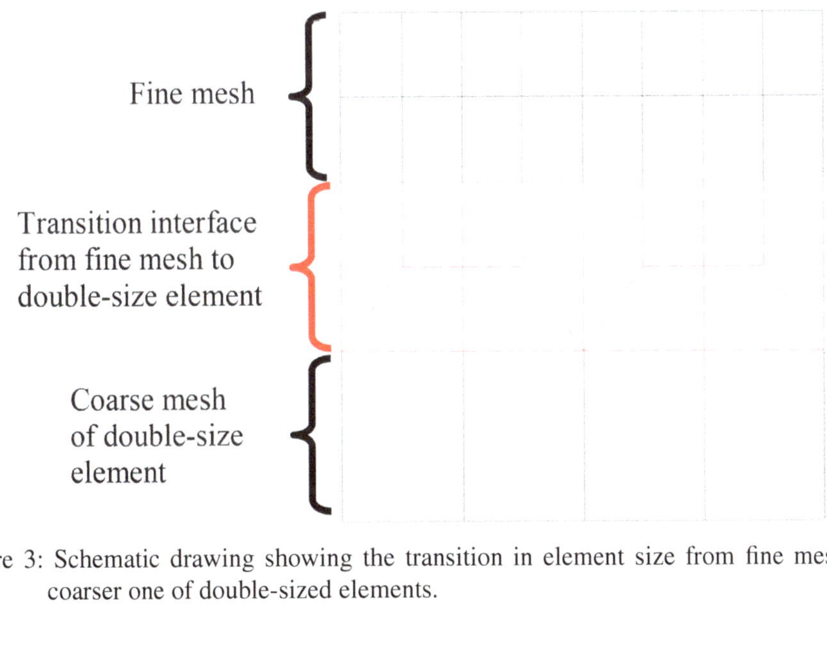

Figure 3: Schematic drawing showing the transition in element size from fine mesh to a coarser one of double-sized elements.

rectangle instead of square. The aspect ratio was taken 2.0 with the longer dimension in the direction of indentation. This helped the model reach the sought penetration depth without any convergence problems, thus this kind of mesh was adopted for all subsequent FE simulations.

4.2 Contact modelling

Contact between the specimen and the indenter was modelled using contact elements generated over the region of possible contact between the two surfaces. The contact element, initially used to represent node-to-surface contact, was a triangle with the base being the target line between two nodes on one of the surfaces and the opposing vertex being the contact node on the other surface. The indenter surface, which should be convex, is considered 'contact' and the flat sample surface is the 'target'.

Contact between two bodies is detected when virtual penetration occurs within a certain limit. A compatibility control method is used to ensure that one surface does not penetrate into another surface by more than an acceptable tolerance. The node-to-surface model uses as control parameter a normal contact stiffness, whose optimum value can be ascertained by trial and error, that is, by testing the performance of the model for a range of its values.

It was noted that, with this contact model, the solution was extremely slow to converge with the number of iterations per load step increasing considerably after only a few microns of simulated indenter penetration. The convergence problem became even worse when sliding friction was included between the two surfaces. This prompted the adoption of an alternative contact modelling, according to which contact is represented by a pair of contact elements, one of them used to represent contact and sliding between the target surface, that of the indenter, and a deformable surface, that of the specimen. The geometric description of the two elements is illustrated in the diagram of Fig. 4.

The specimen top surface was overlaid with the contact element and the indenter surface was meshed with the target element. The penalty method with Lagrange multiplier was used to ensure that penetration compatibility is satisfied. Convergence issues were resolved with this modelling, which was subsequently adopted in all simulations producing the results presented in this paper.

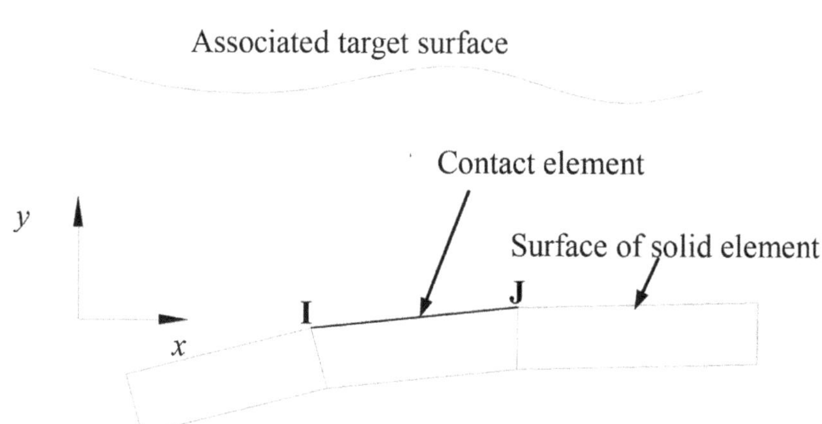

Figure 4: 2-D surface-to-surface contact element (2 nodes).

4.3 Characteristics of the specimen mesh

Initially, the mesh of the indented specimen was generated with its refined part in the indentation zone made up of element size $1.0 \times 2.0\ \mu m^2$. The dimensions of the mesh that covers the modelled half of the cylinder were $672 \times 448\ mm^2$. This is referred to as mesh A and is illustrated in Fig. 5.

To investigate the sensitivity of the results to mesh changes, two other meshes, labelled B and C, were generated by enlarging the specimen model dimensions and the refined zone covering the indentation region. In all three meshes, the indenter is meshed using the free mesh option in ANSYS. Elements on the indenter perimeter from the tip up to the height that corresponds to the expected indentation depth are taken to be 1.0 μm wide, which is the same width as the corresponding elements of the specimen over the contact region. The characteristics of all meshes are summarised in Table 3.

5 RESULTS

5.1 Validation against the elastic solution

The developed FE model was first applied to simulate the indentation process of a perfect sphere with a radius of 150 μm on a linear elastic material. It was thus possible to validate the model against Sneddon's elastic solution of a sphere pressed on an elastic flat surface [1]. The

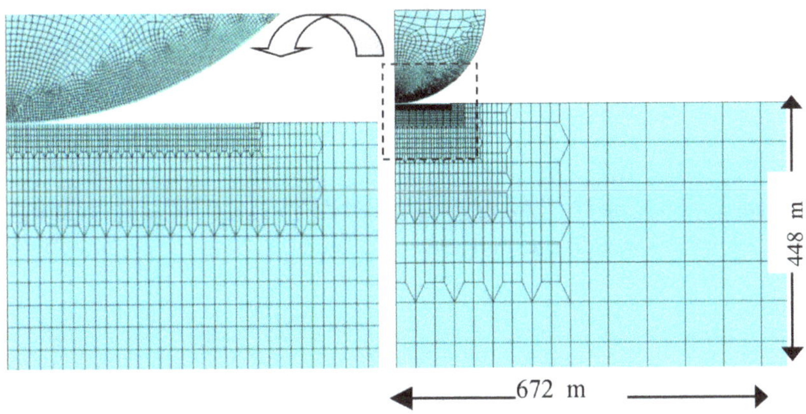

Figure 5: Mesh A of the specimen produced manually using PATRAN.

Table 3: Details of the meshes considered.

Mesh	Specimen dimensions (mm) Width	Height	Number of Nodes	Number of Elements	Size of zone with finest mesh in the contact region (mm) Width	Height
A	672	448	1602	1510	92	8
B	960	1600	4200	4046	112	16
C	1472	1568	8665	8470	127	96

contact was modelled as frictionless. Loading was prescribed as displacement controlled increased by 0.1 μm at every load step. Peak load was attained when maximum depth reached 20 μm. At every load step the indent profile geometry, contact radius and depth were directly measured and compared with the corresponding predictions of Sneddon's solution. The predicted load-indentation depth curve by FE for each mesh A, B and C, described in Table 3, as well as Sneddon's solution are shown in Fig. 6.

As observed in the plots of Fig. 6, best agreement between numerical predictions and Sneddon's elastic solution is achieved with meshes B and C. The same conclusion is reached by plotting the ratio of contact depth, h_c, to indentation depth, h_t, which should be equal to 0.5 according to Sneddon's solution. Since there is no significant difference between the results produced by these meshes and, as can be seen in Table 3, mesh B has less than half the number of elements than mesh C, mesh B was adopted for the FE simulations of contact discussed in this paper.

5.2 Elasto-plastic FE analysis

The initial elasto-plastic indentation simulations were carried out in order to assess the effect of variations in critical input parameters on the results. For this purpose, idealised material properties and a perfect spherical indenter of 150 μm radius were adopted. The stress-strain curve produced by tensile testing of specimen S5 was modified to an idealised curve with a single value for the strain hardening exponent and the strength coefficient. The modified stress-strain curve was based on the actual curve for steel plate S5. It corresponded to a material with Young's modulus $E = 210$ GPa, elastic limit $\sigma_e = 340$ MPa, $\sigma_{0.2} = 375$ MPa, $n = 0.132$, and $K = 782.91$ MPa, satisfying the continuity condition at yield point where

$$\varepsilon_{0.2} = 0.002 + \frac{\sigma_{0.2}}{E} \tag{4}$$

The part on the stress–strain curve joining the elastic limit to the yield stress was created using a quadratic function. The modified stress-strain curve was very close to the original one, characterised by three different regions.

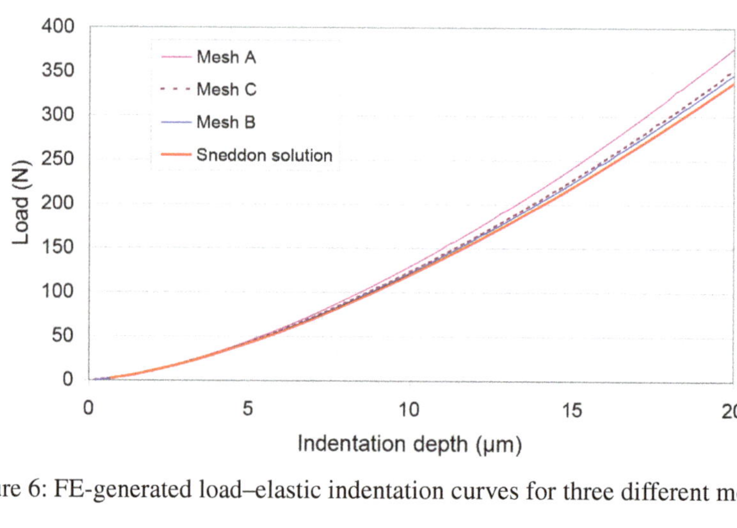

Figure 6: FE-generated load–elastic indentation curves for three different meshes.

Load-indentation curves were obtained for all three meshes detailed in Table 3 and it was noted that the elasto-plastic solutions almost overlap each other. Sensitivity to mesh variation was therefore found negligible. This suggested that mesh B would be satisfactory for further elasto-plastic FE simulations.

Contact tolerance and contact stiffness were two arbitrarily specified numerical parameters associated with contact modelling. Significant changes to their recommended values had no noticeable effect on the indentation output.

Next, the effect of friction between indenter and specimen was examined. Five different values of the friction coefficient of contact were tried, namely 0, 0.05, 0.1, 0.2 and 0.5. The resulting *P–h* curves can be compared by reference to Fig. 7.

As noticed in Fig. 7, the curves start to deviate at around two-thirds of the maximum depth and the maximum difference at the peak load is less than 4%. However, when the pile-up was assessed, the effect of friction was found to be more pronounced. The profiles of the indent for the five friction coefficients can be compared by referring to Fig. 8.

Finally, the effect of changing the elastic modulus of the indenter was assessed. Due to anisotropy of diamond, its Young's modulus, E_i, varies with crystal orientation. E_i can be determined by indirect methods. One such method is by measuring the speed of sound or

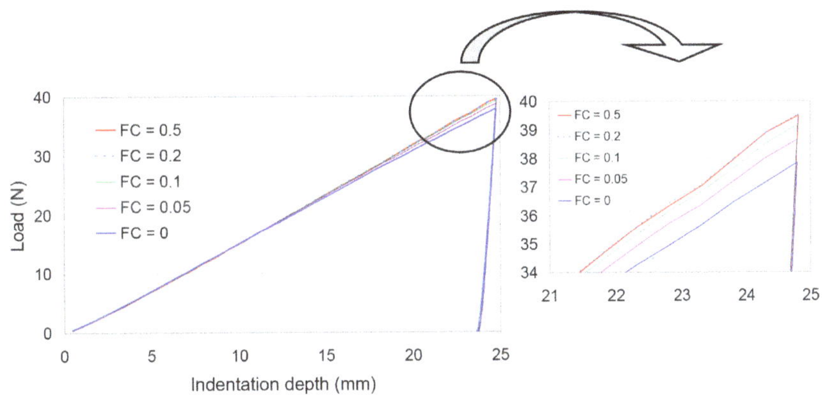

Figure 7: FE-generated load-indentation curves for different assumed friction coefficients.

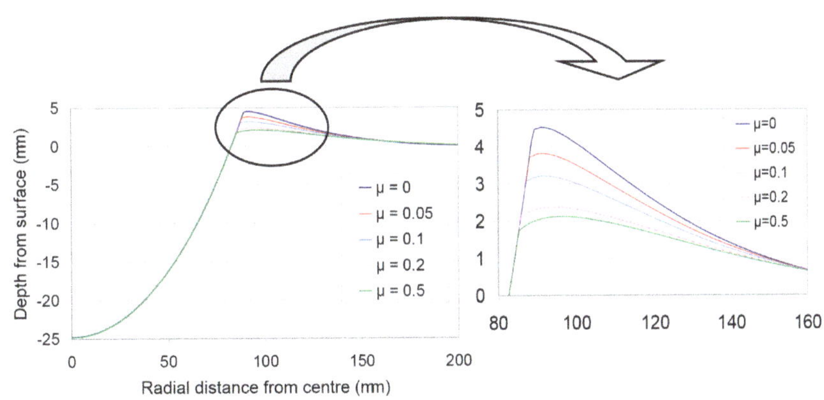

Figure 8: Indentation profiles at peak load using five different friction coefficients.

compression waves in a diamond crystal. An approximate value of 1050 GPa is given for E_i by Wilks and Wilks [16], which was said to vary with the crystal orientation by approximately 10%. This value is very close to that initially adopted.

To assess the effect of any variation of E_i on the *P-h* curves resulting from FE simulation, the latter was run for four values of E_i ranging from 800 to 2000 GPa in addition to the case of an infinitely rigid indenter. Comparison of the resulting load-depth curves showed that there is no significant difference between them for the extreme cases of Young's modulus values, and their difference is even more negligible for the realistic range lying between 800 and 1500 GPa. As a result, the adopted value of 1141 GPa was proved to be acceptable.

5.3 Validation versus experimental results

The performance of the developed FE model in the region of elasto-plastic deformation was validated through comparisons of its predictions with experimental data obtained by instrumented indentation testing [17]. FE simulations were performed on plates S5, S6, S7 and stainless steel. The actual properties of these specimens, described in Section 3, were entered as material input to the elasto-plastic FE analysis. The indenter model was changed from perfect sphere to the real profile determined by digitising and averaging SEM photographs of the actual indenter tip [18]. The program was run with friction coefficient μ = 0.2. Plots of both experimental and numerical results for plates S5 and stainless steel are shown in Figs 9 and 10, respectively.

The good agreement observed in these figures was also noted in similar plots for the other two steel plates. The observed variation in the experimental curves is considered to lie within acceptable limits and it can be understood to be due to variations in material microstructure, surface finishes, and equipment error, among others. The experimental results appear to indicate a stiffer material than that anticipated from the predictions but it should be noted that the FE simulation does not consider any creep effect. Taking that effect into consideration, as discussed in the next section, brings the experimental points closer to the corresponding FE predictions.

5.4 Creep effect

The reliability of any output from indentation experiments depends on ensuring ideal measurement conditions, which help eliminate any undesirable effects. Thus, the recorded

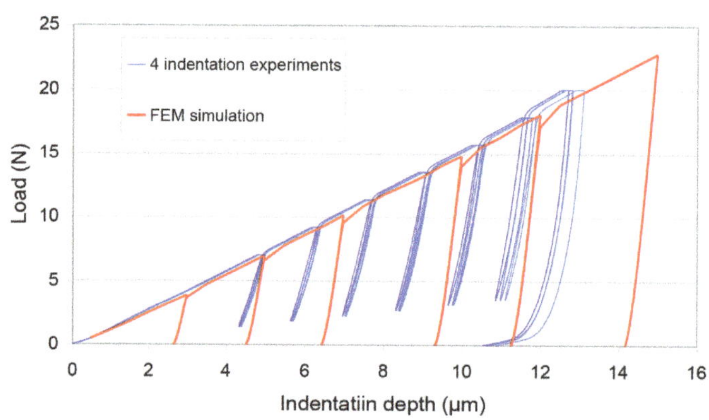

Figure 9: Load-indentation obtained at four locations on S5 and the corresponding FE prediction.

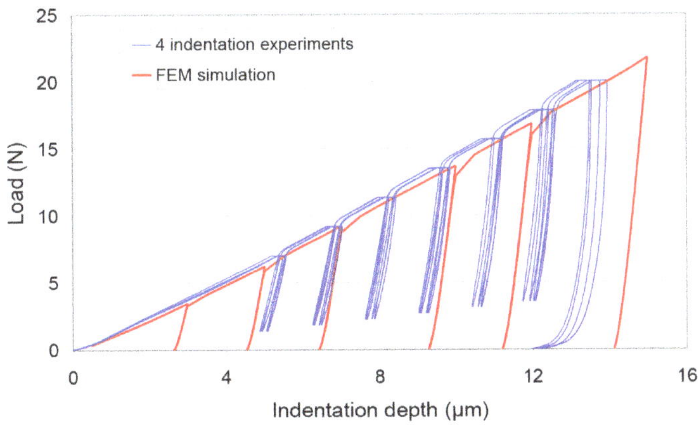

Figure 10: Load-indentation obtained at four locations on stainless steel and the corresponding FE prediction.

indentation data would correspond to the material response predicted by the FE simulation. One effect that cannot be eliminated completely is that of loading rate, which manifests itself as creep; the latter is observed as a plateau in the experimental load-indentation curve at peak load. The extent of creep depends on the material and it is also influenced by the hold/dwell time, which is the duration of maximum load application.

If creep is not considered, this can have a significant effect on the calculation of hardness and the reduced modulus from the unloading curve. It was found by Chudoba and Richter [19] that creep not only affects the measured depth but also the slope of the unloading curve at peak load, which is a key quantity in calculating the reduced modulus. It was reported that when creep was not allowed to take place, the resulting unloading curve showed bowing towards higher depth.

Noticeable creep deformation was reported during indentations on tungsten and aluminium using the Berkovich indenter [19–21]. Hence, it was considered essential to study the effect of creep using a spherical indenter on mild steel in order to eliminate its effect from any subsequently developed material characterisation scheme. For this purpose, spherical indentations were performed at three locations on steel plate S5. The peak load was chosen to be the maximum applicable by the instrument, that is, 20 N.

Since the extent of creep depends on the loading rate, its effect was examined by varying the loading time, that is, the time required to reach the peak load. Three values for the loading time, namely 8, 40, and 200 s, were adopted. When the peak load was reached, it was held constant for a dwell time of 180 s then partial unloading by 80% followed at the same loading rate. The obtained load-indentation curves are shown with the corresponding FE prediction in Fig. 11.

The capability of continuously recording the creep deformation with time made it possible to monitor the development of such deformation over the dwell time. This is shown for the various loading times in Fig. 12. As can be clearly seen in that figure, the creep effect can be removed by holding the peak load constant for a sufficiently long period. Creep deformation takes place at a high rate during the first few seconds then it diminishes within less than a minute. Its value reaches 150 nm/s at the first few seconds of the hold time and reduces to less than 3 nm/s after 15–20 s.

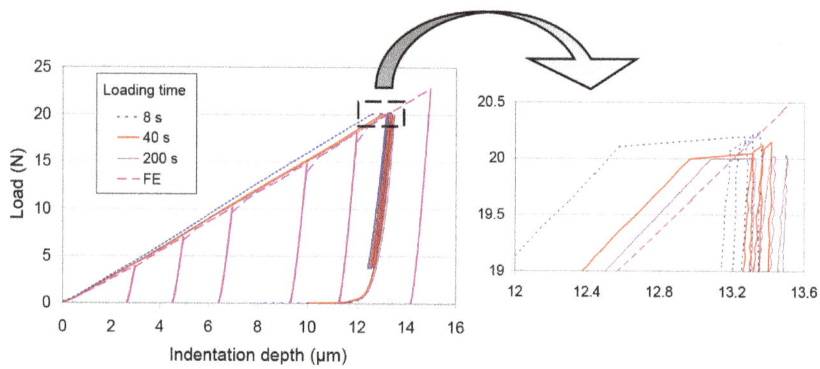

Figure 11: Load–displacement curves for three loading time compared with FE prediction.

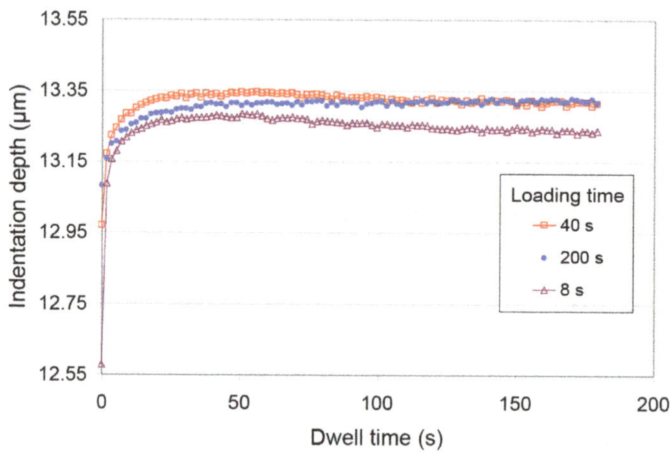

Figure 12: The change of indentation depth over the dwell time, 180 s.

It can also be observed that the loading time has a noticeable effect on the three main segments of the load-displacement curve, namely, the loading curve, hold time at peak load (amount of creep), and unloading curve. In the case of loading time of 8 s, the loading curve appeared stiffer and the amount of creep deformation was large, that is, 1.387 μm. On the other hand, a more compliant loading curve and less creep were noticed when loading time was longer. In the case of loading times of 40 and 200 s, the corresponding measured creep was 0.6894 and 0.4983 μm, respectively.

Even though the long loading time results in less creep, allowing enough dwell time makes the depth at peak load reach the same final value and the unloading curves almost overlap regardless of the loading time as long as it is not very short, that is, at least 40 s.

Applying the required dwell time in the case of the load time of 8 s, will bring the maximum depth close to the corresponding value of the other two cases. Nevertheless, the unloading curve shows bowing at peak load, which will incur an error in calculating the contact stiffness, and thus the modulus. This indicates that short-time loading causes erroneous indentation data, which they seem to include undesired deformations. This explanation of bowing can be supported by the findings of Chudoba and Richter [19] who reported this bowing phenomenon when dwell time is zero and loading time is 92 s. The material behaviour in the case of short-time loading can be understood as a dynamic response to high strain rate loading, whereby the material seems stronger, compared with its response to static loading.

Referring to Figs 9 and 10, it can be observed that the creep value is not only dependent on the loading rate but also on the depth reached, which is in agreement with the findings of Chudoba and Richter [19]. Figure 12 shows that a dwell time of 120 s is enough to allow creep deformation to take place in full and it can compensate for the value of the loading time, as long as the latter is 40 s or more. It also shows, together with Fig. 11, that a loading time of 200 s is adequate to produce reliable indentation data provided enough dwell time is allowed, and that increasing the loading time will not result in noticeable change in the amount of creep. Therefore, a value of 120 s for the dwell time and loading time of 200 s were adopted in the subsequent experiments for material characterisation.

This creep phenomenon clearly implies that since the loading rate effect was not considered in the FE model, only those points corresponding to the end of the creep plateau of *P–h* curves should be considered when comparison with corresponding FE predictions is made, as the other points on the loading curve exclude creep, thus are expected to represent a stiffer response.

6 SUMMARY AND CONCLUSIONS

An axisymmetric FE model was developed for simulating the elasto-plastic spherical indentation of test materials with known, experimentally obtained stress-strain relations. The specimen mesh was generated manually in order to address the high geometric nonlinearity induced by deep indentation. Mesh sensitivity was carried out to ensure the capability of the adopted mesh with regards to accuracy and efficiency.

The FE model was first validated against the Sneddon's elastic solution for a sphere onto a flat surface. Then it was applied to the indentation of idealised elasto-plastic materials, in order to assess the effect of various FE control and physical parameters. These included: the contact tolerance and contact stiffness, friction coefficient between indenter and specimen and the elastic modulus of the diamond indenter. It was found that none of these parameters, within their range of variability and load range of application, had a noticeable effect on the simulated *P–h* curves. However, friction was shown to have a pronounced influence on the predicted pile-up, thus the contact area, which is an important parameter in characterisation procedures based on spherical indentation data.

Time-dependent deformation, so-called creep, was also investigated. The experimental conditions influencing creep were examined. It was found that loading rate and the hold/dwell time at peak load govern not only the extent of creep but also the initial contact stiffness. It was also noticed that the higher the loading rate, the larger the creep effect, thus a longer dwell time will be required to allow for this deformation to take place as long as the full load is not applied abruptly, i.e. in less than 10 s, which can affect the initial unloading contact stiffness. Based on the experimental observations on the test materials, it was found adequate to adopt, for practical purposes, a loading time of not less than 40 s followed by a dwell time of not less than 100 s to account fully for any time-dependent deformation.

Load-indentation curves were simulated using the actual imperfect geometry of the spherical indenter and compared with those obtained from experiments. Indentation experiments were performed in a gradually amplified cyclic loading pattern and the peak load was held for a specified time to remove the effect of creep deformation. The points corresponding to the end of the creep plateau of the experimental *P–h* curves were found in good agreement with the corresponding FE predictions.

REFERENCES

[1] Sneddon, I.N., The relation between load and penetration in the axisymmetric Boussinesq problem for a punch of arbitrary profile. *International Journal of Engineering Science*, **3**(47), pp. 47–57, 1965.
https://doi.org/10.1016/0020-7225(65)90019-4

[2] Taljat, B., Zacharia, T. & Kosel, F., New analytical procedure to determine stress-strain curve from spherical indentation data. *International Journal of Solids and Structures*, **35**(33), pp. 4411–4426, 1998.
https://doi.org/10.1016/s0020-7683(97)00249-7

[3] Cao, Y.P. & Lu, J., A new method to extract the plastic properties of metal materials from an instrumental spherical indentation loading curve. *Acta Materialia*, **52**, pp. 4023–4032, 2004.
https://doi.org/10.1016/j.actamat.2004.05.018

[4] Habbab, H., Mellor, B.G. & Syngellakis, S., Post-yield characterisation of metals with significant pile-up through spherical indentations. *Acta Materialia*, **54**, pp. 1965–1973, 2006.
https://doi.org/10.1016/j.actamat.2005.12.021

[5] Zhao, M., Ogasawara, N., Chiba, N. & Chen X., A new approach to measure the elastic-plastic properties of bulk materials using spherical indentation. *Acta Materialia*, **54**(1), pp. 23–32, 2006.
https://doi.org/10.1016/j.actamat.2005.08.020

[6] Lee, J.H., Kim, T. & Lee, H., A study on robust indentation techniques to evaluate elastic–plastic properties of metals. *International Journal of Solids and Structures*, **47**, pp. 647–664, 2010.
https://doi.org/10.1016/j.ijsolstr.2009.11.003

[7] Donohue, B.R., Ambrus, A. & Kalidindi, S.R., Critical evaluation of the indentation data analyses methods for the extraction of isotropic uniaxial mechanical properties using finite element models. *Acta Materialia*, **60**(9), pp. 3943–3952, 2012.
https://doi.org/10.1016/j.actamat.2012.03.034

[8] Barbadikar, D.R., Ballal, A.R., Peshwe, D.R. & Mathew, M.D., Finite element analysis of deformation due to ball indentation and evaluation of tensile properties of tempered P92 steel. *Metallurgical and Materials Transactions A: Physical Metallurgy and Materials Science*, **46**(8), pp. 3448–3459, 2015.
https://doi.org/10.1007/s11661-015-2974-5

[9] Dean, J. & Clyne, T.W., Extraction of plasticity parameters from a single test using a spherical indenter and FEM modelling. *Mechanics of Materials*, **105**, pp. 112–122, 2017.
https://doi.org/10.1016/j.mechmat.2016.11.014

[10] Yamamoto, T., Kurishita, H. & Matsui, H., Modeling of the cyclic ball indentation test for small specimens using the finite element method. *Journal of Nuclear Materials*, **271–272**, pp. 440–444, 1999.
https://doi.org/10.1016/s0022-3115(98)00799-5

[11] Ma, L., Low, S.R. & Song, J., Finite-element modeling and experimental comparisons of the effects of deformable ball indenters on Rockwell B hardness tests. *Journal of Testing and Evaluation*, **31**(6), pp. 514–523, 2003.

[12] Kogut, L. & Komvopoulos, K., Analysis of the spherical indentation cycle for elastic-perfectly plastic solids. *Journal of Materials Research*, **19**(12), pp. 3641–3653, 2004. https://doi.org/10.1557/jmr.2004.0468
[13] Karthik, V., Visweswaran, P., Bhushan, A., Pawaskar, D.N., Kasiviswanathan, K.V., Jayakumar, T. & Raj, B., Finite element analysis of spherical indentation to study pile-up/sink-in phenomena in steels and experimental validation. *International Journal of Mechanical Sciences*, **54(1)**, pp. 74–83, 2012. https://doi.org/10.1016/j.ijmecsci.2011.09.009
[14] Available at: http://www.ansys.com/en-GB/products/academic.
[15] Available at: http://www.mscsoftware.com/product/patran.
[16] Wilks, J. & Wilks, E., *Properties and Applications of Diamond*, Butterworth-Heineman: Oxford, 1991.
[17] Syngellakis, S., Habbab, H. & Mellor, B.G., Weld zone material characterisation based on spherical indentation data. *International Journal of Computational Methods and Experimental Measurements*, in print, 2017.
[18] Habbab, H., *Post-yield characterisation of welds based on instrumented hardness tester data*, PhD Thesis, University of Southampton, 2005.
[19] Chudoba, T. & Richter, F., Investigation of creep behaviour under load during indentation experiments and its influence on hardness and modulus results. *Surface & Coatings Technology*, **148(2–3)**, pp. 191–198, 2001. https://doi.org/10.1016/s0257-8972(01)01340-8
[20] Pharr, G.M., Oliver, W.C. & Brotzen, F.R., On the generality of the relationship among contact stiffness, contact area, and elastic-modulus during indentation. *Journal of Materials Research*, **7(3)**, pp. 613–617, 1992. https://doi.org/10.1557/jmr.1992.0613
[21] Oliver, W.C. & Pharr, G.M., An improved technique for determining hardness and elastic- modulus using load and displacement sensing indentation experiments. *Journal of Materials Research*, **7(6)**, pp. 1564–1583, 1992. https://doi.org/10.1557/jmr.1992.1564

A CROWD-STRUCTURE INTERACTION MODEL TO ANALYZE THE LATERAL LOCK-IN PHENOMENON ON FOOTBRIDGES

JAVIER FERNANDO JIMÉNEZ-ALONSO[1], ANDRES SÁEZ[2], ELSA CAETANO[3] & ALVARO CUNHA[3]
[1]Department of Building Structures, Universidad de Sevilla, Spain.
[2]Department of Continuum Mechanics and Structural Analysis, Universidad de Sevilla, Spain.
[3]Department of Civil Engineering, FEUP, Porto, Portugal.

ABSTRACT

In this paper a simplified biomechanical crowd-structure interaction model is proposed and validated in order to analyse the lateral lock-in phenomenon on real footbridges. The proposed crowd-structure interaction model is organized in three levels: (i) pedestrian-structure interaction; (ii) interaction among pedestrians in the crowd; and (iii) interaction between the crowd and the structure. To this end, first, the human-structure interaction of each pedestrian is modelled via a simplified two degrees of freedom system. Second, the interaction among pedestrians inside the crowd is simulated using a multi-agent model. The considered model simulates the movement of each pedestrian from the dynamic equilibrium of the different social forces that act on him/her. Finally, the crowd-structure interaction is achieved modifying the behaviour of the pedestrians depending on the comfort level experienced. For this purpose, the recommendations established by the French standards have been considered. The integration of the three levels in an overall model is achieved by the implementation of a predictive–corrective method. The performance of the proposed model is validated correlating the numerical and experimental dynamic response of the Pedro e Inês footbridge during the development of a lateral lock-in pedestrian test. As the first lateral natural frequency of the footbridge is inside the range that characterizes the walking pedestrian step frequency in lateral direction, numerical and experimental studies were performed to analyse its behaviour under pedestrian action. The agreement between the numerical and experimental results is adequate. However, further studies are recommended in order to generalize the proposed approach and facilitate its use during the design project of future footbridges.
Keywords: crowd dynamics, footbridge, human-structure interaction, lateral lock-in, simplified biomechanical model

1 INTRODUCTION

During the last 20 years numerous models have been proposed in order to analyse the lateral lock-in phenomenon on footbridges [1]. Although all these models have shed light in order to simulate numerically this phenomenon, they present the limitation of only taking into simplified account the modification of the modal parameters of the structure due to the pedestrian-structure interaction. The consideration of this fact may improve the numerical estimations of the dynamic response of these structures under pedestrian flows [2, 3].

In order to overcome this limitation, a crowd-structure interaction model is proposed and validated, in this study, to simulate more accurately the lateral lock-in phenomenon on footbridges. The proposed model allows estimating the change of the modal properties of footbridges under pedestrian flows. As consequence of this fact, the model allows a more accurate numerical analysis of the lateral lock-in phenomenon on real footbridges. The performance of the proposal is validated correlating the experimental and numerical lock-in phenomenon on a real footbridge, Pedro e Inês footbridge (Coimbra, Portugal) [4, 5].

The manuscript is organized as follows. The lateral crowd-structure interaction model is discussed in section 2; the Pedro e Inês footbridge is described in section 3; section 4 is dedicated to correlate the experimental and numerical results of a lateral lock-in pedestrian test

DOI: 10.2495/CMEM-V6-N4-764-771

performed on the footbridge; and finally, section 5 presents the main conclusions of this research.

2 CROWD-STRUCTURE INTERACTION MODEL IN LATERAL DIRECTION

The lateral crowd-structure interaction phenomenon is simulated using two individual sub-models: (i) a pedestrian-structure interaction sub-model and (ii) a crowd sub-model. In the first sub-model, each pedestrian is modelled as a TDOF (two degree of freedom) system and the dynamic behaviour of the structure is simulated via its modal parameters. In the second sub-model, the crowd is simulated using a behavioural model that provides a description of the individual pedestrian position, walking pedestrian velocity, step pedestrian frequency and phase between pedestrians. In order to consider the modification of the pedestrian behaviour associated with the vibration level, two additional conditions have been included in this sub-model: (i) a reduction factor, that affects the pedestrian velocity according to the comfort level experienced by each pedestrian; and (ii) a lock-in threshold, that modifies the step pedestrian frequency and the phase among pedestrians in order to simulate the crowd-structure synchronization [2]. In the following subsections, each sub-model and their interactions are described briefly. A more detailed presentation of the model may be found in Ref. [6, 7].

2.1 Pedestrian-structure interaction in the lateral direction

The proposed model follows from the application of dynamic equilibrium equations [8] to a TDOF-system (Fig. 1) with sprung ($m_{a,lat}$) and unsprung masses ($m_{s,lat}$).

The following coupled equation system may be obtained considering the dynamic balance of the proposed system, structure and pedestrian model.

$$M_i\ddot{y}_i + C_i\dot{y}_i + K_i y_i = \varphi_{num\_i}(v_{px}\cdot t)\cdot F_{\text{int}} \tag{1}$$

$$m_{a,lat}\ddot{y}_a + c_{p,lat}\left(\dot{y}_a - \dot{y}_s\right) + k_{p,lat}\left(y_a - y_s\right) = 0 \tag{2}$$

$$m_{s,lat}\ddot{y}_s + c_{p,lat}\left(\dot{y}_s - \dot{y}_a\right) + k_{p,lat}\left(y_s - y_a\right) = F_{p,lat} - F_{\text{int}} \tag{3}$$

where

$m_{a,lat}$, $m_{s,lat}$ and $m = m_{s,lat} + m_{a,lat}$ are the lateral pedestrian sprung mass, unsprung mass and the total pedestrian mass [kg], respectively. A medium pedestrian weight of 70 kg, according to Synpex guidelines [9], was considered in this study.

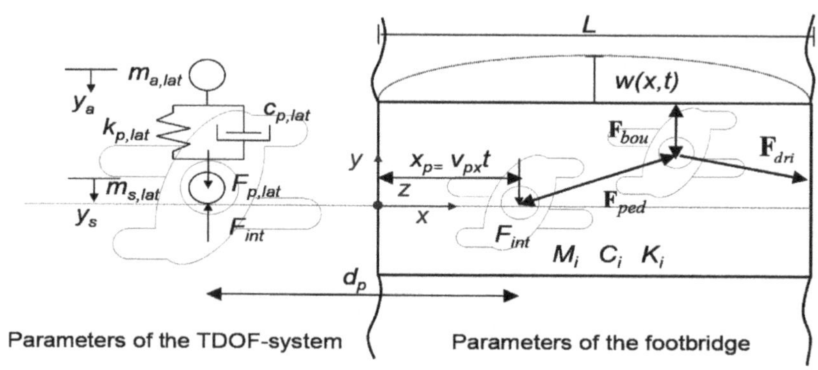

Figure 1: Biomechanical pedestrian-structure interaction model in lateral direction [6, 7].

y_i, y_a and y_s are the lateral displacement of the vibration mode i of the footbridge, the pedestrian sprung mass and the unsprung mass [m].

$k_{p,lat}$ and $c_{p,lat}$ are the lateral pedestrian stiffness [N/m] and damping [sN/m].

$F_{p,lat}$ and F_{int} are the lateral pedestrian walking and pedestrian-structure interaction forces [N].

M_i, C_i and K_i are the modal mass [kg], modal damping [sN/m] and modal stiffness [N/m] of the vibration mode i [kg].

φ_{num_i} is the lateral component of the numerical vibration mode i.

v_{px} is the longitudinal component of the pedestrian velocity vector [m/s].

t is the time [sec.].

d_p is the distance among pedestrians [m].

$w(x,t)$ is the lateral displacement of the footbridge at the position x [m].

L is the length of the footbridge [m].

The lateral pedestrian walking force can be expressed, according to the Synpex guidelines [9], in terms of Fourier series decomposition as:

$$F_{p,lat} = m \cdot g \cdot \sum_{i=1}^{3} \alpha_{i,lat} \cdot \sin\left(\pi \cdot i \cdot f_s \cdot t - \phi_{i,lat} - \varphi_p\right) \tag{4}$$

where

g is the acceleration of gravity.

$\alpha_{i,lat}$ is the Fourier coefficient of the *ith* harmonic for lateral forces or lateral dynamic load factor (LDLF).

f_s is the pedestrian step frequency [Hz].

$\phi_{i,lat}$ and φ_p are the phase shift of the *ith* harmonic of the lateral pedestrian force and among pedestrians respectively [rad].

Applying at the contact point the compatibility conditions of displacements, velocity and acceleration between the footbridge and the TDOF-system that models the pedestrian, the above relations, eqns (1–3), may be organized in a matrix form.

$$\mathbf{M}(t) \cdot \ddot{\mathbf{y}}(t) + \mathbf{C}(t) \cdot \dot{\mathbf{y}}(t) + \mathbf{K}(t) \cdot \mathbf{y}(t) = \mathbf{F}(t) \tag{5}$$

For a group of k pedestrians (Fig. 1), we may further represent each one by the above TDOF-system. Thus, a system of $n+k$ differential equations will need to be solved (being n the number of considered vibration modes). Finally, for the numerical evaluation of the resulting system, the β-Newmark integration family method is used, considering as parameters $\beta = 1/4$ and $\gamma = 1/2$, thus ensuring an unconditionally stable system.

2.2 Parameters of the pedestrian-structure interaction model

The parameters of the pedestrian-structure interaction model (pedestrian modal parameters and walking pedestrian lateral force) have been estimated experimentally via the solution of two inverse problems based on the results of several pedestrian tests performed on a real footbridge [6, 7]. According to the results of this study, the following Gaussian distributions are established in Table 1 for the parameters that characterize the pedestrian-structure interaction model ($N(\mu,\sigma)$, being μ the mean value and σ the standard deviation).

Once determined the pedestrian-structure interaction parameters, the definition of the proposed model is complete.

Table 1: Parameters of the pedestrian-structure interaction model [6, 7].

Modal Parameters		
Definition	**Parameter**	**Value**
Pedestrian lateral sprung mass	$m_{a,lat}$	$N(73.216, 2.736)\%$
Pedestrian lateral damping ratio	$\zeta_{p,lat}$	$N(49.116, 5.405)\%$
Pedestrian lateral natural frequency	$f_{p,lat}$	$N(1.201, 0.178)$Hz
Walking Pedestrian Lateral force		
Definition	**Parameter**	**Value**
First LDLF	$\alpha_{1,lat}$	$N(0.086, 0.017)$
Second LDLF	$\alpha_{2,lat}$	$N(0.094, 0.009)$
Third LDLF	$\alpha_{3,lat}$	$N(0.040, 0.019)$
First lateral phase shift	$\phi_{1,lat}$	0°
Second lateral phase shift	ϕ_{2lat}	0°
Third lateral phase shift	$\phi_{3,lat}$	0°

2.3 Modelling the crowd behaviour

The proposed multi-agent model that simulates the behaviour of the crowd consists in the sum of three partial forces that represent the different influences that the pedestrians undergo when interacting in a crowd [10]. Thus, the resultant social force, $\mathbf{F}_{pci}$, is defined as (Fig. 1):

$$\mathbf{F}_{pci} = \mathbf{F}_{dri} + \mathbf{F}_{ped} + \mathbf{F}_{bou} \quad (6)$$

where

$\mathbf{F}_{dri}$ is the driving force that reflects the motivation of each pedestrian to reach his desired destination [N].

$\mathbf{F}_{ped}$ is the repulsive force, which simulates the interaction among pedestrians [N].

$\mathbf{F}_{bou}$ is the repulsive force, which simulates the interaction among the pedestrians and the boundaries [N].

In order to simulate the pedestrians flow, three parameters must be defined: (i) the pedestrian density [P=Person/m^2], (ii) the value of the desired pedestrian velocity, and (iii) the distance between pedestrians. In the present study the pedestrian density is established depending on the expected pedestrian traffic according to the Synpex guidelines [9]. The values of the desired velocity of each pedestrian are obtained from the pedestrian step frequencies, defined according to the Gaussian distributions proposed by Zivanovic *et al.* [11], $N(1.87, 0.186)$Hz. Subsequently each assigned step frequency is converted into the desired velocity using the empirical relation proposed by Bertram and Ruina [12]. Additionally, the original distance among pedestrians is calculated assuming a rectangular-shaped mesh of pedestrians and the considered pedestrian density. Finally, the phase shift among pedestrians, φ_p, is determined initially adopting a Poisson distribution [8]. Subsequently, this parameter is modified via the crowd-structure interaction.

2.4 Crowd-structure interaction

The maximum lateral acceleration experienced by each pedestrian may be compared against the acceleration threshold values established by the Synpex guidelines [9] in order to modify the individual pedestrian behaviour inside the crowd. Two types of thresholds have been considered in the model for this purpose: (i) comfort thresholds and (ii) a lateral lock-in threshold. First, in order to reflect in the model the change of the behaviour of each pedestrian due to the modification of his comfort level, a behavioural factor has been applied to the pedestrian velocity. According to the results provided by previous studies [3], minimum and maximum comfort thresholds of 0.20 m/s^2 and 2.10 m/s^2, respectively, have been selected in the model. In this manner, if the lateral acceleration of each pedestrian exceeds the minimum value, the pedestrian reduces its velocity, stopping in case the maximum value is overcome. Second, in order to take into account in the model the lateral lock-in phenomenon, the criterion provided by the French standard [13] is considered. If the lateral acceleration experienced by each pedestrian is above 0.15 m/s^2 and his step frequency is within ±10% of the lateral natural frequency of the structure, both his step frequency and phase shift are modified to match the natural frequency of the footbridge, synchronizing the movements of the pedestrian and the structure.

3 DESCRIPTION OF PEDRO E INÊS FOOTBRIDGE

The Pedro e Inês footbridge is located at Coimbra (Portugal). The total length of the structure is 274.5 m, configured by one central arch of 110 m, two lateral semi-arches of 64 m and two transition spans of 30.5 and 6 m (Fig. 2a). The main feature of the footbridge is the anti-symmetrical configuration of the deck and the arches with respect to the longitudinal axis of the structure. The deck is a concrete-steel composite box-girder with a variable width between 4 and 8 m, which generates a panoramic square at mid-span of the footbridge (Fig. 2b). Even

Figure 2: (a) General view of Pedro e Inês footbridge; (b) Perspective of the footbridge and (c) experimental lateral lock-in pedestrian test on this footbridge [4, 5].

during the design phase, numerical studies indicated that the structure was prone to vibrations induced by pedestrians in the lateral direction. This fact motivated the development of a precise and detailed work for the experimental assessment of the dynamic response and the implementation of a control system in order to guarantee an adequate comfort level of the footbridge. This work was performed and reported by Caetano *et al.* [4, 5] and its results have been used in this paper in order to validate the proposed lateral crowd-structure interaction model.

The footbridge presented a first lateral vibration mode with an experimental natural frequency of 0.91 Hz, with an associated damping ratio of 0.55 %, which was easily excited by the pedestrian flows. In order to determine experimentally the number of pedestrians that originates the lateral lock-in phenomenon, an experimental test was performed. The result of this experimental test, reported in the literature [4, 5], has been correlated with the numerical simulation based on the implementation of the proposed lateral crowd-structure interaction model on an updated finite element model [14] of the Pedro e Inês footbridge.

4 LATERAL LOCK-IN PEDESTRIAN TEST ON THE PEDRO E INÊS FOOTBRIDGE

In the experimental lateral lock-in pedestrian test, the lateral acceleration, a_{lat}, at mid-span of the footbridge under the passage of different group of pedestrians was recorded (Fig. 2c). A graphical representation of the maximum lateral acceleration at this position versus the number of pedestrians on the footbridge (Fig. 3) allows for identifying the instability situation associated with the lateral lock-in phenomenon. As it is illustrated in Fig. 3, the number of pedestrians that originates the beginning of the lateral lock-in phenomenon is around 75.

Subsequently, a numerical lateral lock-in pedestrian test based on the proposed lateral crowd-structure interaction model was performed. Each considered group of pedestrians were simulated considering as initial spatial distribution a rectangular-shaped grid with an initial distance among pedestrians of 0.50 m in the longitudinal direction and an equidistant distribution in the lateral direction. During the numerical simulation, each considered group of pedestrians walks freely along the footbridge. The number of pedestrians in each group increases gradually between 15 and 85 in increments of 5. The maximum numerical lateral acceleration at mid-span versus the number of pedestrians on the footbridge is shown in Fig. 3.

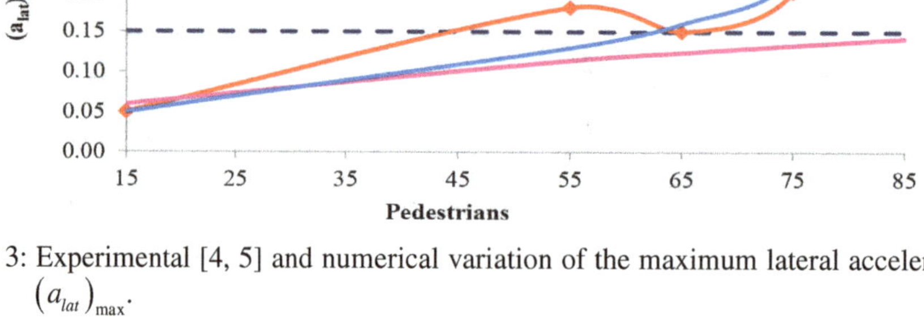

Figure 3: Experimental [4, 5] and numerical variation of the maximum lateral acceleration, $(a_{lat})_{max}$.

As Fig. 3 shows, the correlation between the experimental lateral maximum accelerations and the numerically estimated maximum values is adequate. Additionally, the estimation of the numerical maximum acceleration obtained, applying the methodology proposed by the Synpex guidelines [9], is also shown in Fig. 3. It is clear from Fig. 3 that the proposed model allows obtaining a more accurate numerical analysis of the lateral lock-in phenomenon than these standards. The lateral lock-in criterion established by French standards [13] is also illustrated for reference in Figure 3.

5 CONCLUSIONS

In this paper, a crowd-structure interaction model is proposed and validated to analyse the lateral lock-in phenomenon on footbridges under pedestrian flows. The proposed model is organized in two sub-models: (i) a pedestrian-structure interaction and (ii) a crowd sub-model. The pedestrian-structure interaction sub-model is defined in terms of a TDOF-system. The crowd sub-model is defined in terms of a multi-agent model based on the social force model. The interaction between the two sub-models is achieved by imposing two behavioural conditions: a comfort and lateral lock-in thresholds. According to these requirements if certain acceleration limits are exceeded the affected pedestrians modify their behaviour. The proposed model is formulated under the following hypotheses: (i) constant parameters of the pedestrian-structure interaction model; (ii) dynamic effects associated to the change of the pedestrian velocity neglected, and (iii) the interaction problem only considers the walking action. The proposed model has been implemented in order to analyse the correlation between the experimental and numerical lateral lock-in phenomenon on a real footbridge, the Pedro e Inês footbridge (Coimbra, Portugal). The results obtained from the proposed model are adequate, allowing the accurate prediction of the lateral lock-in phenomenon on this footbridge and consequently the validation of the proposal. The proposed model may be an useful tool to analyse the lateral lock-in phenomenon on footbridges during the design phase. Nevertheless, further studies, additional pedestrian and crowd tests on different footbridges, are recommended in order to better characterize the lateral lock-in phenomenon on real footbridges.

ACKNOWLEDGEMENTS

This work was partially funded by the Spanish Ministry for Science under research project DPI2014-53947-R.

REFERENCES

[1] Ingólfsson, E.T., Georgakis, C.T., Ricciardelli, F. & Jönsson, J., Experimental identification of pedestrian-induced lateral forces on footbridges. *Journal of Sound and Vibration*, **330**, pp. 1265–1284, 2011. https://doi.org/10.1016/j.jsv.2010.09.034

[2] Carrol, S.P., Owen, J.S. & Hussein, M.F.M., Modelling crowd-bridge dynamic interaction with a discretely defined crowd. *Journal of Sound and Vibration*, **331**, pp. 2685–2709, 2012. https://doi.org/10.1016/j.jsv.2012.01.025

[3] Bocian, M., Macdonald, J.H.G. & Burn J.F., Biomechanically inspired modelling of pedestrian-induced forces on laterally oscillating structures. *Journal of Sound and Vibration Sound*, **331**, pp. 3914–3929, 2012. https://doi.org/10.1016/j.jsv.2012.03.023

[4] Caetano, E., Cunha, A., Magalhães, F. & Moutinho, C., Studies for controlling human-induced vibration of the Pedro e Inês footbridge, Portugal. Part 1: Assessment of dynamic behaviour. *Engineering Structures*, **32**, pp. 1069–1081, 2010. https://doi.org/10.1016/j.engstruct.2009.12.034

[5] Caetano, E., Cunha, A., Moutinho, C. & Magalhães, F., Studies for controlling human-induced vibration of the Pedro e Inês footbridge, Portugal. Part 2: Implementation of tuned mass dampers. *Engineering Structures*, **32**, pp. 1082–1091, 2010. https://doi.org/10.1016/j.engstruct.2009.12.033

[6] Jiménez-Alonso, J.F., *Proposal and calibration of a biodynamic model of human-structure interaction by the resolution of the inverse dynamic problem: application to pedestrian bridges*, PhD Thesis, University of Seville, 2015.

[7] Jiménez-Alonso, J.F., Sáez, A., Caetano, E. & Cunha, A., Lateral crowd-structure interaction model to analyse the change of the modal properties of footbridges. Submitted for publication.

[8] Jiménez-Alonso, J.F., Sáez, A., Caetano, E. & Magalhães, F., Vertical crowd-structure interaction model to analyse the change of the modal properties of a footbridge. *Journal of Bridge Engineering ASCE*, **21**(8), pp. C4015004/1-19, 2016. https://doi.org/10.1061/(asce)be.1943-5592.0000828

[9] Butz, C.H., Heinemeyer, C.H., Goldack, A., Keil, A., Lukic, M., Caetano, E. & Cunha, A., *Advanced load models for Synchronous Pedestrian Excitation and Optimised Design Guidelines for Steel Footbridges* (SYNPEX). RFCS-Research Project RFS-CR-03019. 2007.

[10] Helbing, D. & Molnár, P., Social force model for pedestrian dynamics. *Physical Review*, **51**(5), pp. 4282–4286, 1995. https://doi.org/10.1103/physreve.51.4282

[11] Zivanovic, S., Pavic, A. & Ingolfsson, E., Modelling spatially unrestricted pedestrian traffic on footbridges. *Journal of Structural Engineering*, **136**(10), pp. 1296–1308, 2010. https://doi.org/10.1061/(asce)st.1943-541x.0000226

[12] Bertram, J.E.A. & Ruina, A., Multiple walking speed-frequency relations are predicted by constrained optimization. *Journal of Theoretical Biology*, **209**(4), pp. 445–453, 2001. https://doi.org/10.1006/jtbi.2001.2279

[13] SETRA/AFGC. Guide méthodologique passerelles piétonnes (Technical Guide Footbridges: Assessment of vibration behaviour of footbridge under pedestrian loading). SETRA, 2006.

[14] Mottershead J.E., Link M. & Friswell M.I., The sensitivity method in finite element model updating: a tutorial. *Mechanical System and Signal Processing*, **25**, pp. 2275–2296, 2011. https://doi.org/10.1016/j.ymssp.2010.10.012

ACTIVE INTENTION INFERENCE FOR ROBOT-HUMAN COLLABORATION

HSIEN-I LIN, XUAN-ANH NGUYEN & WEI-KAI CHEN
Graduate Institute of Automation Technology National Taipei University of Technology, Taipei, Taiwan.

ABSTRACT
Understanding human intention is an important ability for an intelligent robot to collaborate with a human to accomplish various tasks. During collaboration, a robot with such ability can predict the successive actions that a human partner intends to perform, provide necessary assistance and support, and remind for the missing and failure actions from the human to achieve a desired task purpose. This paper presents a framework that allows a robot to automatically recognize and infer the action intention of a human partner based on visualization, in which an inverse-reinforcement learning (IRL) system is learnt based on the observed human demonstration and used to infer the human successive actions. Compared to other systems based on reinforcement learning, the reward of a Markov-Decision process (MDP) is directly learned from the demonstration. In our experiment, we provide some examples of the proposed framework which yields promising results with coffee-making and pick-and-place tasks. Regarding to the human-intention model based on IRL, the coffee-making experiment indicates that the action is globally predicted because the action of putting down the water pot is selected instead of pouring water when the cup is already filled with water.
Keywords: Human gesture recognition; human-robot collaboration; Markov decision process.

1 INTRODUCTION AND MOTIVATION

Nowadays along with the rapid development of intelligent robots, robots are expected to become a partner of elders, a friend of children, or a co-worker in manufacturing factories. However, these needs are still different from what the robots are programmed manually as usual. Thus, developing a robot which can interact with human partners is one of the major challenges in recent robotic research. Among a variety of problems in human-robot interaction, the importance of human-intention prediction is clear because if a robot understands human intention, it can provide necessary support and assistance at an appropriate moment. Similarly in our society, we observe each other, ask directly or based on the observations to understand and infer correctly the intentions of the others, and then we can work together to accomplish a task successfully. The above implies that endowing a robot with an ability to understand human intention helps the robot become a real helper to humans.

Many works for message exchange between humans and robots have done so far. In these methods, human action plays a key means to deliver information from humans to robots. For example, Bascetta *et al.* [1] used human intention to secure human safety in a coexistence system. On the other hand, Kanno *et al.* [2] categorized the operational definition of intention recognition into three types: keyhole recognition, intended recognition, and obstructed recognition. Intended recognition for human-robot collaboration (HRC) indicates that a human and a robot are both aware of cooperating in the recognition. Thus, this paper focuses the mode of intended recognition, meaning that the robot attempts to recognize human intention during HRC.

The recognition models of human intention have been developed by genetic algorithm (GA), Graphical model, Hidden Markov model (HMM), and Dynamic Bayesian network (DBN), neural-fuzzy network, ontology. Zhou *et al.* [3] used the genetic and ant algorithms and driving data to predict a driver's intention such as lane changing and car following. Wu *et al.* [4] proposed a graphical model to recognize attack intention which was represented by

DOI: 10.2495/CMEM-V6-N4-772-784

a plan or sequence of actions and adopted D-S evidence theory to infer intention by probability. Jin *et al.* [5] used a HMM to predict a driver's intention for assisting lane changing. Gehrig *et al.* [6] used 24 motion units and HMMs to form five complex motion sequences in a real-world kitchen tasks. Tahboub [7] used a DBN inferring user intention to reduce the latency in teleoperating a mobile robot. Jeon *et al.* [8] used ontology-based rules to do user intention inference. In other words, user intentions were described by hierarchical rules and sensor data. Huang *et al.* [9] applied human intention to control a robot walking. The intention was predicted by an adaptive network-based fuzzy inference system (ANFIS) using the force sensor data collected from the handle of the robot. Usually, human explicit actions are implicitly controlled by intention. However, the human-intention models such as HMM, DBN and neural-fuzzy networks lack the control mechanism. On the contrary, A MDP involves the control to infer human intention. In the work of [10] McGhan *et al.* introduced a MDP-based human task-level intent system in the context of a collaborative task. Their system used the human subject experiment data to feed the task model, a robot was hypothesized to predict companion intents by observing and identifying actions as a part of a motion sequence. The approach was based on the fulfillment of the human recurrent targets to find the reward function. In a similar way, our method adopts MDPs to infer human intention in HRC, but we apply a more optimal method to get the reward function.

In a human-intention recognition system, hand gestures are commonly used to predict human intention. For example, Song *et al.* [11] presented a probabilistic graphical method to estimate human intention in grasping tasks. A Kinect camera was used to monitor grasp processes and collect image data. Two major components of this framework are modeling grasping tasks and generative soft vector quantization using self-organizing maps (SOMs) and Gaussian mixture models (GMMs). Similarity, the approach in Ref. [12] employed a Bayesian classifier to recognize human intention based on tracking hands. The results were used in a remote mouse. Alternatively, Ref. [13] introduced continuous dynamic programming (CDP) as a solution for recognizing gesture spotting from human demonstration in video streams. However, most of the previous research mainly considers classifying human hand actions, while human intention should be regarded as an optimization problem of task planning from human demonstration. In this paper, we propose to use IRL [14] to find the optimal reward of a MDP from human demonstration.

The novelty of our research is to develop the human-intention recognition framework in HRC scenarios. Based on the marriage of MDP and IRL, the proposed framework is innovative in human-intention recognition. The MDP complements the probabilistic graphical models with the control power and the IRL provides an optimization process for a MDP from human demonstration. To obtain a correct type of hand gestures, we use a GMM as the robust skin colour model and convolutional neural networks (CNNs) [15] as the hand gesture classifiers. Without loss of generality, the image pre-process *e.g.* poses calibration is implemented before gesture classifiers. For the MDP, how to specify the reward is still a challenging task. However, with the demonstration, the IRL derives the optimal reward function of the MDP and accelerates the learning time for obtaining the MDP policy. All in all, the advantages of our approach include: first, the environmental uncertainty and decision making in the process of human intention are modeled by a MDP. Second, the reward function of a MDP is optimally found from human demonstration and the learning time is significantly reduced.

The rest of this paper is organized as follows: Section 2 briefly describes the problem statement. Section 3 presents the proposed human-intention framework, where the motion layer for recognizing hand gestures and the human-intention model are introduced. Section 4 presents the experiments that validate the feasibility of the proposed framework. Finally, concluding remarks and the future work are presented in Section 5.

2 PROBLEM STATEMENT AND FORMULATION

This paper considers a typical human-robot collaboration scenario as illustrated in Fig. 1 in which a robot and a human execute a daily task together. Some of the work is carried out by the human and the rest of the work is done by the robot. When the human cannot achieve the work by himself, the robot should be capable of detecting human intention for the task and providing appropriate assistance with the approval of the human. Figure 2 shows the collaboration principle with the proposed human-intention system. In this case, the robot is not only a collaborator but also a helper. Consequently, the collaboration performance will also be improved significantly.

As we know, human intention is explicitly regarded as a plan of sequential actions based on a human's desire. In other words, human intention is a process in which a human exploits his knowledge to make decisions about selecting suitable actions at different moments in order to accomplish a task. Thus, the goal of this paper is that a robot has to correctly recognize human intention with some uncertain situations during collaboration. Accordingly, a human-intention model that can tolerate noises and infer a reasonable plan of actions is the solution to this problem. Since a MDP has been known as a great probabilistic framework for

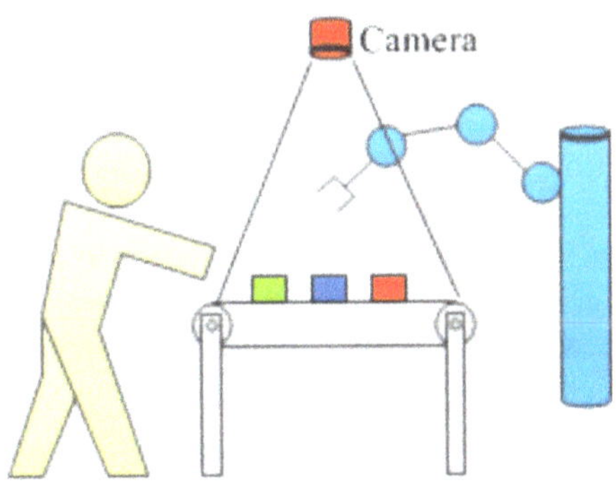

Figure 1: Human-robot collaboration scenario.

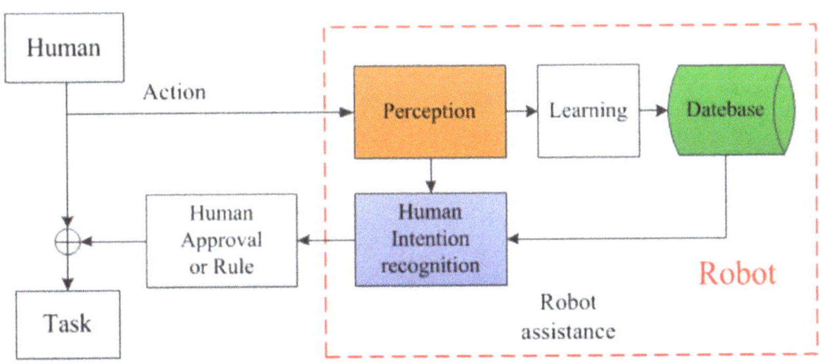

Figure 2: Collaboration principle in the proposed human-intention system.

decision-making tasks, we adopt a MDP to model human intention and then recognizing human intention is to find the policy of the MDP. A MDP (S, A, P, R, γ) consists of a set of states S, a set of actions A, a transitional probability P, and the reward function R that gives a reward for transition from the current state s to the next state s_0 using action a. A discount factor $\gamma \in [0;1]$ to avoid an infinite sum over a potentially infinite horizon. Most importantly, the reward function influences the policy of the MDP significantly but it is manually defined for a task. In this paper, we aim at exploring the reward function by human demonstration. Thus, we propose IRL as a suitable approach because it considers the entire task from scratch to the final step, in order to find a globally optimal policy. That is a unique and most favorable policy. For the proposed system, a human first demonstrates how to manipulate with objects in a certain order to achieve an intended task purpose, and then a robot learns the human intention to provide appropriate assistance.

3 THE HUMAN-INTENTION LEARNING SYSTEM

Figure 3 shows the flowchart of our framework, which consists of two phases: human gesture recognition and human-intention model. In the first phase, when a human and a robot start to collaborate, the captured images of the human gestures and manipulative objects are fed to the motion and object layers. The motion layer is in charge of recognizing human gestures by CNNs and the object layer provides the object attributes such as position and colour. In the second phase, the results from the first phase are encoded as the states in a MDP that serves as a human-intention model. The state and action spaces of the MDP are defined by users. Finally, the IRL is then applied to find the optimal reward function for the MDP using a collection of human demonstrations.

3.1 Hand gesture recognition

In the learning phase, the hand images of human demonstrations for a specific task are collected to train a hand gesture recognition system. For instance, five gesture types for a coffee-making task are shown in Fig. 4. During the execution phase, human actions will be

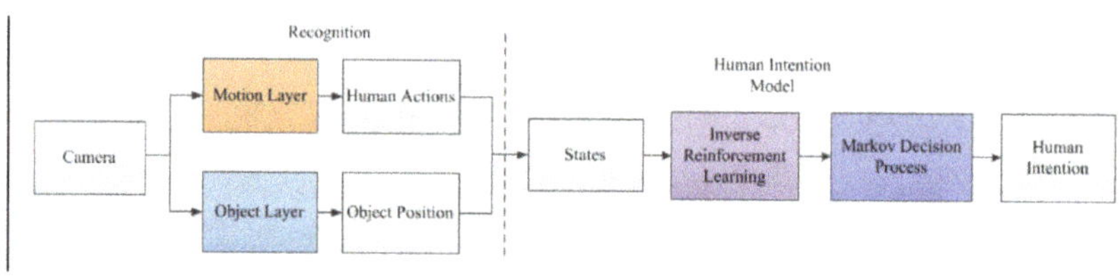

Figure 3: Flowchart of the proposed human-intention learning system.

(a) (b) (c) (d) (e)

Figure 4: Five gesture types to be recognized.

captured and recognized by the classifier above. In order to overcome the changing lighting condition, we adopt a GMM and the YCbCr color space to as the skin model for the images. The details of the implementation were presented in [16]. Using the skin model, the images are binarized into skin and non-skin areas. After image binarization, the images are filtered to remove unexpected noises and sharpen [16].

In addition, since human actions are continuous, the position and orientation of the hand are progressively changing, which makes it difficult to correctly recognize gestures. Thus, a pose calibration is applied to the binarized images to obtain better recognition results. The center position and principal axis of the hand are found by the gravity calculation method and then used to translate and rotate the image to the neutral pose. The pose calibration produces significantly better recognition results. Figure 5 (a) and (b) show an example of original and calibrated images respectively.

We now turn to the problem of recognizing hand gesture types. Recently, convolutional neural networks have been known as a successful tool of multi-class classification. This success is largely attributed to the use of local filtering, max pooling and weight sharing in the CNN architecture, which allows us to save memory size and improve efficiency. Additionally, a CNN uses less pre-processing than conventional neural networks, linear regression and support vector machines (SVMs). From the above mentioned advantages, we adopt CNNs as the approach to classify hand gesture types. However, our CNN is slightly modified to get a better result than the traditional one [17]. Let take coffee-making task as an example, we consider five gesture types, so our CNN output layer here has five units.

Each CNN for a gesture type has nine layers, denoted as I1, C2, S3, C4, S5, C6, S7, H8 and O9 in sequence. In particular, I1 is the input layer, which consists of 50 × 50 pixel calibrated images. C2, C4 and C6 are convolutional layers with 3 × 3, 5 × 5 and 5 × 5 masks, respectively. The masks in C2 layer are designed for sharpen, edge detection, horizontal edge detection, vertical edge detection, horizontal gradient detection, vertical gradient detection, horizontal sobel detection and vertical sobel detection. S3, S5 and S7 are sub-sampling layers. The feature maps in S3 and S5 layers are taken by 2 × 2 pixel sub-sampling but 3 × 3 in layer S7. The outputs of S7 are expanded and concatenated in terms of a 108 × 1 vector. Finally, the hidden layer of the neural nctwork H8 is also fully connected to the five neurons of the output layer O9. Stochastic back-propagation is used to update weights in the CNN. This modified CNN architecture demonstrates good results. Apparently, the modified CNN outperforms the traditional one. For continuous hand motions, transitive hand gestures are also taken into consideration

(a)

(b)

Figure 5: Pose calibration.

the proposed system. The transitive gestures are assigned to the gesture type in the closest preceding or successive frame [16].

3.2 Human-intention model

The human-intention model is very useful to a robot to understand the hidden information behind human motions. The hidden information could be a policy of actions which is a realization of the mental process. To find such a policy for our collaborative tasks, we adopt a MDP as a human-intention model and IRL as the method to find the optimal reward function of the policy.

After the previous stage of hand gesture recognition, the classified hand gestures and the object attributes are encoded as the states of the MDP represented as a 5-tuple $\{S,A,P,R,\gamma\}$. In our framework, let So be the set of possible attributes of the object layer, we consider object features like colour, position, status, *etc*, while S_m be the set of gesture status of the motion layer, for instance gestures are with or without object. S is the set of all possible states in the collaborative task. Consequently, $|S|$ of the MDP is $|S_0|\times|S_m|$. We also define $A=\{a_1,a_2,...,a_n\}$ as a set of all possible actions to accomplish desired tasks where n is the number of necessary actions. The action set is designed to ensure that the desired task can be completed by it and the set size should be kept to a minimum. In addition, the discount factor $\gamma\in[0,1]$ is selected during the optimization process and the square matrix $P(s,a,s')$ is the transition probability matrix that is used to describe all the probabilities over all the state transitions. In order to obtain the transition matrix *P*, we implement each task 100 times, represent each task by states *S*, and analyze the transitions among them. A policy $\pi:S\rightarrow A$ is evaluated by a value function (so-called Bellman equation). The Bellman equation is expressed as follows:

$$V^{\pi}=\sum_{a}\pi(s,a)\sum_{s'}P^{a}_{ss'}\left[R^{a}_{ss'}+\gamma V^{\pi}(s')\right] \tag{1}$$

For each policy π, a corresponding value V_{π} from Equation (1) is found. Furthermore, the goal of the learning agent is to find the optimal policy π^{*}, meaning that V_{π} should be the maximum value V^{*}:

$$V^{*}=\max_{a}\sum_{s'}P^{a}_{ss'}\left[R^{a}_{ss'}+\gamma V^{*}(s')\right] \tag{2}$$

In Equation (1), the reward *R* plays a crucial role in finding the optimal policy π^{*}. In other words, the optimal criterion depends on the definition of the reward function [14]. In fact, the reward function is usually handcrafted, which is non-convincing and can lead us to a non-optimal policy. For example, in our previous research, we have applied the frequency-based method to design the reward function [18], but in some cases, it is not optimal according to the demonstration data. In this paper, we attempt to find a convincing and efficient approach to obtain the reward function. Thus, we apply IRL to the human-intention model to derive the reward function directly by the demonstration data. The recognized states and actions of the task by the hand gesture recognition system are used to compute the reward function. The procedure of IRL attempts to maximize the margin from the optimal value function to others and is written as:

$$P_{a^*}V^{\pi}\geq P_{a}V^{\pi}\quad\forall a\in A\setminus a^{*} \tag{3}$$

where: a^* is the demonstrated action, and P_{a^*} and P_a are the transition matrices of action a^* and a, respectively. The details of the derivations are referred to [14]. In particular, the reward function R can be calculated as a linear programming problem, defined as:

$$\begin{aligned} &\max \sum_{i=1}^{N} \min_{a \in A \backslash a^*} \left(\left(P_{a^*} - P_a \right) \left(I - \gamma P_{a^*} \right)^{-1} R \right) - \lambda \| R \|_1 \\ &s.t. \ \left(P_{a^*} - P_a \right) \left(I - \gamma P_{a^*} \right)^{-1} R \geq 0 \quad \forall a \in A \backslash a^* \\ &\left| R_i \right| \leq R_{\max}, i = 1, \ldots, N \end{aligned} \tag{4}$$

In our tasks, λ is set to 0.65. Usually, robots are pre-programmed by default polices. To meet users' requirements, the policies need to be updated over time. However, in some cases such that gesture recognition error and insufficient amount of demonstration data, it is ineffective to update the human-intention model using demonstration data. As a consequence, in the experiment, when the number of the demonstrated trials is larger than a threshold, the IRL is launched to update the model. In the experiment section, we discuss more about the superiority of IRL with a specific task.

4 EXPERIMENTAL WORK

The feasibility and the efficiency of the proposed learning framework were experimentally evaluated with a coffee-making task and a pick-and-place task. The robot work is to monitor the task process, predict human intention and remind human what action s/he should take by display a text on a screen. In the coffee-making task, the experiments demonstrated that the different reward functions *e.g.* frequency-based and IRL affect the human-intention model. The frequency-based method is a local approach which weights $R^a_{ss'}$ of the Bellman equation by the times of the demonstration. However, the IRL obtains $R^a_{ss'}$ in a sense of globally optimizing the Bellman equation. In the pick-and-place task, the experiment shows that the sequence of human actions was correctly predicted based on the predefined plan learned by the presented learning framework. When the blue, green, and red cubes were randomly grasped and placed, the system suggested a successive action to go back to the learned plan. By doing so, a robot could help the human continue the task with the human's approval.

4.1 Coffee-making task

In the first task, the workspace is illustrated in Fig. 6, which consists of an inside and an outside area. A camera was mounted right above the work-space to monitor the environment. A pot, spoon, cup, and coffee can were placed initially in the outside area, while a human will manipulate objects in inside area. In the experiment, we implemented three different tasks to learn the reward functions. In task 1, we took a pot and pour water in a cup; in task 2 we took a spoon and spooned up coffee in the cup. Finally, task 3 was performed and illustrated in Fig. 8. Table 3 shows the optimal policies that we obtained from the experiments.

For these tasks, the actions were referred to the five classified human gestures in Fig. 4. Table 1 shows the action meaning of each hand gesture, while Table 2 shows the comparison of the recognition rates between the modified and traditional CNN architecture. Figure 7 depicts the cup attributes in the object layer. For example, when the cup was placed in the outside area, it meant that the work had not been started and there was only one 'empty' state for the cup. Similarly, when the cup was located in the inside area, there were three possible

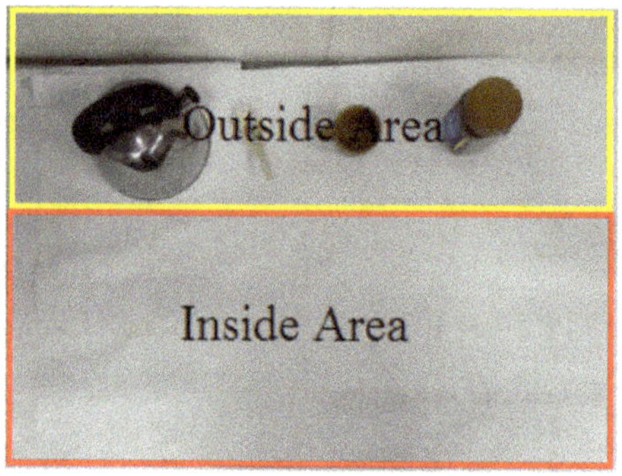

Figure 6: Workspace of the pick-and-place task.

Table 1: Action set in the coffee-making task.

	Gesture	Action
1	Empty	Grasp pot, stop
2	Grip	Place cup
3	GripTrans	Place pot, pour water
4	Spoon	Spoon up and down coffee powder
5	Hold	Hold cup

Table 2: Prediction results of actions performed.

Total (14 Testers)	Empty	Grip	Grip trans	Spoon	Spoon trans	Hold	Average
Modified CNN - Success	1400	1395	1354	1315	1351	1354	
Modified CNN - Rate (%)	100	99.64	96.71	93.93	96.50	96.71	97.24
CNN - Success	1395	1385	1341	1049	1246	1308	
CNN - Rate (%)	99.64	98.93	95.97	74.93	89.00	93.43	91.95

states (empty, with liquid, and with coffee powder). Statuses from the motion layer and the object attributes from the object layer were composed of the set of states S in the MDP of the human-intention model. Here, we have 10 states as illustrated in the set of actions in Fig. 8. Table 1 and object attributes were presented in a state-diagram to reflect the human intention for the task.

Figures 8 (a)–(g) depict the steps of the human demonstration for the coffee-making task 3. The demonstration was used to explore the optimal reward function of the MDP. Figure 9 shows the state diagram of the human-intention model based on the frequency-based method and IRL. It is clear to see that there were two differences between the two approaches at states 4 and 8. When the process was at state 7 with the spoon in the inside area and the empty cup, the action 'spoon up coffee' was selected by both approaches to go to state 8. However, at state 8, we had two different results. The frequency-based approach still stuck at the same

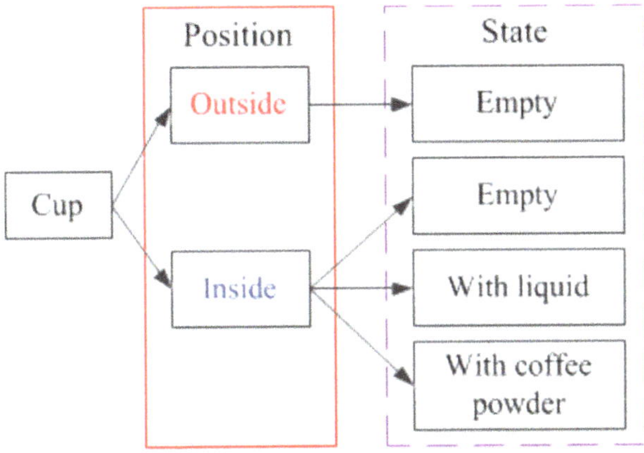

Figure 7: Cup attributes in the object layer.

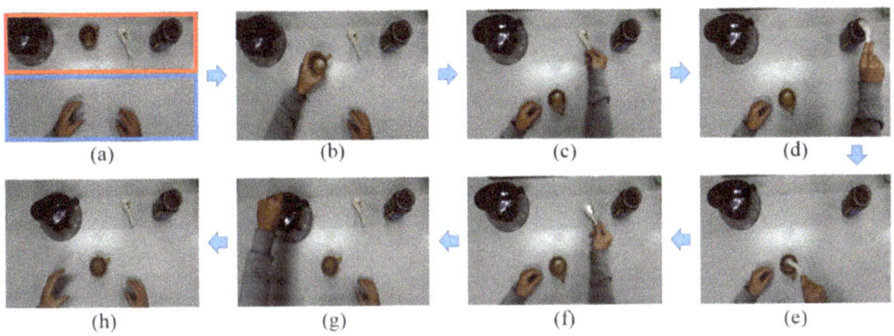

Figure 8: Human demonstration for the coffee-making task. (a) The human and objects were at initial status; (b) grasping and placing the cup; (c) grasping the spoon; (d) spooning up the coffee powder; (e) putting back the spoon; (f) grasping the pot and pouring water into the cup; (g) putting back the pot.

Table 3: Optimal policies.

	Task 1		Task 2		Task 3	
	Frequency	IRL	Frequency	IRL	Frequency	IRL
State 1	Grasp cup	Grasp cup	Grasp cup	Grasp cup	Grasp cup	Grasp cup
State 2	Place cup	Place cup	Place cup	Place cup	Place cup	Place cup
State 3	Grasp pot	Grasp pot	Grasp spoon	Grasp spoon	Grasp spoon	Grasp spoon
State 4	Pour water	Pour water	Pour water	Place pot	Pour water	Place pot
State 5	Place pot	Place pot	Place pot	Place pot	Pour water	Place pot
State 6	Stop	Stop	Stop	Stop	Stop	Stop
State 7	Place spoon	Place spoon	Spoon up coffee	Spoon up coffee	Spoon up coffee	Spoon up coffee
State 8	Place spoon	Place spoon	Place spoon	Place spoon	Spoon up coffee	Place spoon
State 9	Grasp pot	Grasp pot	Grasp pot	Grasp pot	Grasp pot	Grasp pot
State 10	Pour water	Pour water	Pour water	Pour water	Pour water	Pour water

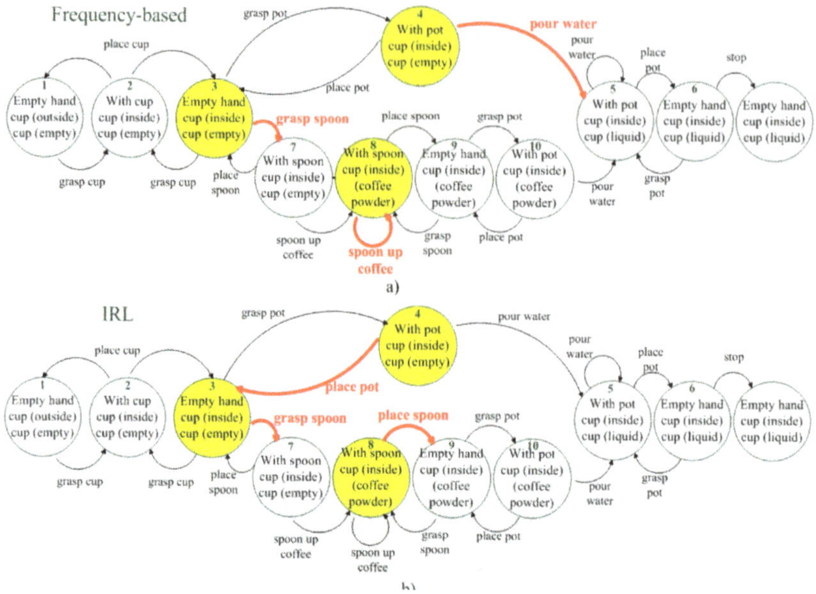

Figure 9: The reproduction results for the two methods. a) The result of frequency-based method. b) The optimization of the reward functions through IRL method.

state by the action 'spoon up coffee' while the IRL approach chose the action 'place spoon' to state 9. Even though the state 8 shows that the coffee powder was in the cup, the frequency-based approach kept doing the action 'spoon up coffee'. Because of such an imperfection of human demonstration at state 8 for a few times, the action 'spoon up coffee' was repeated. However, the IRL approach took an optimal action 'place spoon' to accomplish the task.

Furthermore, another difference between two methods also appeared at state 4 when the action 'grasp pot' was chosen at state 3. Actually, in state 4, when the human already hold the pot, the reward function that we got from the frequency-based method suggested to perform the action 'pour water'. However, before the action 'spoon up coffee' was performed, the action 'pour water' was unnecessary. The process based on the frequency-based approach tended to finish the task without coffee. On the contrary, the IRL approach chose the action 'place pot' before the coffee powder had not been put in the cup.

Evidently, in the coffee-making task, the frequency-based reward function provided a locally optimized solution because it regulated the reward based on the count of the excited action at each state. Alternatively, the IRL explored the optimal reward function from the demonstration and derived the action at each state to achieve the most advantageous performance for task accomplishment.

4.2 Learning a pick-and-place task with a human

This task involved a pick-and-place task and illustrated how the robot can help the human. The goal was to stack three different colour cubes in the same order executed in the human demonstration. When the human stacked the cubes, the robot recognized the human actions and predicted the most possible successive action the human should do. In the task, the set of states was constructed from the object layer. The features including position, colour, and

Table 4: Action set in the pick-and-place task.

	Action	Description
1	Transport empty	Transport without object
2	Transport loaded	Transport with an object
3	Grasp	Grasp an object
4	Release	Release an object
5	Stop	Stop

order of the three cubes were extracted to create the set of statesS_0. There were 16 states. In the further work, more actions could be considered to make the collaboration versatile. Table 4 shows all actions that were used in the task. The reward function and the optimal policy were explored as mentioned in the section 3. In this experimental scenario, to evaluate the obtained policy, some wrong action sequences was purposely performed by the human and we would like to check whether the human-intention model could predict the successive action that made the task being executed in the same action sequence of the demonstration.

The task was demonstrated as the following sequence: grasped the blue cube, placed the blue one in the inside area, grasped the green one on the blue one, pick the red one, and place it on the green one ($blue \rightarrow red \rightarrow green$). The robot watched the human operations and successfully predicted the next actions to implement the task. Figure 10 shows that when the human intentionally executed the wrong task sequence, the robot was able to detect it and suggest the human the correct one (the command in yellow). Figure 10a shows that at the beginning of the task, the robot suggested the human the blue cube. In Fig. 10b, when the human took the cube correctly, the robot continued to suggest the green one. Figure 10c shows that when the human made a mistake to take the red one instead of the green one, the robot suggested the human should remove the red cube. Figure 10d shows that when the human removed the red one, the robot suggested the green one again. Figure 10e shows that when the human took the green one, the red one was suggested in the next step; and finally the human took the red cube to finish the task as shown in the Fig. 10f. The details of the task are shown in the attached video.

The experimental work was the initial attempt to apply MDP and IRL to the problem of modeling collaborative tasks. While the robot was able to recognize human intention to

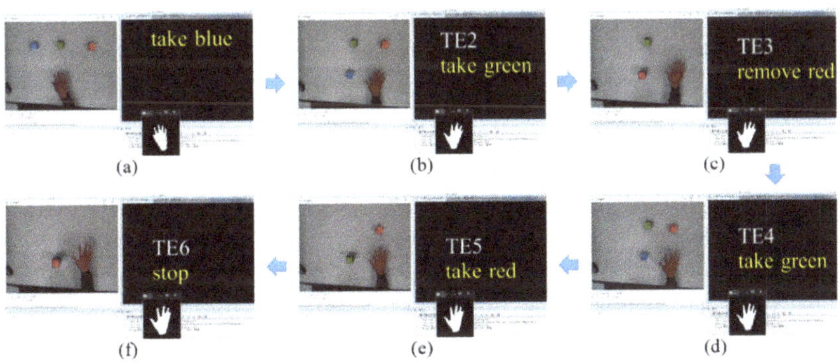

Figure 10: Predicted actions of the pick-and-place task by the human-intention model. (a)–(f) were the steps of the task.

perform the task, it could predict successive actions and suggest them to the human. The following task is about to propose a collaborative model between robot and human according to the obtained human intention. More experiments will be performed on a robot.

5 CONCLUSIONS AND FUTURE WORK

We have introduced a human-intention learning system for HRC. To achieve the learning from human demonstration, we leveraged superiority from the marriage between MDPs and IRL. The MDPs are used to explore decision-making rules and deal with uncertainties from human demonstration and environment. The IRL is used to obtain the optimal reward function based on demonstration and provide a global solution for a given task. Compared to the previous work, our framework involves control actions to infer human intention and provides suggested actions for task accomplishment. Our extensive experimental evaluation with the coffee-making and pick-and-place tasks indicated that the presented human-intention learning system is not only capable of performing collaborative task but also obtaining a globally optimal policy.

ACKNOWLEDGEMENTS

This work was supported in part by the National Science Council under Grant NSC MOST 104-2221-E-027-077. Any opinion, findings, and conclusions or recommendations expressed in this material are those of the authors and do not necessarily reflect the views of the National Science Council.

REFERENCES

[1] Bascetta, L., Ferretti, G., Rocco, P. & Ardo, H., *Towards safe human-robot interaction in robotic cells: An approach based on visual tracking and intention estimation*. IEEE/RSJ IROS, San Francisco, pp. 2971–2978, 2011.
https://doi.org/10.1109/iros.2011.6094642

[2] Kanno, T., Nakata, K. & Furuta, K., A method for team intention inference. *International Journal of Human-Computer Studies*, **58**(4), 393–413, 2003.
https://doi.org/10.1016/s1071-5819(03)00011-9

[3] Zhou, S. & Wu, C. H., *A recognition method for drivers intention based on genetic algorithm and ant colony optimization*. Conference on Natural Computation, Shanghai, pp. 1033–1037, 2011.
https://doi.org/10.1109/icnc.2011.6022185

[4] Wu, P., Wang, Z. & Chen, J. H., Research on attack intention recognition based on graphical model. *Proceedings Of International Conference on Information Assurance and Security*, Shanghai, **1**, pp. 360–363, 2009.
https://doi.org/10.1109/ias.2009.158

[5] Jin, L., Hou, H. & Jiang, Y., Driver intention recognition based on continuous hidden Markov model. *Conference on Transportation, Mechanical, and Electrical Engineering (TMEE)*, **1**, pp. 739–742, 2011.
https://doi.org/10.1109/tmee.2011.6199308

[6] Gehrig, D., Khne, H., Wrner, A. & Schultz, T., *HMM-based human motion recognition with optical flow data*. Proceedings of the 9th IEEE-RAS International Conference. on Humanoid Robots, Paris, pp. 425–430, 2009.

[7] Tahboub, K. A., Intelligent human-machine interaction based on dynamic bayesian networks probabilistic intention recognition. *Journal of Intelligent and Robotic Systems*, **45**(1), 31–52, 2006.

[8] Jeon, H., Kim, T. & Choi, J., *Ontology-based user intention recognition for proactive planning of intelligent robot behavior*. International Conference On Multimedia and Ubiquitous Engineering, Busan, pp. 244–248, 2008.
[9] Huang, Y.C., Young, H.P., Ko, C. H. & Young, K.Y., Design and implementation of a robot control system with traded and shared control capability. *Proceeding Of Asian Control Conference*, pp. 311–316, 2011.
[10] McGhan, C.L.R., Nasir, A. & Atkins, E.M., Human intention prediction using Markov Decision Processes. *Journal of Aerospace Information Systems*, **12**(5), pp. 393–397, 2015.
https://doi.org/10.2514/1.i010090
[11] Song, D., Kyriazis, N., Oikonomidis, I., Papazov, C., Argyros, A., Burschka, A. & Kragic, D., *Predicting human intention in visual observations of hand/object interactions.* IEEE International Conference Robotics and Automation, pp. 1608–1615, 2013.
https://doi.org/10.1109/icra.2013.6630785
[12] Keskinpala, H.K., Adams, J.A. & Kawamura, K., *PDA-based human robotic interface.* IEEE International Conference on SMC, Washington, DC, pp. 1310–1315, 2003.
https://doi.org/10.1109/icsmc.2003.1244502
[13] Breazeal, C., Emotion and social humanoid robots. *International Journal of Human-Computer Studies*, **59**, pp. 119–155, 2003.
https://doi.org/10.1016/s1071-5819(03)00018-1
[14] Ng, A. Y. & Russell, S., *Algorithms for inverse reinforcement learning*. Proceedings of ICML, pp. 663–670, 2000.
[15] Nagi, J., Ducatelle, F., Caro, G., Ciresan, D., Meier, U., Giusti, A., Nagi, F., Schmidhuber, J. & Gambardella, L.M., *Max-pooling convolutional neural networks for vision-based hand gesture recognition*. IEEE International Conference on signal and Image Processing Applications, pp. 342–347, 2011.
[16] Lin, H.I. & Chiang, Y.P., Understanding human hand gestures for learning robot pick-and-place tasks. *International Journal of Advanced Robotic Systems,* **12**(5), p. 49, 2015.
https://doi.org/10.5772/60093
[17] Lecun, Y., Bottou, L., Bengio, Y. & Haffner, P., Gradient-based learning applied to document recognition. *Proceedings of the IEEE*, **86**, pp. 2278–2324, 1998.
https://doi.org/10.1109/5.726791
[18] Lin, H.I. & Chen, *W-K., Human intention recognition using markov decision processes.* Proceedings of International Conference on Automatic Control (CACS), pp. 240–343, 2014.
https://doi.org/10.1109/cacs.2014.7097213

GOAL-ORIENTED ACTIVE LEARNING WITH LOCAL MODEL NETWORKS

JULIAN BELZ[1], KONRAD BAMBERGER[2], OLIVER NELLES[1] & THOMAS CAROLUS[2]
[1]Institute of Mechanics and Control Engineering, University of Siegen, Germany.
[2]Institute of Fluid- and Thermodynamics, University of Siegen, Germany.

ABSTRACT

A methodology for goal-oriented active learning with local model networks (LMNs) is proposed. It is applied for the generation of training data for a computational fluid dynamics (CFD) metamodel. The used metamodel is an LMN trained with data originating from CFD simulations. This metamodel describes the total-to-static efficiency for a given design point, defined by the pressure rise at a specific volume flow rate, depending on geometrical parameters of an impeller of centrifugal fans. The goal-oriented nature originates from three main targets that are addressed simultaneously during the active learning procedure. (I) The concentration on possibly optimal geometries and (II) the focus on areas in the input space where the metamodel's performance is considered to be worst. Additionally, (III) new measurements should differ from already simulated geometries as much as possible. With these goals three important issues in modeling are addressed simultaneously: (I) optimality, (II) model bias, (III) model variance/uniformly space-filling property. In order to fulfill all goals, special properties of LMNs are utilized (embedded approach). Through the structure of LMNs, it is possible to assign *local* model errors to specific areas in the input space. New measurements are preferably placed in such high-error regions, while concentrating on presumably optimal geometries that differ most from the ones already available in the training data. In the field of fluid machinery, the range of achievable design points is usually identified by the Cordier diagram. While the design points obtained in the passive learning phase fairly agree with the standard Cordier diagram, an extension of achievable design points was observed due to the proposed goal-oriented learning strategy. In addition, the total-to-static efficiency could be improved in some areas of the Cordier diagram.

Keywords: active learning, aerodynamic optimization, design of experiments, experimental modeling, impeller of centrifugal fans, metamodeling.

1 INTRODUCTION

Measurements play the key role in experimental modeling. The quality of the data used to build models restricts the performance that can be achieved. Since measurements are always time-consuming and might be very expensive, the aim is often to find the necessary minimum of measurements in order to fulfill some task sufficiently well. Therefore, the design of experiments (DoE) should contain the most informative measurements. In principle there are two ways to obtain such a DoE, which are faced in Fig. 1. If a passive learning strategy is pursued, all points contained in the DoE are known before the first measurement is carried out. This experimental design might be optimized according to some optimality criterion, see Refs. [1–3] for common criteria. In case of an active learning strategy the experimental design is not known completely a priori. Information gathered through already obtained measurements are used to select new *queries*. A query specifies the point in the input space at which a label or the output value should be measured. In the context of metamodeling tasks, a more common expression for active learning is sequential sampling, see Refs. [4, 5].

Active learning strategies have the potential to achieve the same model quality with significant less data compared to passive learning strategies, as stated e.g. in Refs. [6–9]. Active learners can be distinguished by their query strategies, i.e. how they choose the next query. A good survey on possible query strategies for active learners is given in [10]. The proposed

DOI: 10.2495/CMEM-V6-N4-785-796

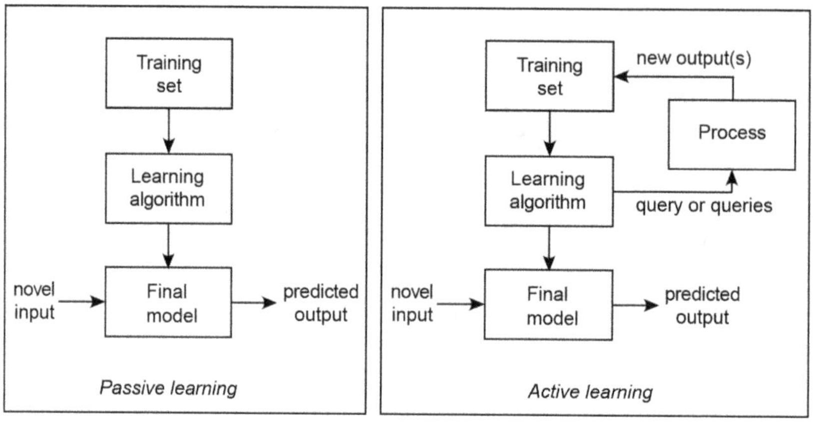

Figure 1: Comparison of passive and active learning strategies.

goal-oriented active learning strategy here is an extension of the hierarchical local model tree for design of experiments (HilomotDoE) proposed in Ref. [11]. The main focus of the query strategy of HilomotDoE is to reduce the error of a local model network (LMN). Since the error of the model can be decomposed into a bias and variance part [12], HilomotDoE addresses and reduces both error parts as described in [13].

The extension of HilomotDoE proposed in this paper is meant for models that should be utilized for optimization tasks and is described further in Section 2. It is successfully used for the generation of training data for a computational fluid dynamics (CFD) metamodel. This metamodel describes the total-to-static efficiency for a given design point depending on geometrical parameters of an impeller and is valid for all typical design points of centrifugal fans according to the Cordier diagram. The optimization task is to find geometrical parameters of the impeller for a given design point such that the efficiency of the centrifugal fan is maximized, see Section 3 for more details. A customization and application of the proposed active learning strategy for the CFD metamodel task together with the corresponding results is described in Section 4 before Section 5 concludes this paper.

2 ACTIVE LEARNING STRATEGY

The pursued active learning strategy deals with models utilized for optimization tasks and it exploits special properties of LMNs. Through the structure of LMNs it is possible to assign *local* model errors to specific areas in the input space. New measurements are preferably placed in such high-error regions, while concentrating on presumably optimal geometries that differ most from the ones already available in the training data. Therefore a lot of *potential* queries are generated, from which only one or a small subset is chosen to be measured. These potential queries will be named *candidate points* from now on. In order to focus on presumably optimal geometries, all existing candidate points are obtained through optimizations. For these optimizations the currently available model is used. In order to concentrate on areas with high local errors, only candidate points lying in these regions are considered during the selection of new queries. The similarity between the leftover candidate points and the already existing training data is evaluated in terms of the Euclidean distance. As a result, the candidate point closest to the largest hole in the training data within the area of the highest model error is chosen as query. More details about LMNs, the active learning strategy, and the new generation method of the candidate points are given in Sections 2.1, 2.2 and 2.3.

2.1 Local model networks

LMNs follow a divide-and-conquer strategy. The whole input space, here spanned by the physical inputs $\underline{u}$, is divided into subregions. In each subregion a local model (LM) $\hat{y}_1$ is estimated. So called validity functions Φ_i define regions in which the LMs are valid and how interpolation between neighboring LMs should be conducted. According to [14] the model output $\hat{y}$ of a LMN with M LMs is calculated by

$$\hat{y} = \sum_{i=1}^{M} \hat{y}_i(\underline{u})\Phi_i(\underline{u}). \tag{1}$$

For a reasonable interpretation the so-called partition of unity has to be fulfilled [12], i.e. the validity functions Φ_i have to sum up to one at any point in the input space.

For the active learning strategy the training algorithm Hilomot (hierarchical local model tree) [15] is used with local affine models. The schematic LMN training procedure of Hilomot can be explained with the help of Fig. 2. The algorithm incrementally grows an LMN, i.e. in each iteration, the number of LMs is increased by one. At the beginning there is only one LM valid in the whole $\underline{u}$-input space. Then all axes-orthogonal splits through the center of the so-called *parent* LM are tested (dashed lines in Fig. 2). Hilomot uses the best orthogonal split as an initialization for a nonlinear split optimization in which the position and orientation of that split are further adjusted. Only the current split is optimized, all already existing splits are kept unchanged. The parameters of the two LMs affected by the split adjustment are obtained through a weighted least squares technique nested in the nonlinear split optimization. After that the LM with the worst *local* error measure (gray shaded regions in Fig. 2) are subdivided further. The resulting validities are utilized to weight the errors of the whole LMN yielding the local error measures. After the first split is performed one additional split is tested before the nonlinear optimization starts, which is the direction of the parent split going through the center of the parent LM. The algorithm stops as soon as a measure for the LMN's generalization performance gets worse. Usually Akaike's information criterion (AIC) [16] or a distinct validation data set is used to evaluate the generalization performance of the LMN. Through the nonlinear split optimization Hilomot allows for an axes-oblique input space partitioning yielding advantages especially for high-dimensional input spaces. For more details on how the validities are constructed and additional advantages of Hilomot compared to other LMN training algorithms please refer to Ref. [12].

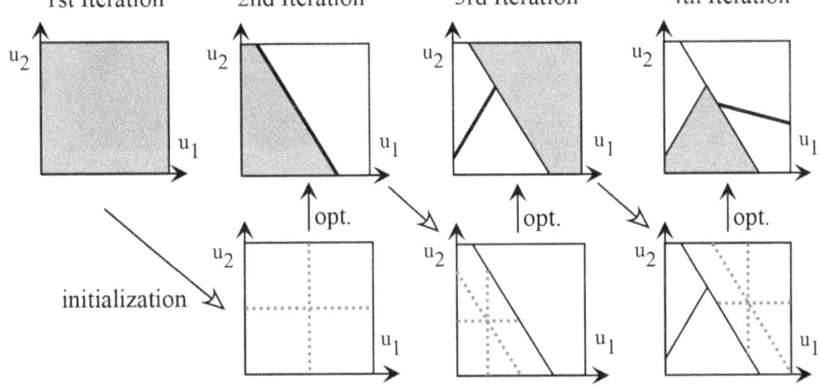

Figure 2: Schematic Hilomot training procedure for a 2-dimensional input space.

2.2 Active learning with local model networks

The active learning algorithm HilomotDoE proposed in Ref. [11] is described in the following. The novelty of this paper regarding the candidate point generation is outlined in Section 2.3. Another innovation is concerned with the generation of more than one query and is explained at the end of this section. As already mentioned the active learning strategy with LMNs aims at reducing the error of the final model. It is assumed that the highest error reduction can be obtained if the queries are placed in the areas of the input space with the highest local error measure $e_{LM,i}$ $i \in \{1,\ldots, M\}$. For the calculation of the local error measures $e_{LM,i}$ the measured outputs $\underline{y}$, the local model outputs $\underline{\hat{y}}_i$, the effective number of local parameters $n_{eff,i}$ and the local validity matrix $\underline{Q}_i = \text{diag}(\Phi_i(\underline{u}))$ are required:

$$e_{LM,i} = \sqrt{\frac{(\underline{y}-\underline{\hat{y}}_i)^{\mathrm{T}} \underline{Q}_i (\underline{y}-\underline{\hat{y}}_i)}{\text{trace}(\underline{Q}_i) - n_{eff,i}}}, i \in \{1,\ldots,M\}. \tag{2}$$

Each squared error between an LM output and the measured output is weighted with the corresponding validity value and is divided by the leftover degrees of freedom. For more details on how to compute the effective number of parameters $n_{eff,i}$ for each LM please refer to Ref. [13].

Figure 3 visualizes the situation during the active learning phase for a two-dimensional input space if only one query is sought. An LMN is trained with Hilomot based on all currently available data. As one result the partitioning of the input space is obtained and for each LM a local error measure according to (2) can be calculated, see Fig. 3a. In the original, not yet extended version of HilomotDoE, candidate points are generated through random sampling from a uniform distribution (dots in Fig. 3b). From all randomly generated candidate points only the ones lying in the LM with the worst local error measure are considered in the following. The candidate point inside the worst LM that fills the greatest hole of the already measured training data is chosen as query. In order to find the greatest hole, the nearest neighboring training data point to each considered candidate is determined. The candidate point

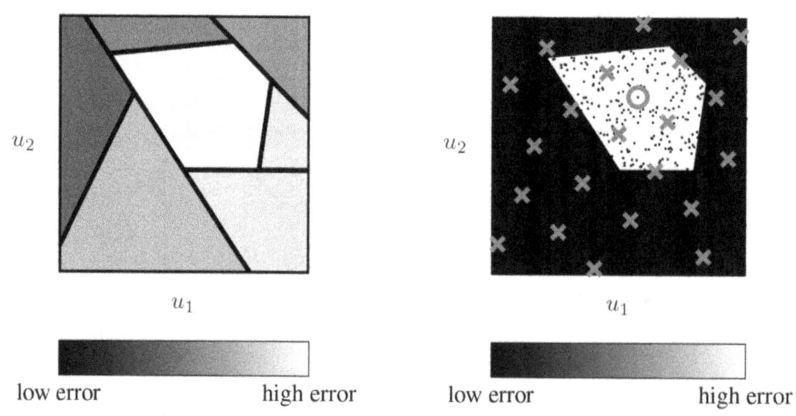

(a) Qualitative local errors

(b) Training data (x), candidates (·) and chosen query (o)

Figure 3: Partitioning of a LMN with local errors (a) and with focus on the worst LM together with training data (x), candidates (·) and the chosen query (o).

with the greatest distance to its nearest neighboring training data point is chosen as query. During the distance calculations, all training data samples are considered, but only candidates within the worst performing LM are taken into account. After the measurement for the current query has finished the training data set is updated, a new LMN is trained with Hilomot and the next query can be determined.

For algorithmic efficiency reasons it might be reasonable to request more than one query at a time. If $n_q > 1$ queries are demanded, a new strategy for HilomotDoE is proposed. Still Hilomot is used to train an LMN and yield a partitioning of the input space. Then, all local error measures are calculated and normalized, such that their sum equals one:

$$\hat{e}_{LM,i} = \frac{e_{LM,i}}{\sum_j^M e_{LM,j}} \tag{3}$$

The number of demanded queries from each LM $n_{qLM,i}$ is obtained by rounding the overall number of demanded queries n_q multiplied with the corresponding normalized local error measure $\hat{e}_{LM,i}$:

$$n_{qLM,i} = [n_q \cdot \hat{e}_{LM,i}]. \tag{4}$$

With this approach more queries are demanded from an LM the bigger its local error measure is in relation to the local error measures of all other LMs.

2.3 Generation of candidate points

Another novelty of this paper lies in the generation of the candidate points, which are subsequently used by HilomotDoE. A block diagram of the whole goal-oriented active learning procedure is shown in Fig. 4. The candidate points are generated through an evolutionary optimization algorithm based on the currently available model, trained with all yet accessible measurements. More details of the used evolutionary optimization algorithm can be found in Ref. [17]. With the help of the model the objective function $J(\underline{u})$ can be determined for each design vector $\underline{u}^T = [u_1 \; u_2 \; \ldots \; u_{nd}]$. In order to distribute the initial values for the optimization runs in a space-filling manner throughout the whole input space, the extended deterministic local search (EDLS) algorithm proposed in Ref. [18] is used to generate a maximin Latin Hypercube (LH) design. To prevent the optimization from generating too similar candidate points, n_c constraints for each initialization point coming from the LH design are applied. If no constraints would be demanded almost all optimizations might lead to the same optimum. More details about the used constraints follow in the next paragraph. All resulting candidate points are provided to HilomotDoE, that determines queries as described in Section 2.2. After the queries are measured, they are added to the available measurements. Through the altered training data the model and therefore the outcome of optimizations with the new model might change, even if neither the constraints nor the LH design is changed.

If no constraints would be applied to the optimization runs, a lot of very similar or identical candidate points could possibly be the result, depending on the properties of the optimization problem. Additionally, these constraints might reflect later usage scenarios, if some design parameters are bounded, e.g. due to limited available construction space. If the number of optimization variables is n_d, there are $n_c = 2^{n_d}$ possible constraint combinations specifying which variables are held fixed. For each point contained in the LH design all possible constraint combinations are applied. Constrained optimization variables are held fixed at the

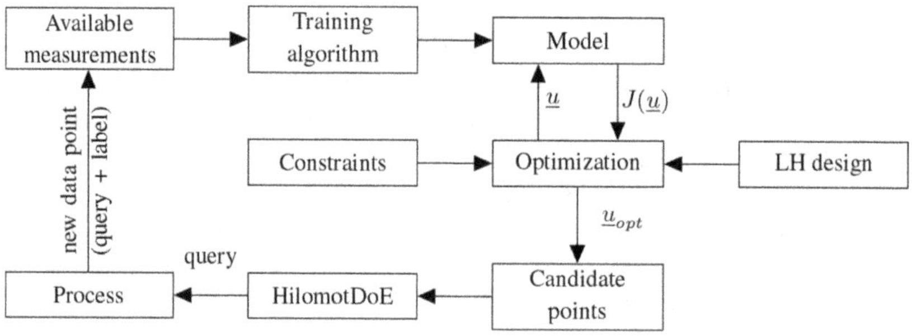

Figure 4: Block diagram of the goal-oriented active learning procedure.

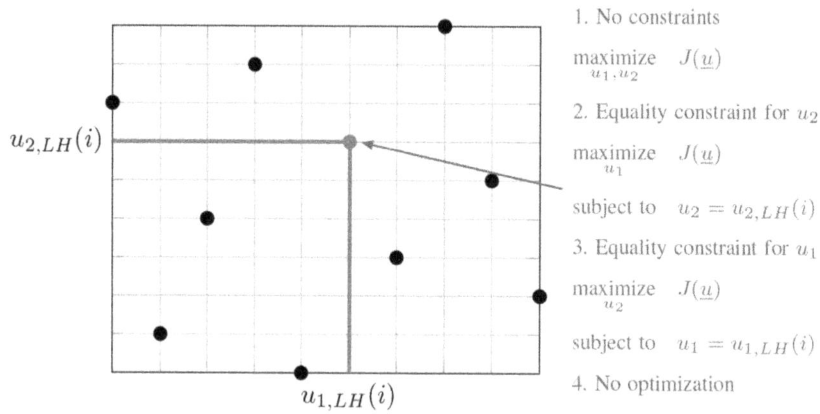

Figure 5: Illustration of all optimization scenarios for point **i** of an LH design with two inputs.

Table 1: Each column contains one constraint for the optimization. An optimization variable u_j might either be constrained to a specific value ($u_{j,\ LH}$ (i)) or can be free, i.e. the value is determined through the optimization (opt.).

Optimization scenario:	1	2	3	4
Opt. variable u_1	opt.	opt.	$u_{1,LH}(i)$	$u_{1,LH}(i)$
Opt. variable u_2	opt.	$u_{2,LH}(i)$	opt.	$u_{2,LH}(i)$

value of the corresponding LH design point. As an example Fig. 5 together with Table 1 illustrate this for $n_d = 2$ optimization variables for the i-th point of the LH design. Four optimization scenarios arise with different constraints; all listed in Fig. 5 and Table 1. The variable $u_{j,LH}$ denotes the value coming from a point in the LH design that is fixed during the corresponding optimization run. In case of opt. in a field of Table 1, the value of the corresponding variable is yielded by the optimization. With this approach $N_c = N_{LH} \cdot n_c$ optimization runs are necessary to obtain all candidate points.

3 APPLICATION: AERODYNAMIC OPTIMIZATION OF CENTRIFUGAL FANS

The proposed goal-oriented active learning strategy is applied in the field of centrifugal fans. The general purpose of fans is to generate a gaseous fluid flow under build-up of pressure.

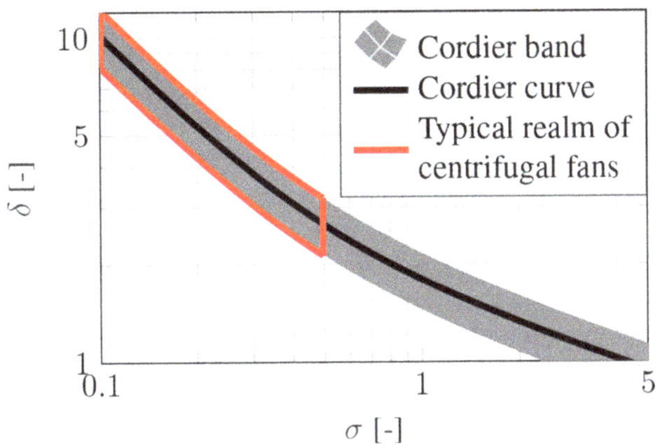

Figure 6: Cordier diagram with indication of the typical realm of centrifugal fans.

Typically, the design of a new fan comprises two main targets. Firstly, the design point (i.e. the desired flow rate Q and pressure rise Δp) must be fulfilled. Secondly, the shaft power P_{shaft} shall be as low as possible. The achievability of the first design target mainly depends on the choice of the outer fan diameter D and the rotational speed n. Cordier [19] found that the specific fan diameter

$$\delta = \frac{D}{\left(\frac{8}{\pi^2}\right)^{1/4}\left(\frac{\Delta p}{\rho}\right)^{-1/4} Q^{1/2}} \tag{5}$$

and the specific fan speed

$$\delta = \frac{n}{\left(2\pi^2\right)^{-1/4}\left(\frac{\Delta p}{\rho}\right)^{3/4} Q^{-1/2}} \tag{6}$$

of all fans and pumps lie in a narrow band around the curve depicted in Fig. 6, which were later known as the Cordier curve and the Cordier diagram, respectively. The original Cordier diagram is based on fan performance data stemming from the 1950s, but its validity was confirmed in numerous more recent studies, see e.g. the work by Willinger *et al.* [20–22]. Figure 6 furthermore indicates the typical realm of centrifugal fans, which is in the area of low specific fan speeds but high specific fan diameters. The rest of the Cordier band is associated with other fan types such as axial or mixed flow. The second design target (the minimization of P_{shaft}) is equivalent to the maximization of the aerodynamic efficiency defined as

$$\eta = \frac{Q.\Delta p}{P_{shaft}}. \tag{7}$$

In the present application, only the impeller as the key component with respect to aerodynamic efficiency is investigated. The impeller geometry is described by nine geometrical parameters including the inner diameter, the inlet width, the outlet width, the inlet blade angle and the outlet blade angle. Those parameters are supposed to be most relevant in order to adapt the fan geometry to a large variety of potential design points [23]. Optimal geometrical parameters are found by the evolutionary algorithm described in Ref. [17]. The objective function is maximization of total-to-static efficiency η_{ts} with penalty terms for the violation of the desired design point. The focus on total-to-static efficiency originates from the assumption that the fan exhausts into an open environment. Under these circumstances, the kinetic energy of the fluid at the fan exit must be considered as loss and Δp must be diminished accordingly when computing the efficiency with eqn. (7). The objective function is evaluated using metamodels of CFD. CFD is a numerical simulation technique, which solves the flow field in a discretized computational domain for given boundary conditions (e.g. flow rate, ambient pressure, velocity of moving walls, etc.). A detailed description of the CFD model can be found in [24]. This reference furthermore contains detailed information about the parameterization of the impeller. Since CFD simulations are time-consuming and require considerable computational resources, there is a strong interest to keep the number of query points to a minimum. At the same time, the metamodels must be extremely precise for optimal geometries to avoid that the optimizer exploits weaknesses of the metamodel instead of finding the real aerodynamic optimum. Given these two requirements, the active learning strategy suggested in this paper is well-suited to be applied in the fan optimization problem.

4 ACTIVE LEARNING FOR THE CFD METAMODEL

In case of the active learning for the CFD metamodel the procedure illustrated in Fig. 4 is applied with minor adjustments to meet problem specific needs. A maximin LH design for an 11-dimensional input space is generated with $N_{LH} = 586$ samples. 2 of the 11 inputs specify a design point defined by the pressure rise and a specific volume flow rate. These two inputs always belong to the set of constrained optimization variables, since the efficiency is maximized for these design points. The remaining $n_d = 9$ inputs correspond to the geometric parameters that should be optimized. It follows that there are $n_c = 2^{n_d} = 512$ possible optimization problems with all constraint combinations. As explained in Section 2.3, all possible combinations of constrained optimization variables are applied to each point contained in the LH design, resulting in $N_c = 512 \cdot 586 = 300032$ candidate points after all optimization runs are finished. Here, 500 queries are demanded in each loop of the active learning strategy. As soon as one loop is finished, i.e. the optimization of all candidate points is accomplished and queries are calculated, the list of queries to be CFD-simulated is updated. This procedure avoids any time lags between CFD simulations due to not in time finished query optimizations.

4.1 Results

The influence of the proposed goal-oriented active learning strategy is evaluated based on three aspects. First, the model quality over the number of samples available for the training is assessed. Second, the extension of achievable design points in the Cordier diagram is shown. And third, the improvements of the achievable total-to-static efficiency are visualized for all achievable points in the Cordier diagram.

For the model quality assessment CFD-based optimizations are completed for 545 different design points. The optimized designs together with the achieved total-to-static efficiencies serve as test data for the models generated with different amounts of training data. The target value of the model is the total-to-static efficiency η_{ts}. Figure 7 shows the curve for the model quality versus the training data amount. Lower root mean squared errors (RMSE) correspond

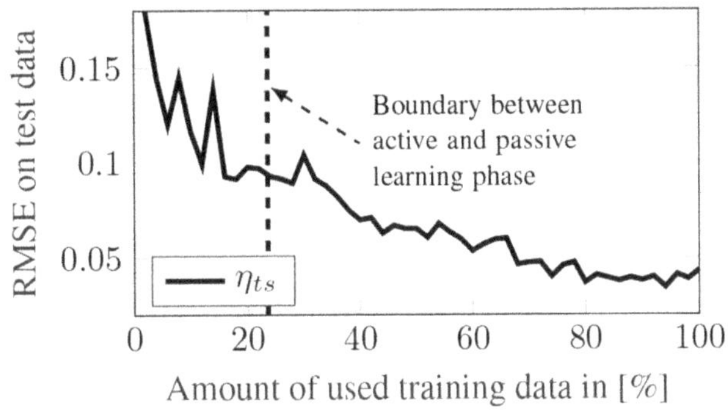

Figure 7: RMSE on test data originating from CFD-based optimization runs vs. amount of training data.

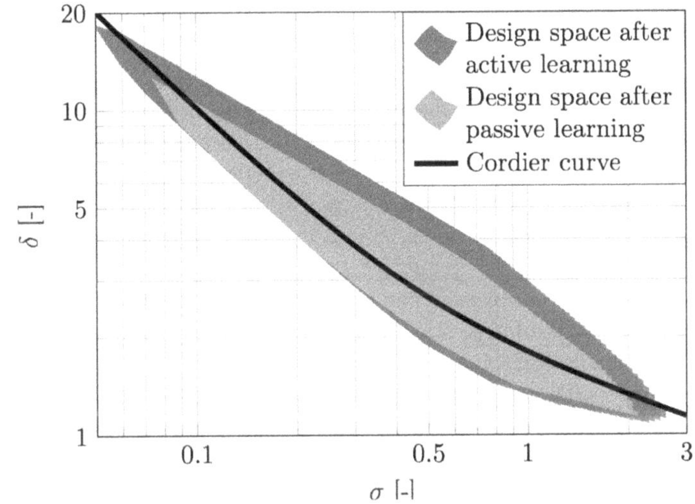

Figure 8: Extension of achievable design point area in the Cordier diagram.

to better generalization performances. The dashed line marks the point, where the passive learning phase has ended and the active learning strategy has started.

Figure 8 shows the extension of achievable regions in the Cordier diagram through the proposed goal-oriented active learning strategy. For almost each viable specific fan speed σ higher specific fan diameters δ are possible. Additionally, lower and higher specific fan speeds could be achieved through the active learning phase compared to the passive one.

The absolute total-to-static efficiencies in the area of possible impeller designs after the active learning strategy and the corresponding improvements are shown in Fig. 9a and b, respectively. The absolute efficiency could be improved up to a value of $\Delta\eta_{ts} = 0.3$ ($\approx 100\%$) by the active learning strategy. The improvements are mostly achieved above the Cordier curve, where the specific fan diameter δ is relatively high. The efficiency improvements are only shown in areas, where data from the passive learning phase is available, compare to Fig. 8. That is the reason, why the area covered by possible designs is less in Fig. 9b compared to Fig. 9a.

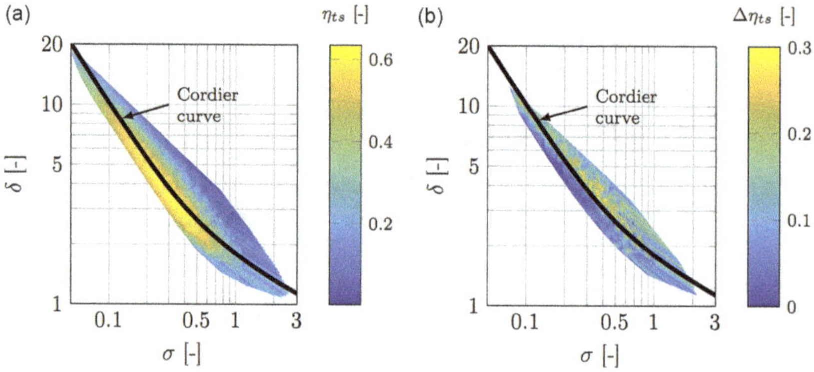

Figure 9: Absolute total-to-static efficiencies (a) and improvements through the active learning strategy (b).

5 CONCLUSION

The proposed goal-oriented active learning strategy with local model networks (LMN) is meant for models that should be utilized for optimization tasks. Active learning strategies allow for an interactive adaption of the design of experiments (DoE) to the process under investigation. Information gathered through already obtained measurements are used to select new *queries*. This adaption has the potential to achieve the same model quality with significant less data compared to a-priori fixed DoEs (passive learning). Through an extension of the already proposed hierarchical local model tree for design of experiments (HilomotDoE) [25] algorithm, a goal-orientation is introduced. These three goals are (I) the concentration on possibly optimal geometries, (II) the focus on areas in the input space with minor generalization performance and (III) a high diversity of training data samples. Therefore, the way *potential* queries or candidate points are generated is adjusted, such that all three goals are met. The proposed methodology is applied to the generation of training data for a computational fluid dynamics (CFD) metamodel. It is shown, that the area of achievable design points is extended, see Fig. 8, and the total-to-static efficiencies could be improved up to 100% in some areas of the Cordier diagram, see Fig. 9. Even though the application incorporates a metamodel, there is no principal restriction to metamodeling tasks of the proposed active learning strategy.

ACKNOWLEDGMENT

This work was funded by the German Ministry for Economic Affairs and Energy (BMWi), the German Federation of Industrial Research Associations (AiF) and the Research Association for Air and Drying Technology (FLT).

REFERENCES

[1] Fedorov, V.V., *Theory of Optimal Experiments,* Elsevier, 1972. https://doi.org/10.2307/2334826

[2] Atkinson, A., The usefulness of optimum experimental designs. *Journal of the Royal Statistical Society Series B (Methodological),* pp. 59–76, 1996. https://doi.org/10.2307/2344398

[3] Fedorov, V.V. & Hackl, R., *Model-Oriented Design of Experiments, Volume 125 of Lecture Notes in Statistics,* Springer Science & Business Media, 2012. https://doi.org/10.1007/978-1-4612-0703-0

[4] Jin, R., Chen, W. & Sudjianto, A., On sequential sampling for global metamodeling in engineering design. *ASME 2002 International Design Engineering Technical Conferences and Computers and Information in Engineering Conference,* American Society of Mechanical Engineers, pp. 539–548, 2002. https://doi.org/10.1115/DETC2002/DAC-34092

[5] Wang, G.G. & Shan, S., Review of metamodeling techniques in support of engineering design optimization. *Journal of Mechanical Design,* **129**(4), pp. 370–380, 2007. https://doi.org/10.1115/1.2429697

[6] Cohn, D.A., *Minimizing statistical bias with queries.* Technical report, DTIC Document, 1995.

[7] Cohn, D.A., Neural network exploration using optimal experiment design. *Neural Networks,* **9**(6), pp. 1071–1083, 1996. https://doi.org/10.1016/0893-6080(95)00137-9

[8] Cohn, D.A., Ghahramani, Z. & Jordan, M.I., Active learning with statistical models. *Journal of Artificial Intelligence Research,* **4**, pp. 129–145, 1996.

[9] MacKay, D.J., Information-based objective functions for active data selection. *Neural Computation,* **4**(4), pp. 590–604, 1992. https://doi.org/10.1162/neco.1992.4.4.590

[10] Settles, B., Active learning literature survey. *University of Wisconsin, Madison,* **52**, pp. 55–66, 2010.

[11] Hartmann, B. & Nelles, O., Adaptive test planning for the calibration of combustion engines - methodology. *Design of Experiments (DoE) in Engine Development,* pp. 1–16, 2013.

[12] Nelles, O., *Nonlinear System Identification: From Classical Approaches to Neural Networks and Fuzzy Models,* Springer, 2001.

[13] Hartmann, B., *Lokale Modellnetze zur Identifikation und Versuchsplanung nichtlinearer Systeme.* Ph.D. thesis, Universitat Siegen, 2014.

[14] Murray-Smith, R. & Johansen, T., Local learning in local model networks. *Artificial Neural Networks, 1995, Fourth International Conference on, IET*, pp. 40–46, 1995. https://doi.org/10.1049/cp:19950526

[15] Nelles, O., Axes-oblique partitioning strategies for local model networks. *IEEE International Symposium on Intelligent Control,* Munich, Germany, pp. 2378–2383, 2006.

[16] Akaike, H., A new look at the statistical model identification. *Automatic Control, IEEE Transactions on,* **19**(6), pp. 716–723, 1974. https://doi.org/10.1109/TAC.1974.1100705

[17] Bamberger, K., *Aerodynamic Optimization of Low-Pressure Axial Fans.* Ph.D. thesis, University of Siegen, 2015.

[18] Ebert, T., Fischer, T., Belz, J., Heinz, T., Kampmann, G. & Nelles, O., Extended deterministic local search algorithm for maximin latin hypercube designs. *IEEE Symposium on Computational Intelligence in Control and Automation (CICA),* Cape Town, South Africa, pp. 375–382, 2015. https://doi.org/10.1109/SSCI.2015.63

[19] Cordier, O., Ahnlichkeitsbedingungen fur Stromungsmaschinen. *Brennstoff-Warme-Kraft (BWK),* **5**(10), pp. 337–340, 1953.

[20] Willinger, R., Das CORDIER-Diagramm fur Stromungsarbeitsmaschinen: Eine theoretische Begriindung mittels Stufenkennlinien. *VDI-Berichte,* **2112**, pp. 17–28, 2010.
[21] Willinger, R., Theoretical interpretation of the CORDIER-lines for squirrel-cage and cross-flow fans. *Proceeding ASME TurboExpo,* Copenhagen, Denmark, pp. 675–684, 2012.
https://doi.org/10.1115/GT2012-68186
[22] Willinger, R. & Kohler, M., Influence of Blade Loading Criteria and Design Limits on the Cordier-Line for Axial Flow Fans. *Proceeding ASME TurboExpo,* Dusseldorf, Germany, 2014.
https://doi.org/10.1115/GT2014-25140
[23] Bommes, L., Fricke, J. & Grundmann, R., *Ventilatoren.* Vulkan Verlag, Essen, 2003.
[24] Bamberger, K., Belz, J., Carolus, T. & Nelles, O., *Aerodynamic optimization of centrifugal fans using CFD-trained meta-models.* 16th International Symposium on Transport Phenomena and Dynamics of Rotating Machinery (ISROMAC), Hawaii, USA, 2016.
[25] Hartmann, B., Ebert, T. & Nelles, O., Model-based design of experiments based on local model networks for nonlinear processes with low noise levels. *American Control Conference (ACC), IEEE,* pp. 5306–5311, 2011.

COMPUTER-AIDED MODEL OF COLONIC PROPULSIVE ACTIVITY

O. AL QATRAWI & R. N. MIFTAHOF
Arabian Gulf University, Manama, Kingdom of Bahrain.

ABSTRACT

A biomechanical model and mathematical formulation of the problem of propulsion of a solid non-deformable pellet by an isolated segment of the colon are presented. The organ is modeled as a soft orthotropic cylindrical biological shell. Its wall is reinforced by transversely isotropic muscle fibers of orthogonal type of weaving embedded in a connective tissue stroma. The mechanical properties of the wall are assumed to be nonlinear, deformations are finite. The longitudinal and circular smooth muscle syncitia possesses electrical properties and are under control of a pacemaker, which is represented by the interstitial cell of Cajal. The model describes the dynamics of the generation and propagation of the mechanical waves of contraction–relaxation along the surface of the bioshell and propulsion of the pellet. The governing system of equations has been solved numerically. The combined finite-difference and finite-element method has been used. The results of numerical experiments demonstrate that pendular movements alone provide a normal transit, without mixing though, of the bolus. Non-propagating segmental contractions show small amplitude librations of the pellet without its visible propulsion. Only the coordinated activity of the longitudinal and circular smooth muscle layers in a form of the peristaltic reflex provides physiologically significant simultaneous propulsion and mixing of the intraluminal content.

Keywords: Bolus, colon, Computational biology, gastrointestinal tract

1 INTRODUCTION

The primary functions of the large intestine (colon) are to store, process and expel fecal mass residues. These require sustained motor activity – the generation of migrating myoelectrical complexes (MMC) that mix and propel the content. Patterns of MMCs produced by the colon under normal physiological conditions are associated with non-propagating and high amplitude propagating mass movements. They occur regularly and assist to push the bulky content ahead of them. The disparity between mechanical and propulsive activities caused by anatomical and/or neuropathological changes in the colon, e.g. Hirschpurg' disease, irritable bowel syndrome, result in idiopathic/slow transit constipation, obstructed defecation or diarrhea. The severity and diversity of clinical symptomatology, relative inaccessibility and the complexities posed by the presence of solid or semi-solid fecal masses, along with the heterogeneity of in *vitro* experimental findings, makes it difficult to unravel underlying pathophysiological mechanisms and thus design effective treatments.

2 AIM

The aim of the study is to investigate numerically the propulsion of the intraluminal content (bolus) and to assess the dynamics of stress-strain distribution, changes in the shape and configuration of the colon.

3 METHODS

Below are the basic assumptions that have been used in model construction [1–3]:

1. The colon is a soft orthotropic tube of cylindrical shape. The wall of the colon has two muscle layers and a network of connective tissue. The outer layer contains muscle fibers

DOI: 10.2495/CMEM-V6-N4-797-803

that are organized in the longitudinal direction of the organ while muscle fibers of the inner layer are arranged in a direction that is specified as orthogonal and circular.

2. The rigid sphere of the radius, R, is included within the tubular segment of length, l, and radius, r0, where isometric contractions take place. Throughout the mechanical reactions, the intraluminal pressure, p, varies in accordance with the adiabatic law.
3. The contact forces that takes place between the two parties: the wall and rigid sphere are assumed to be orthogonal to the sphere surface. Dryness and viscous friction are subject of the motion of the bolus.
4. Muscle layers contract in an independent way with generation of active forces (Ta). The first contractions begin with the longitudinal muscle layer.
5. It is assumed that the muscle layers are syncytia that possess cable electrical properties. As for the electrical properties, the longitudinal layer is characterized to be anisotropic while the circular layer is isotropic.
6. The pacemaker cell that is located at the left boundary controls the mechanical activity of the tube, and hereby generates an excitatory stimulus of given intensity.
7. The dynamics of the colon segment is given by:

$$\gamma_0 \frac{dV}{dt} = \left(T_1 e_{ij}\sqrt{g_{22}}\right)s_1 + \left(T_2 e_{ij}\sqrt{g_{11}}\right)s_2 + p\sqrt{gn_j}, \tag{1}$$

where:

$$T_{1(2)} = k\frac{\partial \varepsilon_{c(1)}}{\partial t} + \varphi_{c(1)} T^a (\lambda_{c(1)} T^p\ (\lambda_c, \lambda_1). \tag{2}$$

Here – γ_0 the linear density of a biomaterial in an undeformed sate; λ_c, λ_1 – the rate of elongation (here after the subscripts l and c refer to the longitudinal and circular muscle layers, respectively; – the direction cosines of the $\varepsilon_{c,1} = \lambda_{c,1} + 1$; e_{ji} outward normal n_j, to the surface with respect to the cylindrical j – axis (I = 1, 2; j = r, s, z); g_{ij}, g – the components and determinant of the fundamental tensors; V (V_r, V_s, V_z) – the velocity vector and its radial circumferential and longitudinal components; $T_{1,2}$ – the components of the tensor of membrane forces; T^p,T^a – the passive and active components of the total force ($T_{c,l}$); k – the rheological parameter; p – intraluminal pressure; s_1,s_2 the Langragian coordinates of the bioshell.

The passive $T^p_{c,1}$ components are derived from:

$$T^p_{(c,1)} = \frac{\partial \gamma_0 W}{\partial(\lambda_{(c,1)} - 1)}, \tag{3}$$

where W – the strain energy density function has the form:

$$\gamma_0 W = \frac{1}{2}[c_1(\lambda_1 - 1)^2 + 2_{c_3}(\lambda_1 - 1)(\lambda_c - 1)^2 + c_2(\lambda_c - 1)^2] + \\ + c_{10} \exp\ [c_4(\lambda_1 - 1)^2 + c_5(\lambda_c - 1)^2 + 2_{c_6}(\lambda_1 - 1)(\lambda_c - 1)]. \tag{4}$$

As for the active components ($T^a_{c,l}$) components we have:

$$T^a_{(c,1)} = c_{7(c,1)}\,\lambda^2_{(c,1)} + c_{8(c,1)}\,\lambda_{(c,1)} + c_{9(c,1)}. \tag{5}$$

The mechanical constant of a biocomposite are $c_{1\text{-}10}$.

8. The dynamics of propagation of the electrical waves of depolarization along the anisotropic longitudinal muscle layer is (φ_1) defined as:

$$C_m(\varphi_1)_t = I_{m1}(s_1, s_2) + I_{m2}(s_1 - s_1, s_2 - s_2) - I_{ion} \tag{6}$$

The transmembrane ion currents per unit volume are I_{m1}, I_{m2}:

$$I_{m1}(s_1, s_2) = M_{vs}\left\{\frac{-2(\mu_{s1} - \mu_{s2})}{(1+\mu_{s1})(1+\mu_{s2})}\operatorname{arctg}\left(\frac{ds_2}{ds_1}\sqrt{\frac{Gs_2}{Gs_1}}\right) + \frac{g_{os_2}}{Gs_2}\right\} \times$$
$$\times\left(\left(\frac{g_{os_1}}{\varphi_c}(\varphi_1)s_1\right)s_1 + \left(\frac{g_{os_1}}{\varphi_1}(\varphi_1)s_2\right)s_2\right), \tag{7}$$

$$I_{m2}(s_1 - s_1, s_2 - s_2) = M_{vs}\iint_x \frac{\mu_{s1} - \mu_{s2}}{2\pi(1+\mu_{s_1})(1+\mu_{s_2})G} \times$$
$$\times \frac{(s_1 - s_1)^2 / G\mu_{s_1} - (s_2 - s_2)^2 / G\mu_{s_2}}{\left[(s_1 - s_1)^2 / G_{s_1} + (s_2 - s_2)^2 / G_{s_2}\right]} \times \tag{8}$$
$$\times\left(\left(\frac{g_{os_1}}{\lambda_c}(\varphi_1)s_1\right)s_1 + \left(\frac{g_{os_2}}{\lambda_c}(\varphi_1)s_2\right)s_2\right)ds_1 ds_2.$$

here:

$$\mu_{s_1} = g_{os_1} / g_{is_1},\ \mu_{s_2} = g_{os_2} / g_{is_2}, \tag{9}$$

$$G_{s_1} = \frac{g_{os_1} + g_{is_1}}{\lambda_c},\ G_{s_2} = \frac{g_{os_2} + g_{is_2}}{\lambda_1},\ G = \sqrt{G_{s_1} G_{s_2}}. \tag{10}$$

The following notations are: C_m – the capacitance of smooth muscle; g_{is1}, g_{is2}, g_{os1}, g_{os2} – the maximal intracellular (the subscript (i) and interstitial space), (the subscript (o) conductivity of the longitudinal and circular muscle layers in the longitudinal and circumferential directions), respectively; M_{vs} – the membrane volume to surface ratio; I_{ion} – the total ionic current.

In the case of propagation along the isotropic circular muscle layer (φ_c):

$$C_m(\varphi_c)_t = \frac{M_{vs}}{(1+\mu_{s_1})}\left(\frac{g_{os_1}}{\lambda_c}(\varphi_1)_{s_1}\right)s_1 + \left(\frac{g_{os_2}}{\lambda_c}(\varphi_1)s_2\right)s_2 - I_{ion}, \tag{11}$$

where the above-mentioned abbreviations are used.

The points of the wall lie on the surface of the sphere throughout all stages of dynamic reaction.

$$(Z_c - u_z)^2 + (r_0 + u_r)^2 + (r_0 + u_s)^2 - R^2 = 0,\ z \in [z_1, z_2]. \tag{12}$$

Here: u (u_z, u_r, u_s) – the displacement vector; Z_c – the position of the center of the sphere at time t; z_1, z_2 – the boundary points of contact of sphere and the wall. The kinematic equation of the motion of the bolus along the tube is governed by:

$$\eta \frac{dZ_c}{dt} + F_d = 2\pi r_0 \int_{z_1}^{z_2} \int_{r_0}^{r} F_c \, d\xi \, d\xi, \tag{13}$$

where F_c, F_d are the contact force and the force of dry friction, respectively, η is the coefficient of viscous friction.

At the initial moment of time, the whole system is in the resting state. It is supposed that the left and right boundaries of the tube are rigidly fixed. Discharges of the pacemaker cell trigger contractions in smooth muscle. When the maximum of the total force in the longitudinal muscle layer is achieved, contraction of the circular muscle layer starts. The complete activation of the circular muscle layer begins as the maximum of the total force in the longitudinal muscle layer is accomplished. The right boundary of the tubular segment remains in the resting state throughout.

Stages and sequences defining the dynamics of peristaltic propulsion of a solid sphere took part, as shown in Fig. 1. The calculations use a set of mechanical and electrical parameters and constants such as: $l = 10.0$ cm, $r_0 = 0.795$cm; $\gamma_0 = 26.14$ (g/cm2); $c_{71} = -40$, $c_{81} = 120$, $c_{91} = -50$, $c_{7c} = -25$, $c_{8c} = 75$, $c_{9c} = -31.2$ (mN/cm·mV); k = 33.3 (mN·sec/cm); $p_o = 1.0$ kPa; $C_m = 1.0$ (μF/cm^2); $M_{vs} = 0.05$.The use of colon tissue prevented us from estimating some of the mechanical and electrical parameters used in the model. Therefore, the missing data have been adjusted during the numerical simulation: $c_1 = c_2 = 10.4$, $c_3 = 2.59$, $c_4 = 3.79$, $c_5 = 12.7$, $c_6 = 0$; the conductivity constants in case of isotropy were: $g_{is1} = 2$, $g_{is2} = g_{os1} = 20$, $g_{os2} = 80$(mSm/cm^2); in a case of isotropy the conductivity constants were: $g_{is1} = 2$, $g_{is2} = g_{os1} = g_{os2} = 20$ (mSm/cm^2).

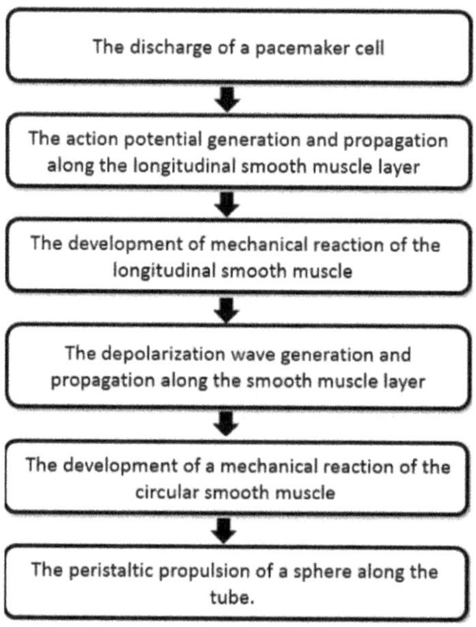

Figure 1: The sequence of events underlying the propulsion of a solid sphere.

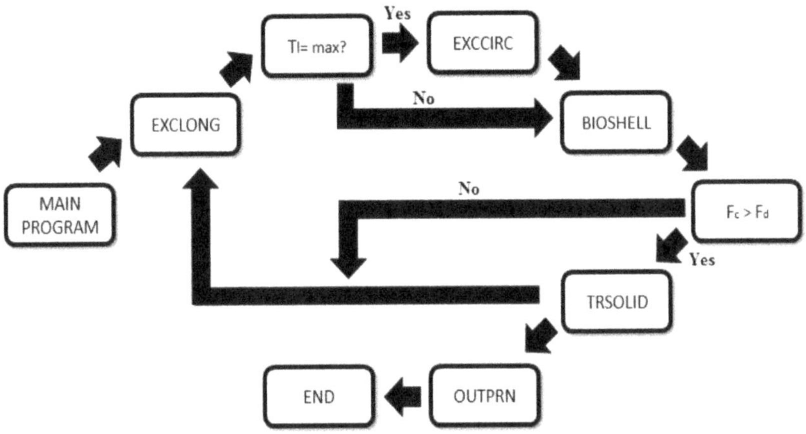

Figure 2: Flow chart of the algorithm.

Numerical algorithm flow chart is shown in Fig. 2. The main program is utilized to define the set of input parameters and constants, the initial configuration and the stress strain state of the tube. It includes the following subroutines:

EXCLONG (EXCCIRC) – Computes the dynamics of the depolarization wave long the longitudinal and circular smooth muscle bi-syncytia; uses the explicit finite-difference scheme of the second order approximation over the spinal and time variables.

BIOSHELL – Calculates the stress-strain states of the intestinal segment; uses the explicit hybrid finite-difference scheme of the second order approximation over the space and time. TRSOLID – Calculates the stress distribution over the enclosed sphere and its propulsion along the tube.

OUTPRN – Prints results.

4 RESULTS

The results of simulations of movement patterns resemble those recorded experimentally. They provide quantitative insights into the spatio-temporal patterns of changes in configuration, the distribution of contact forces over the bolus, and predict the average velocity of colonic transit. Therefore, a reciprocal relationship in the contraction of the longitudinal and circular smooth muscle is needed to guarantee the 'mixing' type of movements. Besides, strong interactive contractions of longitudinal and circular muscle layers are needed to expel the bolus from the gut. The dynamics of stress-strain distribution demonstrate the rise in the intensity of active propulsive forces in the circular smooth muscle layer throughout the entire phase of propulsion of the bolus. Viscous compared to dry friction had a marked effect on the average velocity of colon transit. Thus, the addition of osmotic and rapidly acting lubricant laxatives intraluminally shortened the time required for expulsion of the bolus in a significant way.

In the stimulation, 10 cm long of a part in the colon was constructed. The diameter of it is 0.795 cm, bolus diameter is 1.5 cm, viscous friction is 230 Ns/m, dry friction is 20 N, and intraluminal pressure is 0.1 kPa. Under these physiological circumstances, the bolus moves normally without any irritability within 2.4 seconds. However, whenever viscous friction increases, the irritability is increasing; hence the bolus takes long time to be expelled and as it decreases, the bolus will move normally and faster.

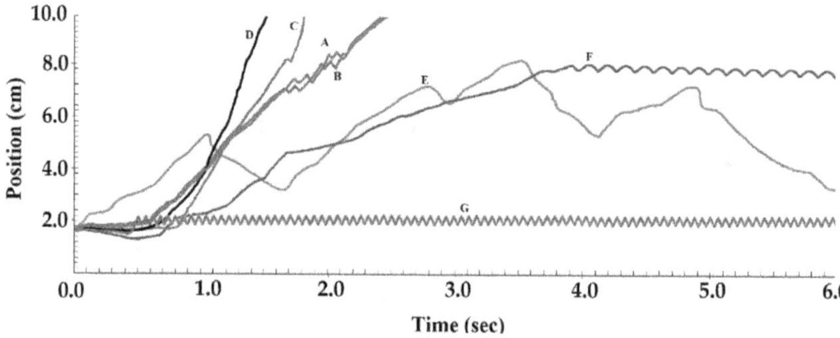

Figure 3: Movements of the bolus as a function of the viscous and dry friction. A: μ = 230 Ns/m; F_d = 20 N. B: μ = 230 Ns/m; F_d = 2 N. C: μ = 200 Ns/m; F_d = 20 N. D: μ = 90 Ns/m; F_d = 450 N. E: μ = 230 Ns/m; F_d = 450 N. F: μ = 1200 Ns/m; F_d = 20 N. G: μ = 230 Ns/m; F_d = 800 N.

Dry friction is a highly effective parameter. The bolus takes long time to be expelled by defecation if the parameter increases but it can be adjusted by decreasing the viscose friction. Hence, the bolus moves in special pattern if the dry friction is between 565N and 2000N. On the other hand, the bolus moves faster if the dry friction decreases, as illustrated in Fig. 3.

Intraluminal pressure is an important parameter. As it increases or decreases, the bolus takes long time to be expelled despite the irritability. It can be stopped only if the intraluminal pressure becomes higher than 15 kPa, as shown in Fig. 4.

Increasing or decreasing the amount of calcium entering the slow and fast calcium channels and potassium – calcium channels will cause a disruption in the bowel movement that will lead to a delay in defecation, as described in Fig. 5.

If there are, no electrical pulses such as ICCs and the motor neurons reach the smooth muscles of the colon, the bolus moves due to the fine motor function of the mechanoreceptors, but this movement will be in an irregular pattern, which leads to a delay defecation. In addition, if there are no electrical impulses and mechanoreceptors, the bolus moves due to the elastic forces but in a slow manner and with disruption. This fault can be adjusted by decreasing the viscous friction, as it is shown in Fig. 6.

Finally, there is a strong relationship between diameter of the colon, diameter of the bolus and velocity. If the diameter of the bolus increases, the bolus moves very fast. On the other

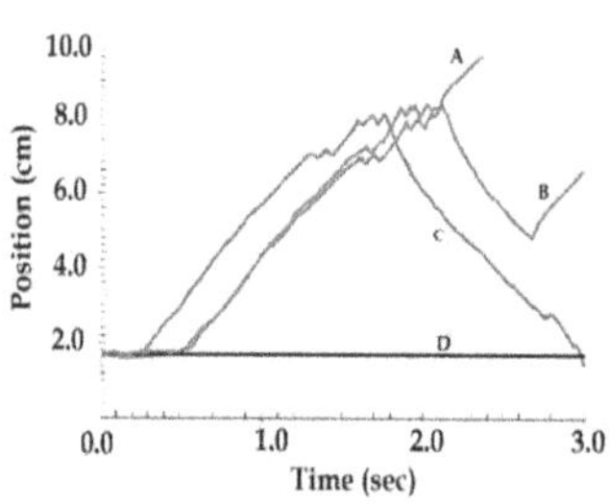

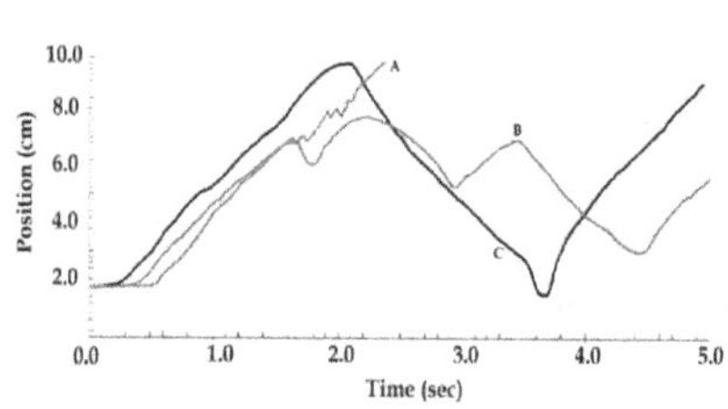

Figure 4: Effects of intraluminal pressure on propulsion. A: p = 0.1 kPa. B: p = 0.15 kPa. C: p = 0.075 kPa. D: p = 20 kPa.

Figure 5: Effects of ion channel, K^+ - Ca^{2+}, fast Ca^{2+}, and slow Ca^{2+} permeability on bolus movements. A – norm; B – decreased permeability; C - increased permeability.

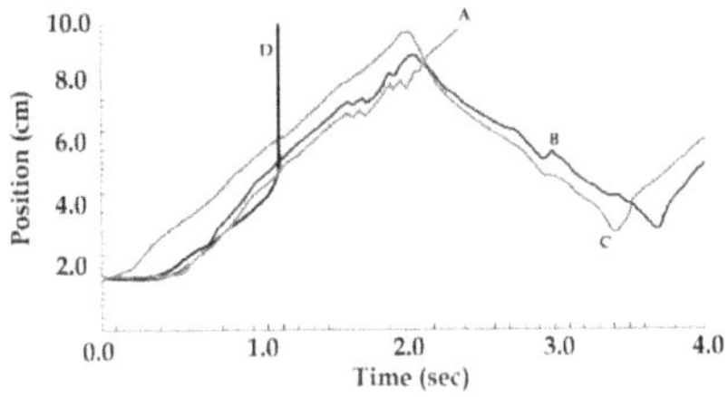

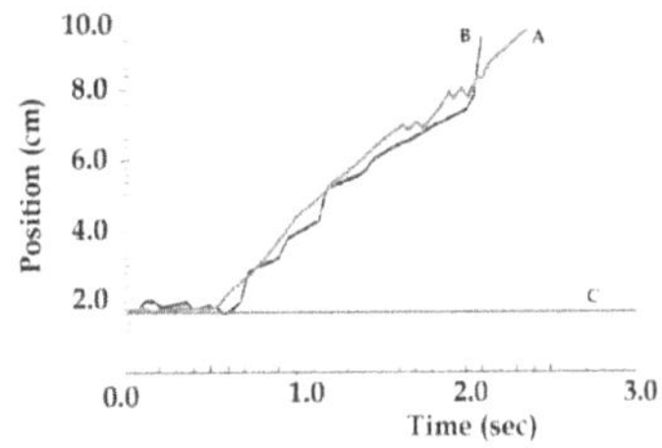

Figure 6: Effects of stimulation and viscous friction on bolus propulsion. A: $\mu = 230$ Ns/m, with stimulation. B: $\mu = 230$ Ns/m, no stimulation. C: $\mu = 230$ Ns/m, with stimulation; D: $\mu = 100$ Ns/m, no stimuli.

Figure 7: The effect of the radius on bolus movements. A: $r_0 = 0.795$ cm. B: $r_0 = 0.9$ cm. C: $r_0 = 0.6$ cm.

hand, if it decreases below 105% of the diameter of the colon, the bolus stops moving because mechanoreceptors will not fire and muscles of the colon cannot affect the bolus, as there are no stretching forces (Fig. 7).

5 CONCLUSIONS

The mathematical model of a segment of the gut reproduces qualitatively and quantitatively the dynamics of colonic transit. Viscous, and not dry, friction is the dominating parameter in the stability of propulsion. The reciprocal relationship of coordinated contractions of the longitudinal and circular smooth muscle layers is essential for the movement of the intraluminal content. Slow and fast calcium agonist/antagonist drugs must not be given to patients who suffer from constipation. Doctors should prescribe a combination of laxative lubricant drugs (mineral oil) and laxative bulk-formation (psyllium, methylcellulose) to have effective response.

REFERENCES

[1] Miftahof, R. & Akhmadeev, N., Dynamics of intestinal propulsion. *Journal of Theoretical Biology*, **246**, pp. 377–393, 2007. https://doi.org/10.1016/j.jtbi.2007.01.006

[2] Miftahof, R. & Fedotov, E.M., Intestinal propulsion of a solid deformable bolus. *Journal of Theoretical Biology*, **235**, pp. 57–70, 2005. https://doi.org/10.1016/j.jtbi.2004.12.019

[3] Miftahof, R. & Abdusheva, G., Small bowel propulsion: Transit of a solid bolus. In *Biophysics and Life Science*, ed. D. Ghista. Munich: Springer, pp. 253–260, 2007.

NOISE FILTRATION OF SHOCKWAVE PROPAGATION IN MULTI-LAYERED SOILS

LAITH I. NAMIQ & YOUSOF Q. ABDALJALIL
Retired Professor, Geotechnical Department, AKRF, Inc., USA.

ABSTRACT

Numerical methods, and especially the finite-element method (FEM), are usually adopted for the analyses of shockwave propagation in nonlinear inelastic media. Noise or spurious oscillations, in the calculated stresses and displacements, frequently appear in the FEM solutions. This article introduces and describes a numeric filter based on least-square analysis that can smooth out such fictitious noise. The sliced least-square method (SLSM) filter is implemented in a finite elements program that solves 1D time integration of dynamic equilibrium sets of equations that simulate shockwave propagation in multi-layered soils supported by a hard stratum. Elastic and elasto-viscoplastic material models with dynamic yield surface constitutive relations are invoked to model sand, clay, and concrete materials in the analyses. Results of the analyses of shockwave propagation in layers of soil and concrete without the filter are compared with identical conditions with the inclusion of the new filter in the finite-element program. Oscillations in calculated stresses and displacements were observed in the results when no filter was included in the solution program. Solution results showed little or no noise with the application of the new filter. The predicted FEM analyses results were compared with physical test results with very good to excellent comparisons obtained.

Keywords: elasto-viscoplastic material model, finite elements, implicit time integration, spatial filter, spurious oscillations, wave propagation.

1 INTRODUCTION

Analyses of wave propagation in multi-layered soil profiles are required in many geotechnical engineering problems. Among these problems is prediction of magnitudes and frequencies of shock-generated body and surface waves at the location of nearby structures. Closed-form solutions do exist, and utilize assumptions of simple boundary conditions and elastic or simplified elastic-plastic soil behaviour. However, closed form solutions become intractable if more representative constitutive relations for earth materials are utilized, and if large strains and/or cycles of loading and unloading are to be accounted for. Hence, numerical methods, and especially the finite-element method (FEM), are usually adopted for the analysis of these and similar problems. The problem is that noise or spurious oscillations in the calculated stresses and displacements frequently appear in the FEM solutions. Calculations with such imprecision have been recognized, and attempts to circumvent these spurious oscillations and improve upon the solutions have been and are being carried out in several ways. Modifications to the time integration operator adopted in the FEM solution technique were proposed[1]. The incorporation of additional sinusoidal terms in the solution method was attempted [2]. A cut-off wave length was suggested for any given FEM mesh [3]. The addition of harmonic functions to the conventional interpolation function to enrich the FEM formulation was introduced [4]. Calculations of critical element size to be avoided in FEM mesh construction were explored [5]. The basic idea in all these attempts is the modification of one or more phase(s) of the solution technique, i.e. interpolation function, element size, integration scheme, etc. To summarize, all proposed solutions dealt with a pre-result output. A different approach was to improve upon the FEM post-result solution by use of digital [6] or spatial filters [7]. A simple, efficient, new numerical spatial filter that smoothens out FEM-generated fictitious noise is introduced and detailed herein.

DOI: 10.2495/CMEM-V6-N4-804-813

2 SLICED LEAST-SQUARE METHOD FILTER (SLSM)

Least-square analysis forms the mathematical basis of the introduced filter. The least-square method can be expressed mathematically as follows: Let $\psi(x)$ be defined on the interval $a \le x \le b$ and let $\varphi(x, C_1, C_2, \ldots, C_n)$ be the function that is to approximate ψ over the interval, given the selection of values for the parameters $C_i, i = 1,2,\ldots,n$. The basic concept is that the weighted squared error over the whole interval should be a minimum. The error ε at a given point in the above-defined interval is given by eqn (1):

$$\varepsilon = \psi(x) - \phi(x, C_{1,} C_2, \ldots, C_n) \tag{1}$$

and the integral, Γ_2 in eqn (2), is to be minimized:

$$\Gamma_2(C_1, C_2, \ldots, C_n) = \int_a^b w(x)\{\psi(x) - \phi(x, C_1, C_2, \ldots, C_n)\} \cdot dx \tag{2}$$

where $w(x)$ is a weighting function. The target function, ψ, is available at a given number of points, m. Thus, the least-square error procedure is defined as follows:

$$\begin{aligned} &C_1 \sum_{k=1}^{m} w(x_k) \cdot \phi_i(x_k) \cdot \phi_1(x_k) + \ldots + \\ &C_i \sum_{k=1}^{m} w(x_k) \cdot (\phi_i(x_k))^2 + \ldots + \\ &C_n \sum_{k=1}^{m} w(x_k) \cdot \phi_i(x_k) \cdot \phi_n(x_k) = \sum_{k=1}^{m} w(x_k) \cdot \phi_i(x_k) \cdot \psi(x_k). \end{aligned} \tag{3}$$

Defining the elements of the matrix **S** and vector **T** as follows:

$$S_{i,j} = \sum_{k=1}^{m} w(x_k) \cdot \phi_i(x_k) \cdot \phi_j(x_k) \tag{4}$$

$$T_i = \sum_{k=1}^{m} w(x_k) \cdot \phi_i(x_k) \cdot \psi(x_k) \tag{5}$$

one can write n equations, where n is the order of the polynomial fit plus one, in matrix notation as in eqn (6):

$$\boldsymbol{S} \cdot \boldsymbol{C} = \boldsymbol{T} \tag{6}$$

where **C** is a vector containing e coefficients $C_1, C_2, \ldots, C_n$. Solving eqn (6) yields the values of C_1 through C_n. And then the approximating function, $\phi(x)$, can be computed, at any point in the interval (a,b), using eqn (7):

$$\phi(x) = \sum_{i=1}^{n} C_i \cdot x^{i-1} \tag{7}$$

The weight function, w (x_k) is assigned a value of 1.0 in SLSM as no given Gauss integration point is believed to be more accurate than other points.

The least-square regression analysis method (LSM) may be applied to smooth FEM-calculated output (e.g., stresses) of wave propagation throughout the domain of interest of a solid body. Noise-free response is then obtained. However, illusory smoothening and flattening of sharp peaks in computed stresses, as well as erratic response ahead of the wave front, is produced by application of LSM to the FEM solution output. To circumvent this problem, application of LSM is preceded by an additional procedure designed to eliminate the introduction of further erratic results in the FEM output. This procedure is developed herein, and together with LSM, defines SLSM.

SLSM is achieved by sorting in linear order, in vector **V**, the magnitude of generated dynamic loading stress components, at element Gauss integration points along the wave propagation axis. This is followed by searching, per a pre-set criterion, the FEM output for 'jumps' in the calculated stress' magnitudes between arranged consecutive Gauss integration points. The complete set of Gauss integration points is then divided into subsets delimited by consecutive stress 'jumps'. A slice that is the set of Gauss integration points between two 'jumps' is thus generated. LSM is subsequently operated on each slice individually. Finally, displacements, velocities, and acceleration vectors are modified according to the new stress value (i.e. after LSM application) at each Gaussian integration point.

2.1 Stress 'jump' identification criterion

A stress 'jump' is identified at each time step when any of the following conditions is true:

2.1.1 Condition (a)

$D <= R$ as shown in Fig. 1, other variables shown in the figure are given by eqn (8) through (11).

$$A_1 = \sigma_1 / \sigma_{max} \tag{8}$$

$$A_2 = \left(\left(\frac{\sigma_2}{\sigma_{max}} \right) + \left(\frac{\sigma_3}{\sigma_{max}} \right) \right) / 2 \tag{9}$$

$$R = \left| A_2 - \alpha \left(A_1 + A_2 \right) \right| \tag{10}$$

where α is a dimensionless factor. D is calculated using eqn (11):

$$D = \left| A_2 - (\sigma_3 / \sigma_{\max}) \right| \tag{11}$$

In Fig. 1, d is the distance along the selected search direction, and L is the cumulative length of finite elements (FE) along that search direction.

2.1.2 Condition (b)

$D <= R$ as shown in Fig. 2. The variables shown in Fig. 2 are given by eqn (12) through (15):

$$A_1 = \left(\left(\frac{\sigma_{j-3}}{\sigma_{max}} \right) + \frac{\sigma_{j-2}}{\sigma_{max}} \right) / 2 \tag{12}$$

$$A_2 = \left(\left(\frac{\sigma_{j-1}}{\sigma_{max}} \right) + \frac{\sigma_j}{\sigma_{max}} \right) / 2 \tag{13}$$

$$R = \left| A_2 - \alpha \left(A_1 + A_2 \right) \right| \tag{14}$$

$$D = \left| A_2 - (\sigma_j / \sigma_{\max}) \right| \tag{15}$$

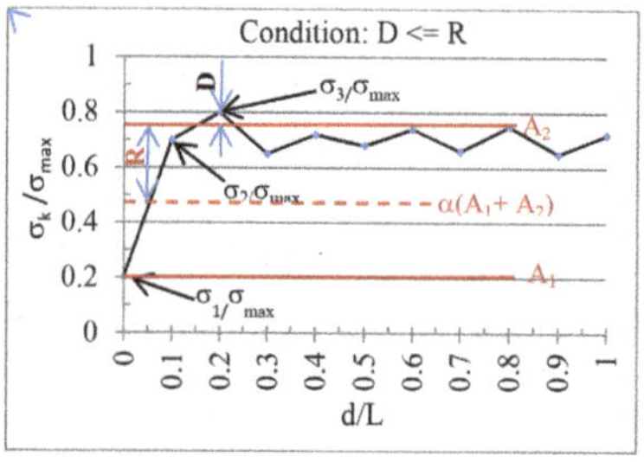

Figure 1: Condition (a) for stress 'jump' identification.

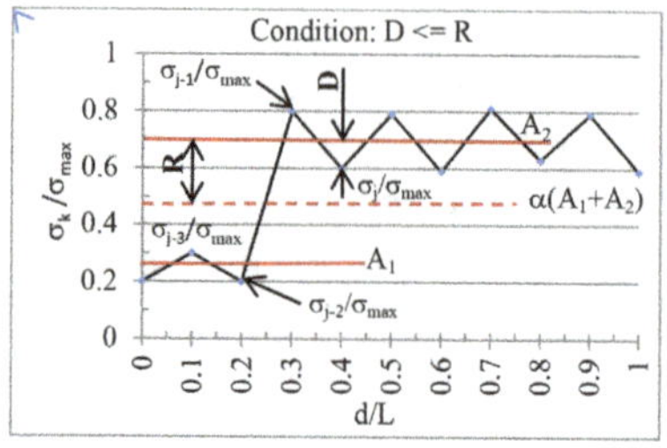

Figure 2: Condition (b) for stress "jump" identification.

2.1.3 Condition (c)

$D <= R$ as shown in Fig. 3, where R is given by eqn (16), and the rest of the variables are as defined for condition (b).

$$R = \left| A_2 - \left(\frac{\sigma_{j-3}}{\sigma_{max}} \right) \right| \tag{16}$$

If none of the conditions (a) through (c) are applicable, then LSM is applied to the stresses calculated at the end of current time step, Fig. 4.

3 EXAMPLES

3.1 Analysis of an elastic rod

Linear-elastic slender column was analysed under a variety of applied loadings as given by eqns (17) through (20):

(a) Initial conditions:

$$\boldsymbol{U}_o = \dot{\boldsymbol{U}}_o = 0 \tag{17}$$

where $\boldsymbol{U}$ is the vector of nodal displacements in the z-direction. Vectors and matrices are in bold type. Note that 'u' is the displacement in the z-direction and 'v' is the displacement in the radial direction; see Fig. 5.

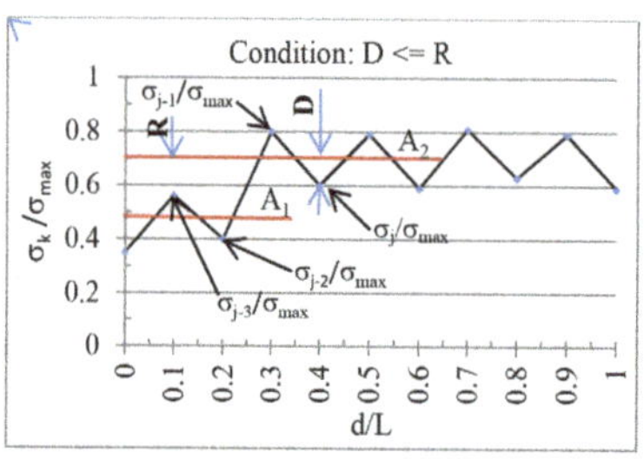

Figure 3: Condition (c) for stress "jump" identification.

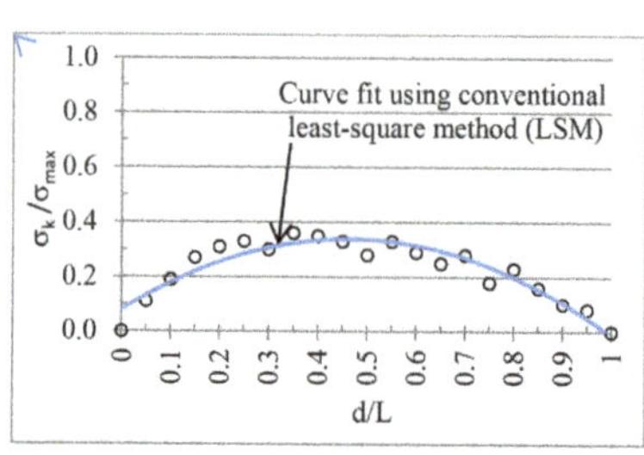

Figure 4: Curve fit where no stress "jump" is identified.

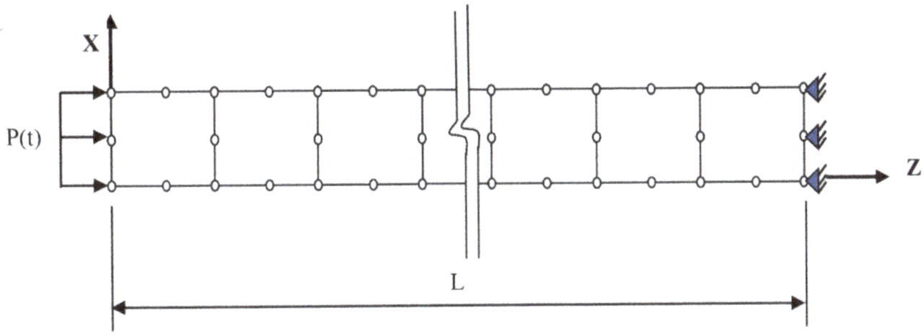

Figure 5: Finite element mesh for slender rod. All side nodes are restrained in the X-direction.

(b) Boundary conditions:

$$v = 0; z = [0, L] \tag{18}$$

$$u = 0; z = L \tag{19}$$

(c) Applied load:

$$P(t) = \begin{cases} 0, t \leq 0 \\ 2070, t = 0.006\,s \\ < 2070, t > 0.006\,s \end{cases} \tag{20}$$

where t is time in seconds, and P(t) is in kPa. The variation of applied pressure with time is given in [9].

The pressure P(t) is uniformly distributed on the element surface where z = 0. Axisymmetric 8-node isoparametric elements were used in that analysis. A cubic polynomial fit in the SLSM was implemented, eqn (21):

$$\sigma(z) = A + B \cdot z + C \cdot z^2 + D \cdot z^3 \tag{21}$$

where A, B, C, and D are coefficients that were calculated using the standard least-square procedure. The exact and filtered solutions are shown on Fig. 6. The filtered solution (i.e. calculated axial stress versus distance along the slender rod) closely matches the exact solution.

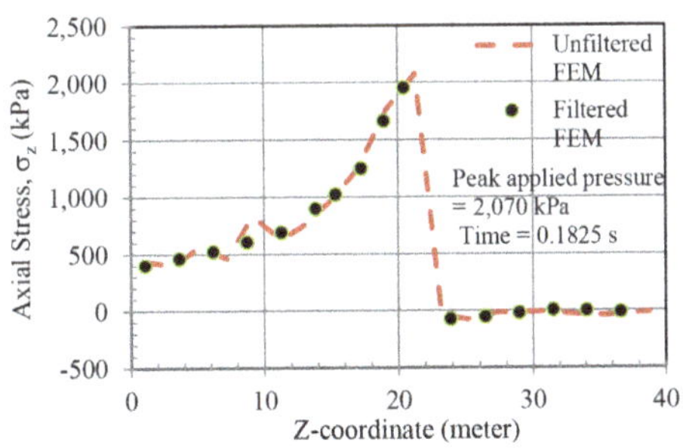

Figure 6: Elastic analysis of a slender load under dynamic load.

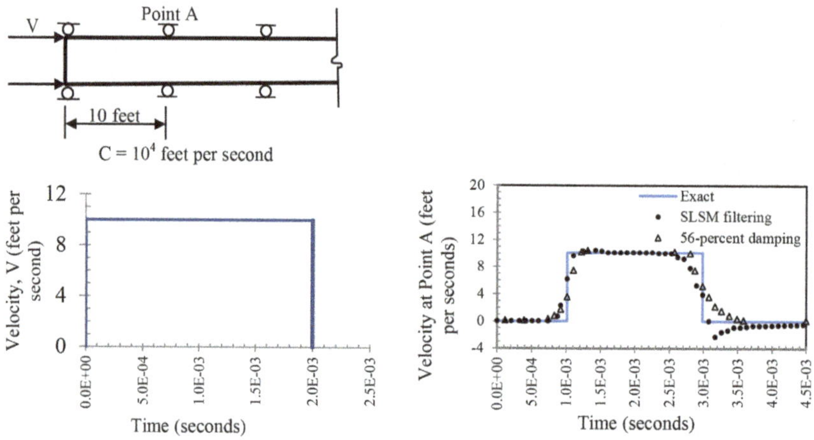

Figure 7: (a) Model of a slender rod and Variation of applied velocity *versus* time. (b) Comparison of the SLSM filtered solution to artificial damping method cited by Belytschko *et al.* [10].

The problem of a slender column under axial strep velocity applied at the free end was solved, Fig. 7a. The problem was presented in Belytschko *et al.* [10], where artificial damping in the FEM solution was employed. The exact solution, FEM solution with artificial damping, and the SLSM filtered FEM solution, are shown in Fig. 7b. As can be seen, the SLSM solution is close to the exact one.

3.2 Non-linear analysis of a soil column

The problem of a soil column subjected to time-varying, vertical pressure was solved. The material model is an elasto-viscoplastic modified Cam clay model [11]. Configuration of FE and boundary conditions are the same as for the case of elastic rod (Section 3.1). The soil column consists of normally consolidated clay with decreasing void ratio with depth. The applied load is as given by eqns (17) through (20). However, in this case, the peak applied pressure, P_o, is 207 kPa.

Attenuation factor, α, is given by eqn (22):

$$\alpha = \Delta\sigma_z / P_o \tag{22}$$

where $\Delta\sigma_z$ is the change in axial stress and is calculated using the output of the FEM program. Variation of the attenuation factor with non-dimensional depth (i.e. along the axis of the soil column) is shown in Fig. 8. As a check on the validity of the solution, the calculated variation of attenuation factor (with depth) is compared to that published in [9]. The solution shown in [9] is presented as Fig. 9. As can be seen from Figs 8 and 9, the variation of the attenuation factor with depth has the same characteristics.

Change in axial (vertical) stress, using an elasto-viscoplastic material model for the clay column with peak overpressure of 2,070 kPa is calculated. Propagation of the stress wave is shown in Fig. 10. SLSM is applied to the calculated change in axial stress and one result is shown in Fig. 11. The filtered solution shows larger compressive and smaller tensile stresses compared with the unfiltered solution.

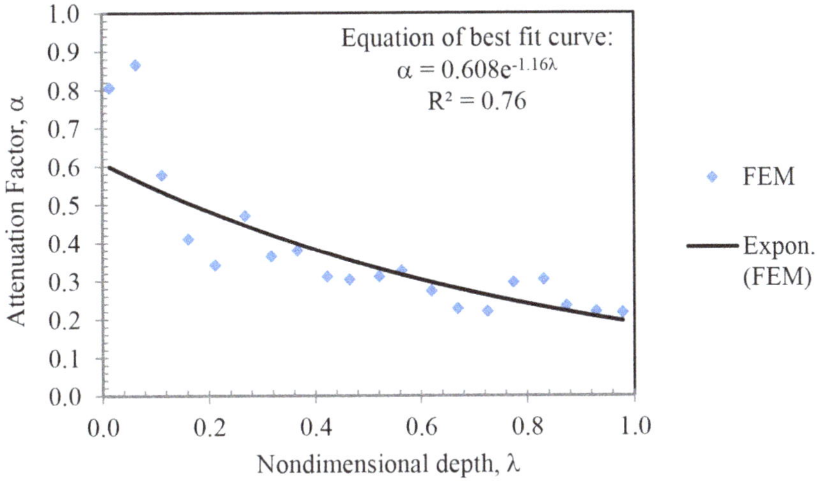

Figure 8: Variation of the attenuation factor *versus* non-dimensional depth.

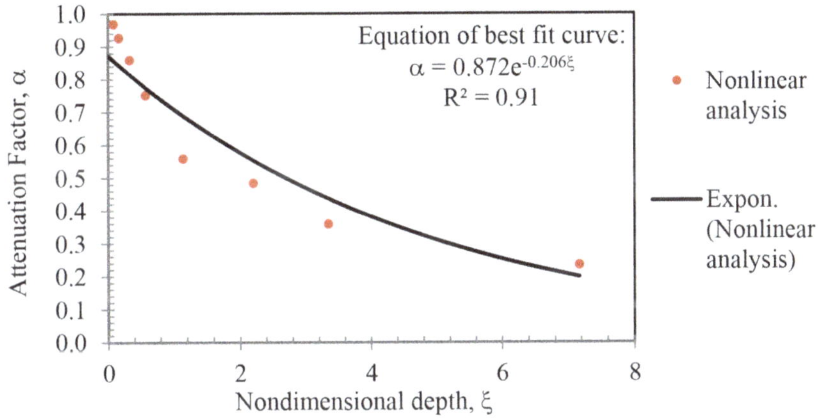

Figure 9: Variation of attenuation factor *versus* depth using bilinear 1-D stress–strain relationship (after Hendron and Auld [9]).

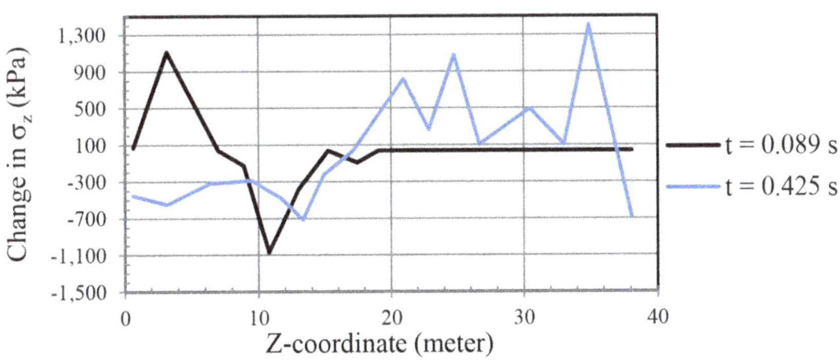

Figure 10: Stress wave propagation in a clay column using elasto-viscoplastic analysis.

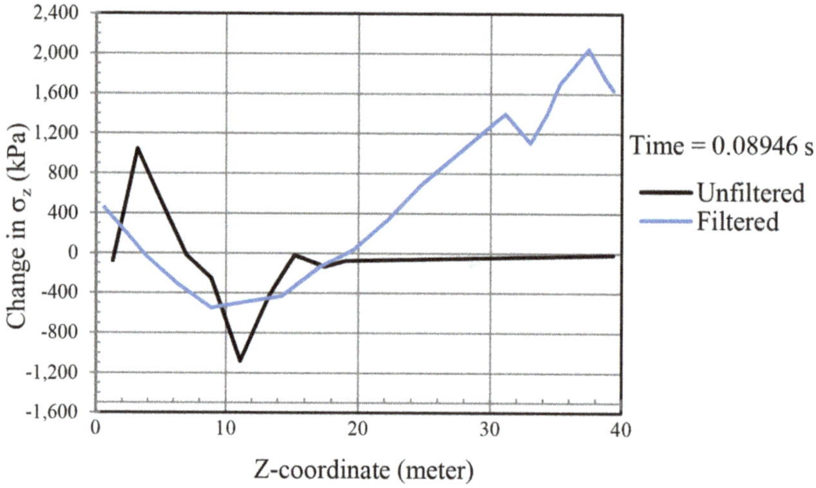

Figure 11: Effect of stress smoothening on predicated change of stress in clay column (peak overpressure = 2,070 kPa).

3.3 Non-linear analysis of sand and concrete strata

A two-strata column with plain concrete on top of sand and with an interface element between the two materials was analysed. The thickness of each stratum was 19.97 m, and the interface element was 0.06 m thick. Plain concrete was modelled using a non-linear stress–strain relationship. Sand was modelled using an elasto-viscoplastic version of modified Cam clay. The applied load, at the top of the plain concrete stratum, was the same as for the clay column, as shown in Figs 10 and 11.

The change in axial stress calculated using a linear, elastic material model was compared with that using the elasto-viscoplastic material model, as shown in Fig. 12. A reduction in peak axial stress was noticed in the case of the elasto-viscoplastic material model. The reduction can be attributed to energy dissipation caused by plastic strain in the concrete and sand, given the high pressure applied with a peak of 2,070 kPa.

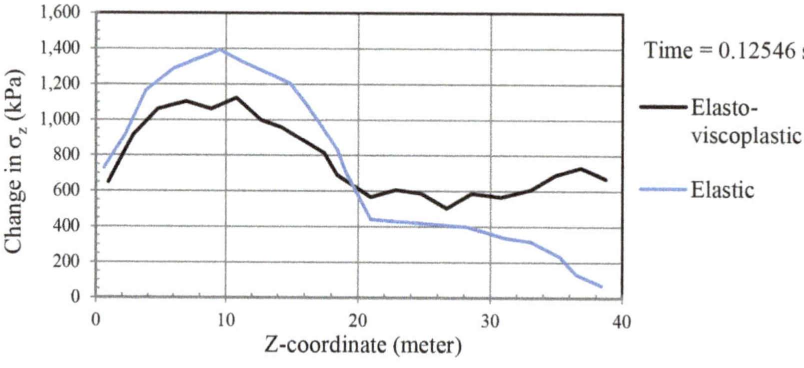

Figure 12: Effect of analysis type on predicted change in axial stress for the case of plain concrete on sand stratum.

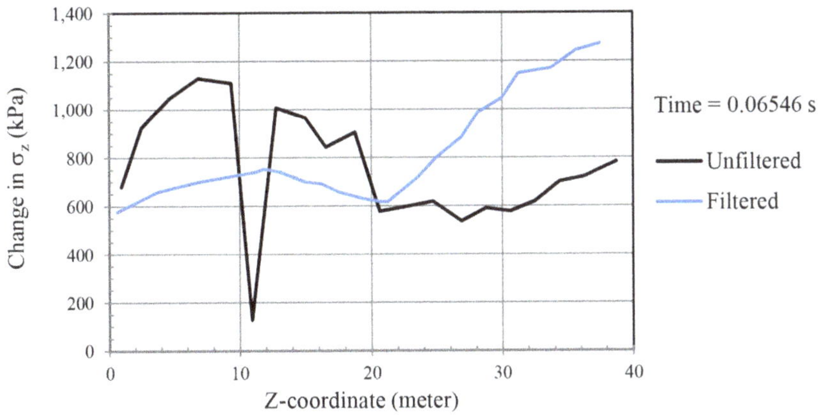

Figure 13: Effect of stress smoothening on predicted change in stress for the case of plain concrete on sand stratum (elasto-viscoplastic analysis).

An example of the output of SLSM filtered and unfiltered change in stress is shown in Fig. 13. The filtered solution indicates a smooth variation of the change in axial stress *versus* depth. It also indicates that a peak in stress has been reached near the fixed end of the concrete-and-sand column. The unfiltered solution indicates a tensile stress jump near a depth of 11 m, which is near the middle of the concrete stratum. From the mechanics of stress wave propagation, such tensile stress jump cannot be explained, given that one end of the concrete-and-sand column is free and the other is fixed (in the direction of wave propagation).

4 CONCLUSIONS

The application of the FEM to the problem of shock wave propagation in multi-layer continua, whose constitutive laws are complex, provides results that cannot be obtained using existing closed-formed solutions. Spurious oscillations in FE solutions are inherent because of the limited cut-off frequency in FE meshes. Spurious oscillations observed in the case of material non-linearity are higher than those in the elastic medium.

A spatial filter, the SLSM, based on the least-squares method, is developed herein. The SLSM filter proved effective in linear elastic analyses in reducing the magnitude of the spurious oscillations in the calculated axial stress in a slender column, while preserving the shape of the advancing stress wave.

The elasto-viscoplastic solution showed attenuation trend in the axial stress, which matches with the published data very well. The SLSM filter provides smoothening of the calculated stress in elasto-viscoplastic slender columns composed of one or more materials.

REFERENCES

[1] Belytschko, T. & Mullen, R., Mesh partitions of explicit-implicit time integration. *Formulations and Computational Algorithm in Finite Element Analysis*, ed. K.J. Bathe, T. Oden & W. Wunderlich, MIT Press: Cambridge, Massachusetts, U.S.A., pp. 672–690, 1978.

[2] Pain, H.G., *The Physics of Vibrations and Waves*, John Wiley and Sons Ltd.: London, 1976.

[3] Kreig, R.D. & Key, S.W., Transient shell response by numerical time integration. *Advances in Computational Methods in Structural Mechanics and Design*, eds J.T. Oden, W.W. Clough & Y. Yamamoto, UAH Press: Huntsville, Alabama, U.S.A., 1972.

[4] Ham, S. & Bathe, K.J., A finite element method enriched for wave propagation problems. *Computers and Structures*, **94–95**, pp. 1–12, 2012. https://doi.org/10.1016/j.compstruc.2012.01.001
[5] Schmicker, D., Duczek, S., Liefold, S. & Gabbert, U., Wave propagation analysis using high-order finite element methods: spurious oscillations excited by internal element eigenfrequencies. *Technische Mechanik*, **34**(2), pp. 51–71, 2014.
[6] Okrouhlik, M., Ptak, S. & Valdek, U., Self-assessment of finite element solutions applied to transient phenomena in solid continuum mechanics. *Engineering Mechancis*, **16**(2), pp. 103–121, 2009.
[7] Martel, L., Love wave propagation across a step by finite elements and spatial filtering. *Geophysical Journal of the Royal Astronomical Society*, **61**, pp. 659–677, 1980. https://doi.org/10.1111/j.1365-246X.1980.tb04837.x
[8] Cope, R.J., Sawko, F. & Tickell, R.G., *Computer Methods for Civil Engineers*, McGraw-Hill Book Company, Ltd.: U.K., 1982.
[9] Hendron, A.J. & Auld, H.E., The effect of soil properties on the attenuation of air-blast-induced ground motions. *Proceedings of Symposium on Wave Propagation*, The University of New Mexico, Albuquerque, New Mexico, U.S.A., 1978.
[10] Belytschko, T., Chiapetta, R.L. & Bartel, H.D., Efficient large scale non-linear transient analysis by finite elements. *International Journal of Numerical Methods in Engineering*, **10**, pp. 579–596, 1976. https://doi.org/10.1002/nme.1620100308
[11] Atkinson, J.H. & Bransby, P.L., *The Mechanics of Soils: An Introduction to Critical State Soil Mechanics*, McGraw-Hill Ltd.: U.K., 1978.

VISUALISATION OF ACOUSTIC STREAMING USING PIV IN NEWTONIAN AND NON-NEWTONIAN LIQUIDS

MADURANGA AMARATUNGA & RUNE W.TIME
Department of Petroleum Engineering, University of Stavanger, Norway.

ABSTRACT

The effect of fluid rheology on acoustic streaming was studied experimentally using a low frequency (600Hz–15kHz) underwater acoustic transducer. The fluid rheology was compared with deionized water and non-Newtonian fluid polyanionic cellulose (PAC). Streaming effect generated by the transducer in a static liquid medium was visualized by particle image velocimetry (PIV) method. The motion of fluid was optically visualized using light scattering 'seeding' particles. Velocity profiles induced by the acoustic streaming have different shapes and range of magnitudes. First, the acoustic streaming in deionized water was visualized for different frequencies and pressure amplitudes (voltages). A maximum of 1 g/L PAC was then introduced in smaller steps for some selected frequency and voltage settings. The streaming disappeared completely when the total concentration of the fluid medium reached 0.19 g/L PAC. The measured streaming velocities are found to be in the range of 2.1 to 9.9 cm/s for water and it is proportional to the applied voltage and the operating frequency of the transducer. When introducing PAC, the streaming velocity within water gradually decreased until zero due to the attenuation of acoustic waves by viscous effects. This confirms that the streaming velocity is approximately inversely proportional to the bulk viscosity of the medium. The velocity vectors and the streaming velocity maps illustrate the induced non-linearities of the fluid medium due to the acoustic propagation. The results are part of a comprehensive study aimed at investigating the influence of acoustic vibration on particle settling in non-Newtonian fluids.

Keywords: Acoustic streaming, flow visualization, particle image velocimetry, rheology.

1 INTRODUCTION

An acoustic wave is pressure oscillation in a medium, propagation with the sound velocity. For small amplitude waves there is no net measurable mass transport, since there is no net molecular motion, only a net transport of energy and momentum. However, when a sufficiently strong acoustic pressure field is applied into a viscous liquid, non-linear effects set in. In such cases the liquid can start to flow in the same direction as propagation of acoustic waves. This fact is explained as momentum transfer from the wave to the fluid by viscous attenuation of a sound (pressure). Due to spatial variation in pressure amplitude around the acoustic beam and the degree of sound attenuation, shear induced eddies and circulation currents can be generated. The liquid motion or the steady bulk flow within any fluid medium under these influences of acoustic waves is called acoustic streaming [1]. Simply stated, acoustic streaming is a dissipation of the acoustic energy in the form of change of the gradients in the momentum flux [2, 3]. Streaming is only one effect [4] of the propagation of the high intensity acoustic waves. Other effects are cavitation, agitation and oscillatory fluid motion, etc. Acoustic streaming is a second-order steady flow [5], which is superimposed on the dominant acoustic velocity and it is induced by the non-linearities of the acoustic propagation. The absorption of the energy of the sound wave by the fluid is responsible for this induced flow [3, 6]. Acoustic streaming is used widely for non-invasive intervention purposes, both in industry and medicine. Among them, detection of ovarian cysts, detection of blood clotting via ultrasound, convective transport in microfluidic applications, removal of non-specifically bound proteins to allow reuse of biosensors, drug delivery, etc. [7, 8] are popular. One of the trending utilization of acoustic streaming is to generate high efficient

DOI: 10.2495/CMEM-V6-N4-814-826

mixing and heat transfer in different applications [1, 9, 10]. This has become increasingly important when some process conditions and the materials are risky to reach. In literature, the studies that investigated the phenomena of acoustic streaming are limited to qualitative description of the phenomena and very few researchers have looked at acoustic streaming in a quantitative way. Almost all of them have reported about the streaming caused by ultrasound sources. Ref [11], who used a lithotripsy pulse to generate acoustic waves have found almost a linear relationship between the streaming velocity and the generated peak-negative pressures. Ref [6] have developed a CFD model in order to predict the acoustic streaming assuming that the hydrodynamic momentum rate of the incoming jet is equal to the total acoustic momentum rate emitted by the transducer. The laser doppler anemometry (LDA) measurements performed by Ref. [1] validate their CFD model for the visualization of acoustic streaming induced by an ultrasonic horn. References [4, 11] state that the streaming velocity due to any acoustic source within a fluid, is proportional to the intensity and the surface area of the acoustic source (an ultrasound source), the attenuation coefficient of the medium and inversely proportional to the bulk viscosity and the speed of sound in the medium. The numerical study performed by Ref. [12] visualized the cavitation and acoustic streaming as a result of periodic changes in pressure within a liquid medium. The application of low frequency ultrasound is limited to studies that have been done using air as the medium to visualize the streaming patterns and velocity. Studies of acoustic streaming within non-Newtonian liquids are very limited except in medical field with regarding to the biological applications. From the study performed by Ref. [10], they have experienced the acoustic streaming produced by ultrasonic transducers in highly viscous fluid. Ref. [8] have presented the effects of non-Newtonian viscosity and the power law model on non-linear acoustofluidics by claiming that streaming velocity is highly dependent on the viscosity model of the blood and increasing ultrasonic frequency or intensity, amplifies the streaming velocity Ref. [13], who used LDA to study the acoustic and streaming velocity inside a cylindrical standing-wave resonator filled with air mention that the magnitude of the streaming is dependent on fluid viscosity with temperature. However, Ref. [5] have performed a similar type of experiment using LDA, but in isothermal conditions to prove the non-linear phenomena of oscillatory flows and the efficiency degradation of thermoacoustic devices. According to the study performed by Ref. [7] to predict the acoustic streaming phenomenon in Newtonian and non-Newtonian fluids, the streaming velocities predicted for a non-Newtonian fluid are smaller than that in a Newtonian fluid. Visualization of acoustic streaming becomes much more representative by using a non-intrusive technique such as LDA or PIV if turbulent flow is dominating. The intrusive measurement techniques or visual observations techniques like streak photography to estimate the velocity magnitude is not accurate enough [1] in this regard. PIV is based on video images of the illuminated fluid flow plane using seeding particles and produces an instantaneous two dimensional velocity vector field [14]. Ref.[4] have used PIV technique to visualize acoustic streaming and established a positive linear relationship between the ultrasonic intensity and the peak streaming velocity while Ref. [11] suggest that PIV is a good non-intrusive technique to estimate the peak-pressures in streaming applications. References [15, 16] have used synchronized PIV technique to measure the acoustic and streaming velocity fields simultaneously. The experimental values of the mean acoustic velocity and streaming velocities obtained from PIV are in good agreement with the theoretical values confirming the applicability of synchronized PIV technique.

The background for these preliminary acoustic experiments at low frequency (lower part of the acoustic range) is to investigate whether they can make any considerable influence to the rheology of drilling fluids and subsequently the particle settling within non-Newtonian drilling fluids. It is obvious that, scaling up of such acoustic processes has shown to be difficult due to the interactions between the sound field and the subsequently generated flow. For that, better analytical and numerical studies of this type of low frequency acoustic streaming are important and valuable. However, a comprehensive experimental study is necessary to get a better insight into the streaming velocity fields within Newtonian and non-Newtonian fluids to think and design further experiments confining to the aim. Therefore, this study considers the acoustic streaming in both Newtonian and non-Newtonian liquids at a very low frequency range and describes how the streaming velocity is affected in both qualitative and quantitative manner when the frequency, input power as well as the fluid rheology is changed.

2 MATERIALS AND METHODS

In this work acoustic streaming was studied with acoustic waves generated from a low frequency underwater acoustic transducer in an open liquid container. The experimental setup is shown in Fig. 1. This rectangular channel is made of glass with the dimensions of 0.97 m × 0.13 m × 0.28 m. De-ionized water and 1 g/L PAC were used as the Newtonian and non-Newtonian fluids, respectively. The white/off-white colour regular PAC granular powder (with a bulk density of 400–880 kg/m^3) provided by M–I Swaco, England was used to prepare the polymer solution. A volume of 20 L liquid was used for all the test runs. After performing the test runs with deionized water, 1 g/L PAC solution was introduced gradually to the same liquid channel in small quantities to identify the PAC concentration that does not show any streaming or any induced flow. A low frequency underwater acoustic transducer BII-7531 (D × H = 60 × 40 mm) provided by Benthowave Instrument Inc. Canada, with an output frequency ranging from 600 Hz to 15 kHz was utilized. Sonication was provided as continuous sinusoidal waves in a range of frequencies and amplitudes as mentioned in Table 1. The transducer was positioned at the centre of the liquid body 12 cm away from the left wall of the open channel as shown in Fig. 1. The two inside ends of the rectangular channel were covered using sponge type sound absorbing material to avoid any sound reflection from the end walls. A function generator model 33220A from Agilent

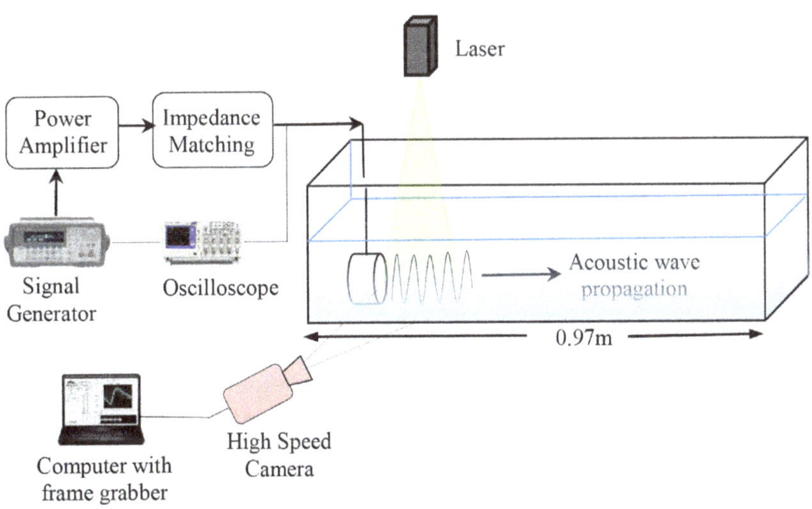

Figure 1: Sketch of the experimental set up.

Table 1: Test matrix used for the experiments.

Frequency (kHz)	Amplitude in terms of supply voltage (mV) from Signal Generator	Input voltage to the acoustic transducer after amplification (V_{RMS})	Input power to the acoustic transducer after amplification (W_{RMS})
Effect of frequency			
4	500	251	11.5
5	500	256	13.4
6	500	263	16.6
Effect of power input			
6	400	213	10.8
6	500	263	16.6
6	600	301	20.3
6	700	328	25.1
6	800	342	27.1

Technologies was used to generate the sinusoidal waves of different frequencies and amplitudes. The accuracy of the generated frequency and amplitude are 1 μHz and 0.1 mV, respectively. A linear power amplifier (BII-5061), which has a maximum power output of 200 W at +48 VDC amplified the signal from the function generator. The amplified signal was fed to an impedance matching unit (BII-6010) in order to match the high acoustic impedance of the transducer. Voltage and the current of the amplified signal just upstream of the transducer was read from an Oscilloscope (Tektronix TDS 2024C), where the input power to the transducer was also computed using the MATH function of it. The effect of input electrical power and frequency of an acoustic wave on the streaming velocity in a Newtonian fluid was studied in axial and radial directions. The effect of the fluid rheology on acoustic streaming was studied by gradually adding some small quantities of 1g/L PAC solution to the deionized water at a constant frequency of 6 kHz and amplitude of 700 mV. All experiments were conducted in room temperature at 21^{0}C. The rheological measurements were taken from Anton Paar MCR 302 viscometer at constant temperature and pressure.

2.1 The PIV system

The two-dimensional velocity fields inside the channel were measured using PIV technique. A Basler A800-510 μm colour camera (500 fps in full resolution 800 × 600) equipped with a 532 nm green laser (Photon DPGL-2200) was used to capture the images. All the images for the test series were captured at 100 fps with a resolution of 688 × 400 pixels and the measurements were made in a plane parallel to the channel length. The laser plane was illuminated vertically through the centre of the fluid filled tank. Polyamide seeding particles (PSP-50 from DANTEC Dynamics) with a mean diameter of 50 μm were added to the liquids as light scattering tracer ('seeding') particles for measuring the field velocity profile. Data acquisition was continued for sufficiently long time in order to allow the streaming flow to achieve its steady state profiles.

2.2 Image analysis

To calculate the streaming velocity field, acquired images were processed using PIVlab [17], which is a Digital Particle Image Velocimetry (DPIV) tool developed in MATLAB. The tool calculated the particle displacement for groups of particles by evaluating the cross-correlation of many small sub images (interrogation areas). The camera was set so that the image contained the whole region with the transducer and the streaming flow approximately up to 18 cm downstream the transducer. For image processing, a partial section that illustrates the streaming flow within the liquid medium was selected as the region of interest rather than the entire image captured. All the images were pre-processed using contrast limited adaptive histogram equalization (CLAHE) and some background noise was removed using high pass filter. For the PIV settings FFT window deformation was adopted with interrogation area with 3 passes. Gaussian 2 x 3 point was used as the sub-pixel estimator. Regarding post-treatment of correlated frames and for data validation, a vector velocity threshold with a high standard deviation value was set.

3 RESULTS AND DISCUSSION

The results of the experiments are presented in this section. The PIV technique was used for the quantification of streaming velocity and different features of the streaming velocity maps are discussed within this section. The effect of applied frequency, input voltages to the transducer and also rheology of the fluid medium are investigated with different visual models. All the velocity profiles presented in this section are regressed using Gaussian model with an accuracy of $R^2 > 90\%$.

3.1 Transient flow development

The transient flow development in pure deionized water due to the application of acoustic waves from the low frequency underwater transducer is shown in Fig. 2. Results shown here are from the test carried out at a frequency of 6 kHz and an input voltage of 700 mV to the signal generator, which provided an electrical input power of $25.1W_{RMS}$. Figure 2 shows a typical map of an acoustically induced flow with its temporal propagation in front of the transducer face for 8 seconds. The flow originates from the transducer as a result of sonication. The streaming is highest along the beam centreline. A maximum flow velocity was reached at a certain distance from the transducer, and beyond this distance the streaming was gradually diffused. This velocity 'smearing' is attributed to the attenuation effects within the liquid medium. Strong streaming was formed in the direction of the beam-axis immediately after the transducer was switched on (at t = 0 s) and the spatial distribution of the velocity vectors is complex with several local peaks and vortices. The temporal movement of the acoustic wave front is similar to a streaming jet confirming the establishment of Ref. [2], where the acoustic streaming takes the form of an inertially dominated turbulent jet at powers above 4×10^{-4} W. The jet-like flow movement sets up large-scale circulations within the liquid body and eventually interact back and modify the streaming jet. The shape of the visual model of the induced flow is much distorted and vigorous as it propagates and that drives away the particles from the transducer along the channel.

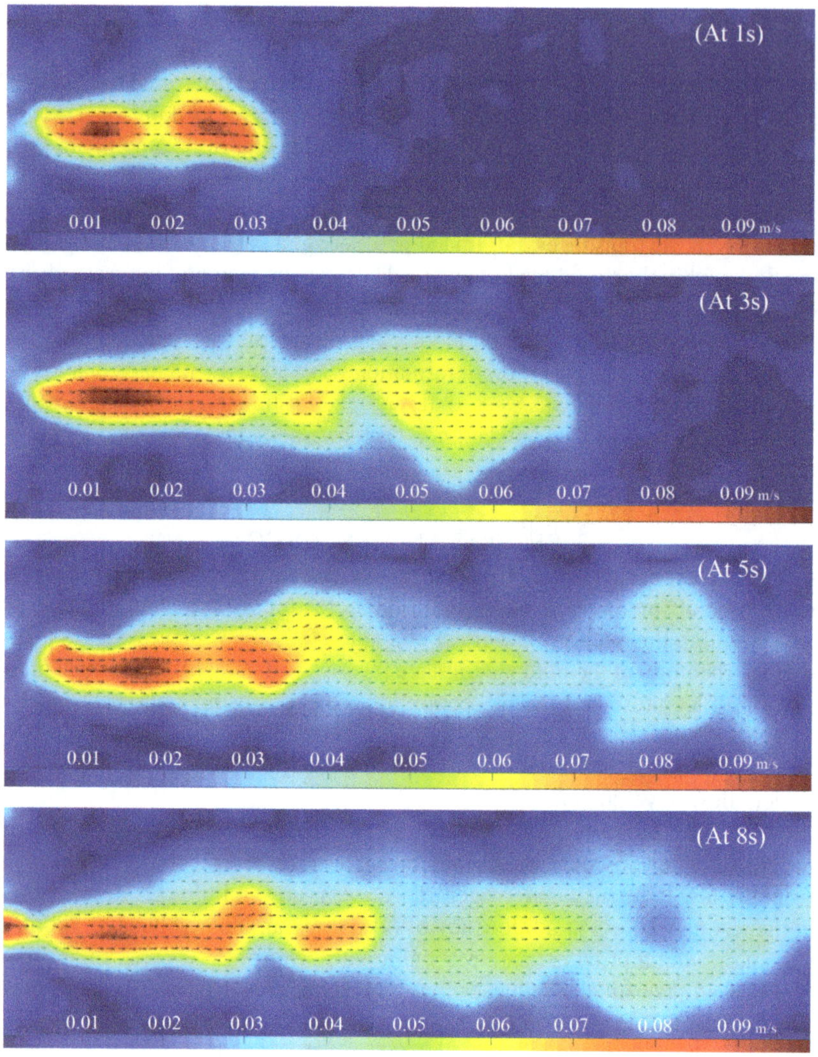

Figure 2: Transient flow development in deionized water for 8s.

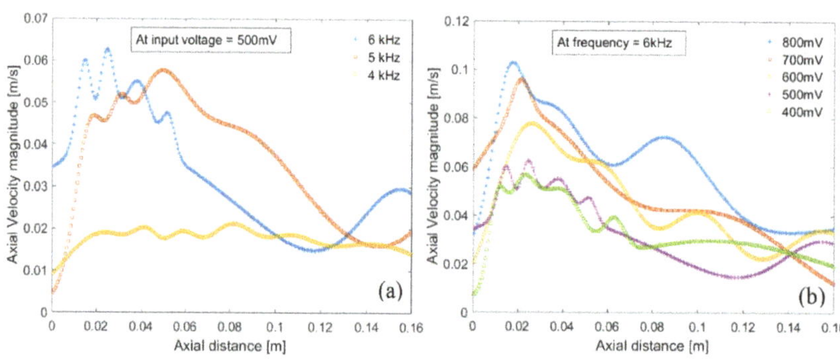

Figure 3: Axial velocity distribution in pure water at 30s. (a) At constant input voltage; (b) At constant frequency. (*Voltage mentioned here is the input from the signal generator)

3.2 Velocity magnitude in axial direction at steady state

Figure 3 shows how the streaming velocity varies within pure water along a horizontal axis which contains the peak streaming velocity. The acoustic transducer is placed at the left side and the results shown here are taken when the flow is assumed to have achieved its steady state at 30 s. Figure 3(a) shows the variation of axial streaming velocity for different frequencies when the input power to the transducer is kept constant at 500 mV and Fig. 3(b) shows the variation of axial streaming velocity when the input power to the transducer varies at a frequency of 6 kHz. As shown in the transient flow development in Fig. 2, the streaming velocity is dying out in the fluid medium when it is moving away from the transducer. It is observed that the peak velocity is obtained only one time along the beam axis closer to the acoustic source and some other minor peaks that are not so strong can also present. Therefore, it can be said that the acoustic streaming is significant in the proximity of the transducer and decreases as the flow moves away from the transducer. This is due do the attenuation of the acoustic wave and this attenuation increases with the distance from the transducer. The shape of the curves is far from being sinusoidal and is actually much distorted. There can be seen a strong asymmetry of the axial velocity, which drives away the particles from the transducer which can be explained due to the non-linear phenomenon and probably the presence of superior harmonics, distorting the sinus. This asymmetric pattern of axial streaming velocity distribution is same for all the frequencies and input power values investigated. Even so, according to Fig. 3(a), it can be seen that the peak streaming velocity increases when the applied frequency of the transducer is increased. That is same for the axial velocity curves shown in Fig. 3(b), where the peak streaming velocity increases when the input electrical power to the acoustic source is increased. The maximum peak axial velocity detected was 9.94 cm/s at a frequency of 6 kHz and at an input voltage of 800 mV to the signal generator, which resulted in a input electrical power of $27.1W_{RMS}$.

3.3 Velocity magnitude in radial direction at steady state

Figure 4 shows how the radial streaming velocity varies within pure water along a line normal to the beam axis where the peak velocity appeared. The results shown here are taken when the flow is assumed to have achieved its steady state at 30s. The velocity profile in radial

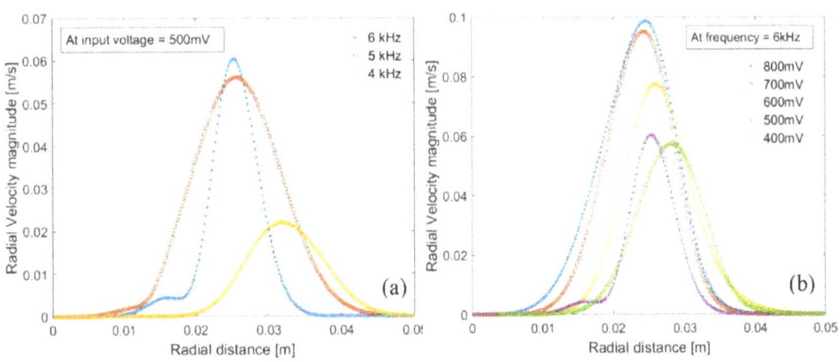

Figure 4: Radial velocity distribution in pure water at 30s. (a) At constant input voltage; (b) At constant frequency. (*Voltage mentioned here is the input from the signal generator)

direction, perpendicular to the beam-axis, is relatively symmetrical and has its maximum at the centre. According to Fig. 4(a), it can be seen that the radial velocity of the acoustically induced flow within the water channel increases as the applied frequency of the acoustic source is increased. The bell-shape of the three curves in Figure 4(a) illustrates the effect of frequency on the velocity distribution in such a way that higher frequencies provide more intensified push up at the centreline of the beam. When the frequency is lowered, the velocity tends to distribute to either sides of the beam axis since they are not having much intensified effect as it was at higher frequencies. Variation of the radial streaming velocity with the input electrical power to the transducer is shown in Fig. 4(b) and, this clearly shows how the tip of the bell-shaped velocity profile moves towards the right direction. This also confirms the axial velocity distributions presented in Fig. 3 in a way that when the intensity of the input signal is reduced, the forcing effect of the bulk liquid flow is also reduced. As a result of this, the peak velocity may not occur exactly at the centre line or the beam axis, but at places little lower than the beam axis.

3.4 Peak velocity variation

Variation of the peak streaming velocity (V_p) with input frequency and power is shown in Fig. 5. This is the same peak velocity displayed in Fig. 3 after the flow is assumed to have obtained its steady state and the two velocity components; u-velocity (axial component) and v-velocity (radial component) are also incorporated to describe the dominant direction of the flow. The peak streaming velocity (V_p) is composed of the two velocity components u and v as described in eqn (1).

$$V_p = \sqrt{u^2 + v^2} \tag{1}$$

It is seen that V_p increases with input frequency and also with the input power to the transducer as it was already discussed in the previous section and already stated by Refs. [4, 18]. The most important fact here is, the radial component of the velocity (v-velocity) is far less than the axial component (u-velocity). This depicts a very firm idea and confirms the visual observations that the flow is mainly driven towards the horizontal direction due to the momentum transfer from the acoustic wave. According to Fig. 5(a), the v-velocity component has been increased with the frequency which can be described as the flow is developing towards

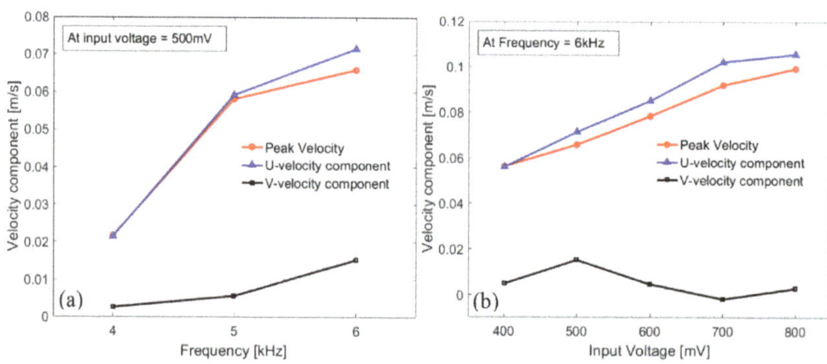

Figure 5: Variation of peak streaming velocity in pure water at 30s. (a) At constant input voltage; (b) At constant frequency. (*Voltage mentioned here is the input from the signal generator)

more vortex type patterns. However, according to Fig. 5(b), the v-velocity component is not much responsive to the input power of the acoustic source. This assures that the flow is more or less intensified with the power input and focusing towards the bulk axial movement away from the transducer. It is important to mention that, the irradiated acoustical energy can differ from absorbed energy in several times and this difference depends on many factors. Here also, in this study, authors are talking about the input electrical power to the acoustic transducer and not the power of sonication process within the liquid medium. To determine the efficiency of acoustic field, it is necessary to know acoustical energy introduced and absorbed in the liquid volume.

3.5 Velocity vectors and streamlines

The length of the vector components (shown in Fig. 6) quantifies the acoustic streaming velocity in pure water for the frequency of 6 kHz and input voltage of 700 mV to the signal generator. Here also, the results are taken when the flow is assumed to have achieved its steady state at 30 s. Figure 6 shows only one particular case of the experiments and vector plots for other cases are not presented here for brevity. When analysing these vector plots, it is observed in all of the cases that maximum flow velocity is observed in the region closer to the acoustic source. At low frequency and low input power, most of the velocity changes occur near the transducer without much distortions on the other side of the liquid channel. That means, the main mixing is observed in the regions near the transducer. However, for the higher frequencies, almost, all of the fluid achieved high velocity and the violent mixing was

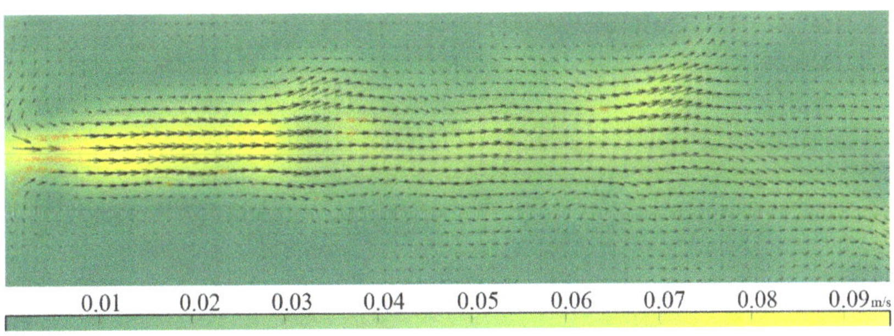

Figure 6: Velocity vector plot in pure water for 6 kHz & 700 mV (steady state at 30s)

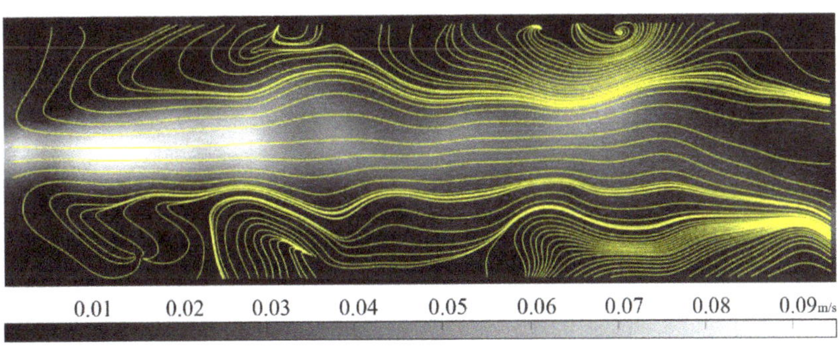

Figure 7: Streamline pattern in pure water for 6 kHz & 700 mV (steady state at 30s)

generated within the channel up to some considerable distance. Velocity vectors give a great insight about the mixing phenomena within the liquid.

Figure 7 shows the stream line pattern for the same case explained above in Fig. 6. The spatial distribution of the velocity vector field is confirmed by the stream line plot which illustrates the complexity of the motion of fluid with several local peaks and vortices. According to Fig. 7, it can be seen that for higher power inputs, the flow reversal takes place near the transducer and flow is in the inward direction. It indicates that there is an entrainment of liquid towards the centre of the transducer due to the high velocity near the source and high-pressure gradient. This phenomenon is evident from the vector plot of the velocity magnitude shown in Fig. 6 also. Some distinct vortices are visible in either sides of the beam axis forming the motion more complex and confirming the statement of [16] which says when the amplitude of the acoustic source increases, circulatory streaming develops and is then distorted to a complicated and irregular structure.

3.6 Effect of fluid rheology on acoustic streaming.

As described in the methodology, polyanionic cellulose (PAC) was used to investigate how the fluid rheology influences the acoustic streaming. That is one of the main objectives of this preliminary study to use acoustic streaming effect on non-Newtonian liquids where particle settling needs to be altered. Known quantities of 1g/L PAC were added gradually in small portions. The streaming flow was the checked for each case as done for the pure water. Figure 8(a) shows the influence of PAC concentration on axial streaming velocity only for some selected cases for brevity. Results shown here are from the test carried out at a frequency of 6 kHz and an input voltage of 700 mV to the signal generator which provided an electrical input power of 25.1 W_{RMS}. And also, the results are taken when the flow is assumed to have achieved its steady state at 30 s.

It can be seen from Fig. 8(a) that the axial streaming velocity is reduced with the added PAC concentration. When the overall PAC concentration within the liquid channel reached 0.19 g/L, the streaming flow was completely disappeared. Similar to the axial streaming velocity profiles seen for pure water in Fig. 3, here also the streaming velocity is dying out in the non-Newtonian fluid medium further away from the transducer. However, it is important to notice that the flow induced by the acoustic waves within the non-Newtonian medium is smearing out quickly at a lower axial distance than that we observed for pure water. This

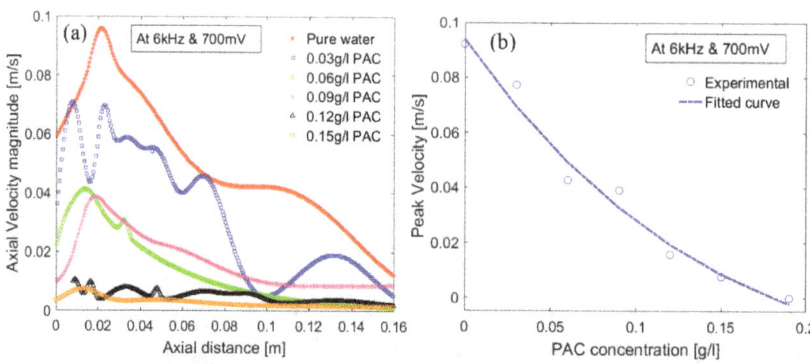

Figure 8: (a) Axial velocity distribution in PAC; (b) Peak velocity variation with PAC concentration. (*Voltage mentioned here is the input from the signal generator)

phenomena is attributed to the strong attenuation property of non-Newtonian liquid. It is observed that the peak velocity is obtained only one time along the beam axis closer to the acoustic source before it smears out. Figure 8(b) illustrates how the peak velocity magnitude is reducing with the added PAC concentration. The peak velocity started with a value of 9.21 cm/s for pure water until it reaches zero when the PAC concentration became 0.19 g/L. The experimental peak velocity magnitudes at different concentrations were well fitted by an exponential curve shown in eqn (2) with an accuracy of $R^2 = 97.76\%$.

$$V_p = a * \exp\left(b * C_{PAC}\right) + c * \exp\left(d * C_{PAC}\right) \tag{2}$$

Where, C_{PAC} is the concentration of PAC in g/L and a, b, c and d are constants with the values of 6.992, -4.107, -6.898 and -4.032 respectively. The rheological properties of the resulted PAC solution with the concentration of 0.19 g/L were measured after the experiments and the parameters of that solution according to the power law are n = 0.97 and $K=3.3 \times 10^{-3}$ Pa.s^n, where n is the behaviour index and K is the consistency index of the fluid.

4 CONCLUSION

The set of experiments were conducted to visualize the acoustic streaming produced by a low frequency underwater acoustic transducer in both Newtonian and non-Newtonian fluids. PIV technique was used for quantification of the streaming velocity. The features of the streaming velocity maps, vector plot and streamline pattern were presented for some selected cases. It was observed that, the streaming is highest along the beam centreline and the maximum peak velocity along the transducer beam is in the axial position. Acoustic streaming is significant in the proximity of the transducer and decreases as the flow moves away from the transducer due to the attenuation of the acoustic wave and this attenuation increases with the distance from the transducer. The temporal movement of the acoustic wave front is similar to a streaming jet. Streaming depends on the applied frequency and the input voltage (power). Streaming velocity vectors give a great insight about the mixing phenomena within the liquid and the vortices visible in streamline plots confirm the complexity and the non-linearity of the fluid flow induced by the acoustic pressure field. The peak streaming velocity (V_p) measured was in the range of 2.1–9.9cm/s for pure water and that depends on the supply frequency and the input voltage settings to the transducer. The streaming velocities observed for non-Newtonian fluid is smaller than that for the Newtonian fluid confirming the fact that the wave attenuation is much higher in non-Newtonian fluids.

ACKNOWLEDGEMENTS

The authors would like to acknowledge the valuable support provided by H.A. Rabenjafimanantsoa, Milad Khatibi from Department of Petroleum Engineering & Jon fidjeland, Romuald K. Bernacki from department of computer and electrical engineering in university of stavanger. Authors gratefully acknowledge the Norwegian Research Council for funding this study under the project "NFR Improved Model Support in Drilling Automation".

REFERENCES

[1] Kumar, A., Kumaresan, T., Pandit, A.B. & Joshi, J.B., Characterization of flow phenomena induced by ultrasonic horn. *Chemical Engineering Science,* **61**(22), pp. 7410–7420, 2006.
https://doi.org/10.1016/j.ces.2006.08.038

[2] Lighthill, S.J., Acoustic streaming. *Journal of Sound and Vibration,* **61**(3), pp. 391–418, 1978.
https://doi.org/10.1016/0022-460x(78)90388-7

[3] Nyborg, W.L., Acoustic streaming due to attenuated plane waves. *The Journal of the Acoustical Society of America,* **25**(1), pp. 68–75, 1953.
https://doi.org/10.1121/1.1907010

[4] Frenkel, V., Gurka, R., Liberzon, A., Shavit, U. & Kimmel, E., Preliminary investigations of ultrasound induced acoustic streaming using particle image velocimetry. *Ultrasonics,* **39**(3), pp. 153–156, 2001.
https://doi.org/10.1016/s0041-624x(00)00064-0

[5] Ellier, E.S., Bailly, Y., Girardot, L., Ramel, D. & Nika, P., *Acoustic streaming measurements by PIV*, presented at the 15th International Symposium on Flow Visualization, Minsk, Belarus, June 25–28, 2012.

[6] Trujillo, F.J. & ,Knoerzer, K., *CFD modelling of the acoustic streaming induced by an ultrasonic horn reactor*, Seventh international conference on CFD in the minerals and process industries, CSIRO, Melbourne, Australia, 2009.

[7] Sankaranarayanan, S.K., Singh, R. & Bhethanabotla, V.R., Influence of non-newtonian fluid dynamics on SAW induced acoustic streaming in view of biological applications, in *Sensors, 2011 IEEE*, pp. 1546–1549, 2011.

[8] Aayani, R., Shahidian, A. & Ghassemi, M., Numerical investigation of non-newtonian blood effect on acoustic streaming. *Journal of Applied Fluid Mechanics,* **9**, pp. 173–176, 2016.

[9] Groznova, A., *Modelling of ultrasound assisted mixing in Newtonian and non-Newtonian liquids*, 2014.

[10] Rabenjafimanantsoa, H.A., Wrobel, B.M. & Time R.W., PIV Visualization of acoustic streaming in Non-Newtonian fluid. *Annuals Transactions of the Nordic Rheology Society,* **17**, 2009.

[11] Choi, M., Doh, D.H., Cho, C.H., Kang, K.S., Paeng, D.G., Ko, N.H., Kim, K.S., Rim, G.H. & Coleman, A.J., Visualization of acoustic streaming produced by lithotripsy field using a PIV method. *Journal of Physics: Conference Series*, **1**(1), pp. 217–223, 2004.
https://doi.org/10.1088/1742-6596/1/1/048

[12] Abolhasani, M., Rahimi, M., Dehbani, M. & Shabanian, S.R., *CFD modeling of low, medium and high frequency ultrasound waves propagation inside a liquid medium*. Presented at the 4rd National Conference on CFD Applications in Chemical & Petroleum Industries, Ahwaz, Iran, 2012.

[13] Thompson, M.W. & Atchley, A.A., Simultaneous measurement of acoustic and streaming velocities in a standing wave using laser Doppler anemometry. *The Journal of the Acoustical Society of America,* **117**(4), pp. 1828–1838, 2005.
https://doi.org/10.1121/1.1861233

[14] Adrian, R.J., Particle-imaging techniques for experimental fluid mechanics. *Annual Review of Fluid Mechanics,* **23**(1), pp. 261–304, 1991.
https://doi.org/10.1146/annurev.fluid.23.1.261

[15] Nabavi, M., Siddiqui, M.H.K. & Dargahi, J., Simultaneous measurement of acoustic and streaming velocities using synchronized PIV technique. *Measurement Science and Technology,* **18**(7), pp. 1811–1817, 2007.
https://doi.org/10.1088/0957-0233/18/7/003

[16] Nabavi, M., Siddiqui, M.H.K. & Dargahi, J., Experimental investigation of the formation of acoustic streaming in a rectangular enclosure using a synchronized PIV technique. *Measurement Science and Technology,* **19**(6), p. 065405, 2008. https://doi.org/10.1088/0957-0233/19/6/065405
[17] Thielicke, W. & Stamhuis, E., PIVlab–Time-Resolved Digital Particle Image Velocimetry Tool for MATLAB, version: 1.32. *Journal of Open Research Software,* **10**, p. m9, 2014.
[18] Chouvellon, M., Largillier, A., Fournel, T., Boldo, P. & Gonthier, Y., Velocity study in an ultrasonic reactor. *Ultrasonics Sonochemistry,* **7**(4), pp. 207–211, 2000. https://doi.org/10.1016/s1350-4177(00)00060-2

VEHICLE OCCUPANT RESTRAINT SYSTEM DESIGN UNDER UNCERTAINTY BY USING MULTI-OBJECTIVE ROBUST DESIGN OPTIMIZATION

HIROSUKE HORII
Department of Mechatronics, University of Yamanashi, Japan.

ABSTRACT

This research reports a vehicle occupant restraint system design that takes account of uncertainties of crash conditions and situations by using a multi-objective robust design optimization method called MORDO. The vehicle occupant restraint system is composed of restraint equipment, such as an airbag, a seatbelt and a knee bolster. The optimization aims to improve the safety performance of the system and its robustness simultaneously. The safety of the system is evaluated by some indexes based on some safety regulations, which are calculated by response surface model of an occupant at a crash. In addition, its robustness is evaluated by the mean value and the standard deviation of objective functions, which are calculated by using Monte Carlo simulation based on a certain probabilistic distribution in space of design variables around each design candidate. Some helpful information for designing the restraint systems, such as trade-off information of safety performance and its robustness, are provided by visualizing and analysing the Pareto optimal solutions.

Keywords: evolutionary algorithm, machine learning, multi-objective optimization, occupant safety, robust optimization

1 INTRODUCTION

Performance of a vehicle occupant restraint system is affected by interaction among occupant restraint equipment, such as an airbag and a seat belt. Furthermore, situations of crashes, such as speed at the crash and posture of occupants, are various. Therefore, performance of the system requires robustness under the interactions and the uncertainties of the design, the condition and the situation. In automotive engineering field, summation of the mean value and the standard deviation is used as the objective function at the robust design optimization of the performance and the stability so far (e.g. Fu and Abramoski [1]).

On the other hand, in the aerospace and aeronautics field, some multi-objective robust design approaches have been proposed, such as the multi-objective robust design optimization, MORDO by Padovan *et al.* [2] and Parussini *et al.* [3], and the design for multi-objective six sigma, DFMOSS by Shimoyama *et al.* [4] The features of those methods are (1) the performance and the stability under uncertainty are set as independent objective functions, and then (2) trade-off information of the performance and the stability are obtained in one time optimization.

The author has been working on improving the vehicle occupant restraint system design so far. Firstly, evaluation time of the safety performance of the design was reduced by using response surface models of the injury criteria by Horii [5]. The response surface models were constructed by using some machine learning methods whose training data set was CAE sampling results of an occupant behaviour simulation. The simulation model was constructed by using a multi-body dynamics simulation. Then significant reduction of the evaluation time in multi-objective optimization by using evolutionary algorithm was derived by applying the response surface models by Horii [6].

In this research, with incorporating the above outcomes, the MORDO is employed for the vehicle occupant restraint system design. This design system evaluates some injury criteria

DOI: 10.2495/CMEM-V6-N4-827-834

and their stabilities under uncertain conditions, and then trade-off information of the performances and the stabilities are provided to users. Subject of the design is an occupant restraint system at a frontal crash of a vehicle.

2 MULTI-OBJECTIVE ROBUST DESIGN OPTIMZATION, MORDO

2.1 Robustness in engineering design

In engineering design, robustness against perturbation is necessary. The perturbation includes tolerance in manufacturing, uncertainty of operating condition, error at operation and so on. On the other hand, usual optimization seeks the best point of an objective function. A conceptual diagram of the above idea is shown in Fig. 1. The point A is the optimal point and the point B is the suboptimal point, but the response of the point A is steeper than that of the point B. If the design has fine precision, the point A may be the best solution. But if the design includes a perturbation, the performance of the point A may be unstable and it may not be suitable. So the engineering design requires taking account of the perturbation and the precision of the design variables, and their effect to the objective function.

2.2 Robustness evaluation with MORDO

In MORDO, each design variable is set any probabilistic distribution of perturbation. Robustness of a design candidate is evaluated by mean value and standard deviation of objective functions, which are calculated by using Monte Carlo simulation based on the above probabilistic distribution in space of design variables around each candidate. A conceptual diagram of the above idea is shown in Fig. 2. Each design variable x_i is set a certain normal distribution. Then the mean

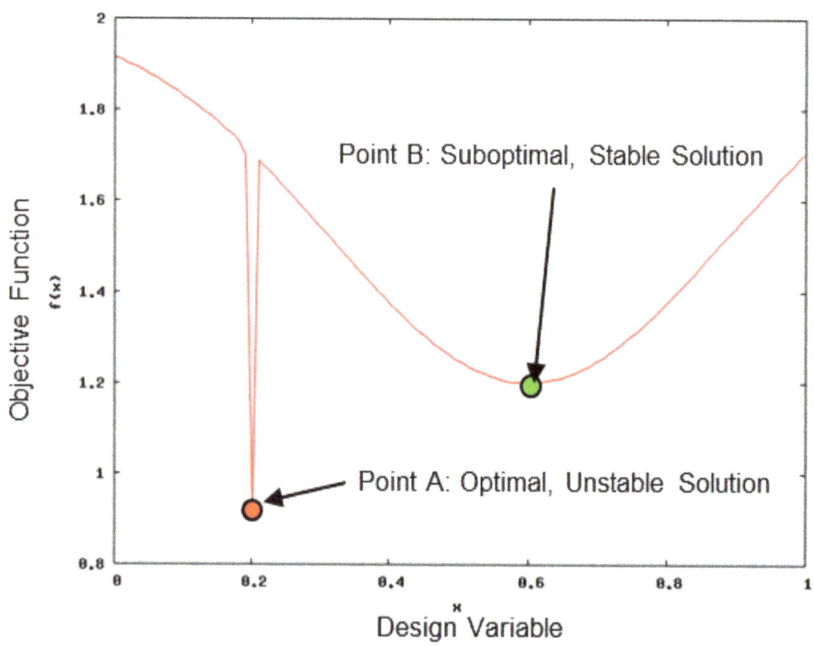

Figure 1: Conceptual diagram of stability of design (at minimization).

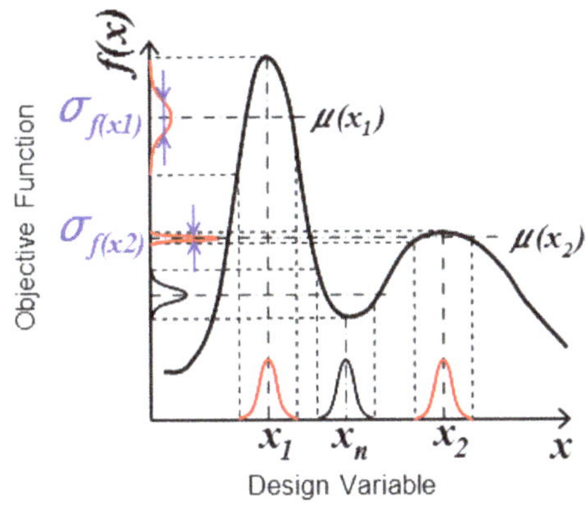

Figure 2: Conceptual diagram of robustness evaluation at MORDO.

value $\mu(x_i)$ and the standard deviation $\sigma_{f(x_i)}$ of the objective function are calculated. Those mean value and standard deviation are also set as another objective function in MORDO. Decision maker of a design is able to understand the balance of performance and robustness of many design candidates all at once after the optimization.

3 OCCUPANT RESTRAINT SYSTEM DESIGN BY USING MORDO

3.1 Response surface model of vehicle occupant restraint system

In this research, since huge number of function evaluation is required for optimization and robustness evaluation, it is impossible to employ direct simulation. Firstly, response surface models are constructed by using limited samplings of direct simulation. Then, the response surface models are employed for optimization and robustness evaluation.

Injury criteria of an occupant at a frontal crash are estimated by using a machine learning method, Gaussian Process. The Gaussian process is a Bayesian approach, based upon the expression of knowledge in terms of probabilistic distribution (Rasmussen and Williams [7]). This method is a powerful regression model specified by parameterized mean and covariance functions and suitable for estimating non-polynomial responses.

An occupant's behaviour model of a full-frontal crash testing shown in Fig. 3 is constructed by using a multi-body dynamics tool, MADYMO. The model is composed of a

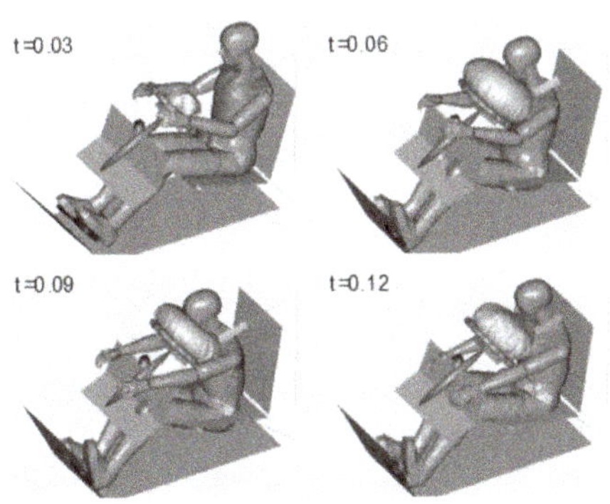

Figure 3: Occupant behaviour model at frontal crash.

Table 1: Definition of design variables.

	Variable	Range and name
Airbag:	Time to fire (sec.)	0.015≦AB_TTF≦0.035
	Mass flow rate	0.5≦AB_MFR≦2.0
	Vent hole factor	0.5≦AB_VHF≦2.0
Seatbelt:	Time to fire (sec.)	0.01≦Belt_Preten_TTF≦0.03
	Load limit (N)	2000≦Belt_Preten_LL≦6000
Knee bolster:	Stiffness factor	0.5≦SF_KB≦2.0

Hybrid-III dummy, surrounding equipment such as a seat and a steering wheel, and restraint equipment such as an airbag, a seatbelt and a knee bolster. The model simulates the occupant's behaviour at the crash for 0.12 sec.

Input variables for controlling the behaviour of the model, consists of six design variables regarding the restraint equipment, which strongly affect safety indexes, an airbag, a seatbelt and a knee bolster. Definition of the design variables are shown in Table 1.

Output variables are three safety indexes based on the Japan NCAP, head injury criterion, thoracic resultant acceleration in 3 msec. and femur load. The head injury criterion, HIC is an index of head injury risk. The thoracic resultant acceleration in 3 msec., T3MS is measured by an accelerometer mounted on centre of mass of a crash dummy's chest. The femur load, FL is measured by load cells mounted on the dummy's left and right femurs. FL_L is the left femur's value, and FL_R is the right femur's value.

After good agreement of actual simulation and response surface models were confirmed by Horii [5], the response surface models, which are constructed by 500 of the number of sampling of the training data set, are employed for following optimization.

3.2 Implementation of MORDO for vehicle occupant restraint system design

Here, the above four response surface models of HIC, T3MS, FL_L and FL_R are employed for optimization of the vehicle occupant restraint system. The optimization problem is defined as four-objective minimization problem. The objective functions are the mean value and the standard deviation of the HIC and the T3MS. The FL_L and the FL_R are set as constraints. Summary of the objective functions and the constraints are shown in Table 2.

Table 2: Summary of the objective functions and constraints.

	Name
Minimize:	Mean of HIC, $\mu(HIC)$
	Mean of T3MS, $\mu(T3MS)$
	Std. Dev. of HIC, $\sigma(HIC)$
	Std. Dev. of T3MS, $\sigma(T3MS)$
Subject to:	$\mu(HIC)+3\sigma(HIC)\leq 1000$
(Regal regulation)	$\mu(T3MS)+3\sigma(T3MS)\leq 588$
	FL_L ≤ 1000
	FL_R ≤ 1000

An evolutionary multi-objective optimization algorithm, the Adaptive Range Multi-Objective Genetic Algorithm, ARMOGA by Sasaki and Obayashi [8] is employed as an optimizer. The flowchart of the ARMOGA is shown in Fig. 4. In order to taking account of perturbation of detecting time of a crash, AB_TTF and Belt_Preten_TTF are selected as robust parameter. The AB_TTF and the Belt_Preten_TTF are the parameters that define the operation timing of the airbag and the seatbelt pretensioner. Normal distribution is selected as the probabilistic distribution for Monte Carlo simulation. At each evaluation point, 100 samplings are generated under the above distribution for assessing the robustness. Summary of the MORDO setting is shown in Table 3.

3.3 Optimization result

All design candidates, which were explored and evaluated by the MORDO, are shown in Fig. 5. Pareto optimal solutions, which were extracted from the above all design candidates,

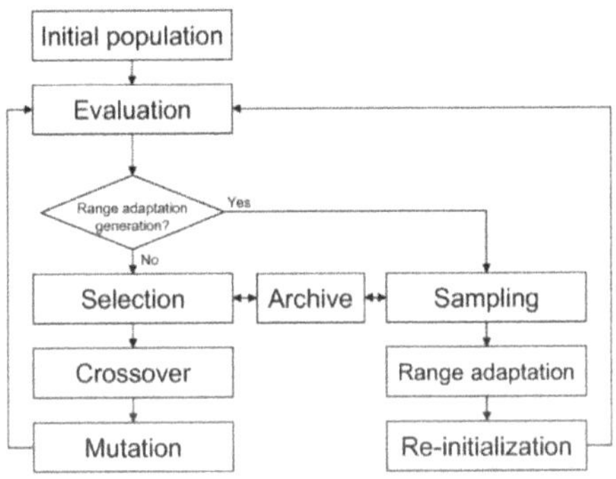

Figure 4: Flowchart of ARMOGA.

Table 3: Summary of the MORDO setting.

Optimizer setting	
Optimizer	ARMOGA
Population size	40
Generation	50
Crossover method	BLX-0.5
Crossover rate	1
Mutation rate	0.1
Range adaptive operation:	
Starting generation	20
AR interval	5
Robust setting	
Robust Parameter	AB_TTF Belt_Preten_TTF
Probabilistic distribution	Normal distribution
Std. Dev. of perturbation	$\sigma = 5.0\times10^{-4} sec.$
Number of sampling	100

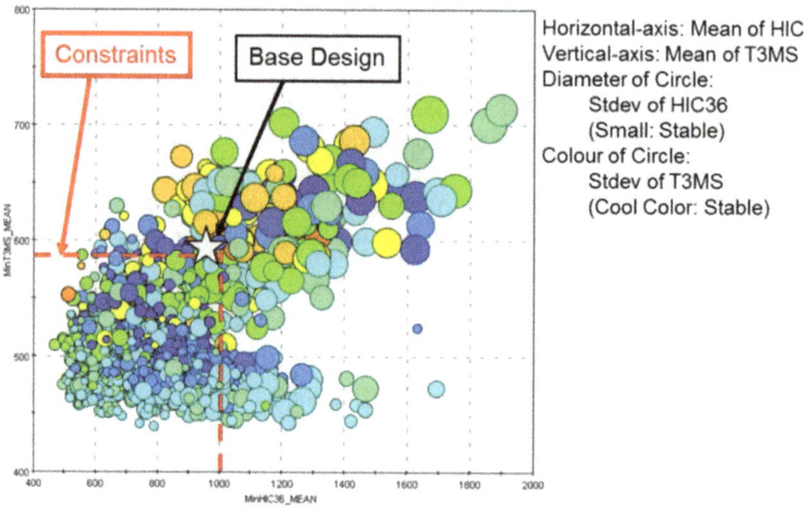

Figure 5: All design candidates that were explored and evaluated by the MORDO.

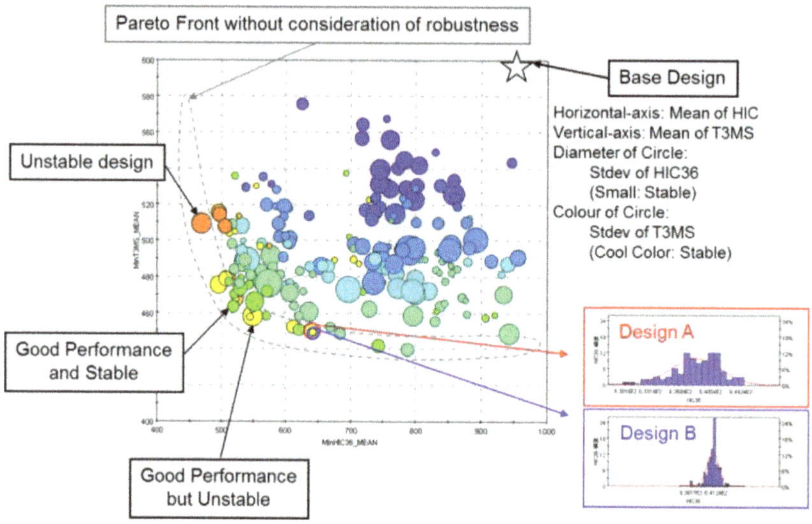

Figure 6: Pareto optimal solutions that were extracted from all design candidates.

are shown in Fig. 6. The horizontal axis represents the mean of the HIC. The vertical axis represents the mean of the T3MS. The diameter of each circle represents the standard deviation of the HIC. The small circle means stable and large circle means unstable. The colour of each circle indicates the standard deviation of the T3MS. Cool colour means stable and warm colour means unstable. These charts help the decision maker to understand the trade-off balance of injury risks between head and thoraces, and stability of each design candidate simultaneously.

As an example of showing an effectiveness of the MORDO, here we focus on two similar design candidates, the design A and the design B in Fig. 6. The frequency distribution charts

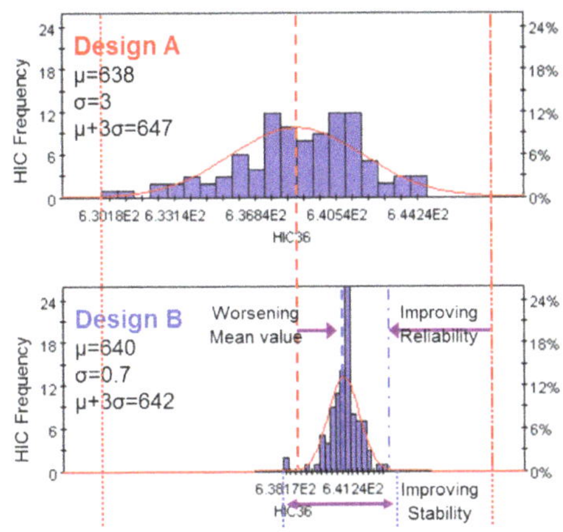

Figure 7: Frequency distribution of Monte Carlo sampling of HIC.

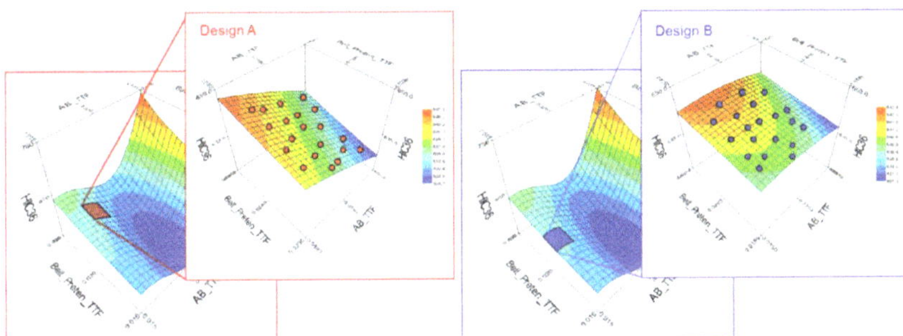

Figure 8: Response surface of HIC

of Monte Carlo sampling of the HIC of the two design candidates are shown in Fig. 7. The response surface of the HIC for two robust parameters, AB_TTF and Belt_Preten_TTF are shown in Fig. 8.

Figure 7 shows that though the mean value of the HIC of the design A is slight superior to that of the design B, the standard deviation of the HIC of the design A is larger than that of the design B. Therefore, the design A is unstable and the statistical worst case, that is, $\mu(HIC)+3\sigma(HIC)$ is inferior to the design B. Figure 8 also shows that the design A is on the slope of the response surface and it seems unstable. On the other hand, the design B is on a relatively flat surface and it seems more stable. Accordingly, if a slight worsening of the mean value is accepted, a robust design candidate can be obtained.

In this way, the MORDO provides a design environment, which is able to decide a final design with taking account of robustness at a vehicle occupant restraint system design.

4 CONCLUDING SUMMARY

In this research, a multi-objective robust design optimization method, the MORDO was applied for a vehicle occupant restraint system design. The MORDO provided a design environment, which was able to decide a final design with taking account of robustness. The effectiveness of MORDO was shown by visualizing and analysing with some results, such as scatter plot, frequency distribution of Monte Carlo sampling designs and shape of response surface of an objective function.

ACKNOWLEDGEMENTS

This work was supported by JSPS KAKENHI Grant Number 26330274. The calculate data of MADYMO was provided by TASS International K.K. The development environment, modeFRONTIER was provided by IDAJ Co. Ltd. and ESTECO S.p.A.

REFERENCES

[1] Fu, Y. & Abramoski, E., *Robust design for occupant restraint system*. Proceedings of Reliability and robust design in automotive engineering 2005 (SAE Technical Paper 2005-01-0814), 2005.
https://doi.org/10.4271/2005-01-0814

[2] Padvan, L., Pediroda, V. & Poloni, C., *Multi objective robust design optimization of airfoils in transonic field*. Proceedings of International Congress on Evolutionary Methods for Design, Optimization and Control with Applications to Industrial Problems, EUROGEN2003, 2003.

[3] Parussini, L., Pedirda, V. & Obayashi, S., Design under uncertainties of wings in transonic field. *JSME International Journal, Series B*, **48**(2), pp. 218–223, 2005.
https://doi.org/10.1299/jsmeb.48.218

[4] Simoyama, K., Oyama, A. & Fujii, K., A new efficient and useful robust optimization approach -- design for multi-objective six sigma, *Proceedings of the 2005 IEEE Congress on Evolutionary Computation,* **1**, pp. 950–957, 2005.
https://doi.org/10.1109/cec.2005.1554785

[5] Horii, H., Estimate modelling for assessing the safety performance of occupant restraint systems. *WIT Transactions on the Built Environment*, **134**, pp. 627–635, 2014.
http://dx.doi.org/10.2495/SAFE130561

[6] Horii, H., Multi-objective optimization of vehicle occupant restraint system by using evolutionary algorithm with response surface model. *International Journal of Computational Methods and Experimental Measurements*, **5**(2), pp. 163–173, 2017.
http://dx.doi.org/ 10.2495/CMEM-V5-N2-163-170

[7] Rasmussen, C.E. & Williams, C.K.I, *Gaussian Processes for Machine Learning*, MIT Press, Cambridge, 2006.

[8] Sasaki, D. & Obayashi, S., Efficient search for trade-offs by adaptive range multi-objective genetic algorithms. *AIAA Journal of Aerospace Computing, Information and Communication*, **2**, pp. 44–64, 2005.
https://doi.org/10.2514/1.12909

IMAGE ANALYSIS APPLICATIONS FOR THE STUDY OF SEGREGATION IN LIGHTWEIGHT CONCRETES

AFONSO M. SOLAK, ANTONIO J. TENZA-ABRIL & FRANCISCO BAEZA-BROTONS
Department of Civil Engineering, University of Alicante, Spain.

ABSTRACT

The use of lightweight concrete allows great flexibility and cost savings when it is used in building construction having a positive impact on the energy consumption of buildings due to its good thermal characteristics. However, it is also known that the differences between the densities of the materials used to produce these concretes make it highly susceptible to the segregation phenomenon. The main objective of the present work is to present a method to quantify this phenomenon using techniques of image analysis. In this work, a lightweight concrete produced was molded in cylindrical molds using different times of internal vibration and causing different degrees of segregation. The samples were cured, vertically saw-cut in two pieces (halves) and the sections were photographed. Subsequently, the halves were saw-cut horizontally in four equal parts and posteriorly their densities were determined experimentally. The densities obtained were used to calculate the segregation index of each sample (experimental method). Furthermore, the photographed sections were processed using image analysis software in order to determine the volumetric proportions of aggregates in each sample (noise reduction, threshold adjustment, binarization and fill holes). The processed images were used to calculate the densities and segregation index of the lightweight concrete produced through image analysis. In addition, using the photographed sections, a vertical density profile was programmed to analyze the distribution of the lightweight concrete components (mortar and aggregate). Finally, the results obtained experimentally and through image analysis were compared. This study demonstrates that the image analysis allows a deeper knowledge of the behavior of segregated concrete

Keywords: density, image analysis, lightweight concrete, segregation index, segregation

1 INTRODUCTION

The use of lightweight concrete allows great flexibility in design and cost savings due to the decrease in dead loads, lowering of foundation costs, etc [1–3]. In addition, the reduction of the density of the concrete produces an increase in the thermal resistance, improving the energy efficiency of the buildings constructed with this type of material.

Lightweight concretes usually used for structural applications are concretes with lightweight aggregates (LWAC), in general obtained by the total or partial substitution of conventional aggregates by lightweight aggregates and characterized by having densities lower than 2000 kg/m^3 [4].

The vibration is an industrial practice to compact fresh concrete into formwork and around reinforcement. During the vibration process, the yield stress of concrete is reduced or removed so the concrete flows by its weight [5] for releasing air bubbles and producing concrete of the highest density, strength and durability [6]. Usually, a concrete that has been vibrated carefully may contain 1 to 2% of entrapped air [7].

A homogeneous and randomly oriented aggregate distribution can improve the mechanical properties, durability, stability and impermeability of concrete [8]. The ability of fresh concrete to remain uniform during consolidation is a critical issue in the mixture design [9]. During the mixing of lightweight aggregate concrete (LWAC), due to the low density of the aggregates used and the longer mixing times [10], LWAC is susceptible to segregation of the aggregates as a result of the differences between the densities of their components [11]. In fact, during the vibration of the concrete, lightweight aggregates tend to float.

DOI: 10.2495/CMEM-V6-N4-835-846

As the mechanical strength of the mortar is considerably higher than the strength of lightweight aggregates [12–14], a non-uniform distribution of the aggregates in the concrete mixture can strongly affect the overall characteristics, which are commonly considered as homogeneous values for design purposes.

All aspects presented above justify the experimental evaluation of segregation in concrete using indexes for it quantification [15].

What follows are some methods to quantify the phenomenon of segregation proposed by other authors.

1.1 Method proposed by Ke *et al.*

Ke proposes a method for determining the segregation index (I_s) [16, 17], dividing the specimens into four equal sections and using the densities obtained from the upper (ρ_{top}) and lower (ρ_{botton}) slices of a cylinder. A possible segregation tends to reduce the density in the upper section because the lightweight aggregates tend to float in the mortar matrix.

The index is calculated according to the eqn. (1).

$$I_s = \frac{\rho_{top}}{\rho_{botton}} \tag{1}$$

If I_s=1, it can be considered that the sample shows perfect uniformity. An index of less than 0.95 indicates a start of segregation [17]. However, experimental results indicate that this segregation index does not always reflect the real conditions of the specimen. Another fact is the difficulty in locating concentrations of aggregates, which could demand the weighing and the comparison of many specimens [18].

1.2 Method proposed by Lopez – Navarrete – Esmaeilkhanian

Using a particular case of the method proposed by Esmaeilkhanian *et al.* [19] and using an unbiased stereology technique based on count pointing [20, 21], Navarrete-Lopez have calculated a segregation index based on the volumetric fraction of aggregates at different heights of a specimen [9]. Each specimen was divided into three equals sections (top, middle and bottom). For the top and bottom sections, the volume of coarse aggregate was calculated using the eqn. (2),

$$V_{ai} = \frac{P_{ai}}{P_{refi}}.100\% \tag{2}$$

where P_{ai} is the sum of the points intersecting the aggregate in section i; P_{refi} is the sum of the points intersecting section i and V_{ai} is the aggregate volume fraction of section i. To evaluate segregation, the volumetric index (VI), proposed by Esmaeilkhanian *et al.* [19] was calculated:

$$VI(\%) = 2 * \frac{|V_{at} - V_{ab}|}{V_{at} + V_{AB}} * 100\% \tag{3}$$

where V_{at} and V_{ab} are the coarse aggregate volume fraction of the top and bottom sections, respectively.

The results of the studies of Kwasny *et al.* [20] suggest that lightweight concrete may be considered as non-segregated when VI values are below 20%. Esmaeilkhanian *et al.* [22]

Table 1: VI range of segregation levels Navarrete-Lopez [11].

Segregation level	VI range (%)
None to slight	0–40
Moderate	40–80
Severe	80–120
Slightly stratified	120–160
Highly stratified	160–200

have studied the dynamic segregation of self-compacting concrete and have proposed the value of VI = 25% as the limit for segregation. Navarrete-Lopez proposes a scale with five degrees of segregation to classify the results obtained (Table 1).

1.3 The use of 2D images to represent 3D phenomenon

When the distribution and location of the aggregates is evaluated based on a 2D image (cross-section of the specimen), it is important to understand its representation in the three-dimensional (3D). According to the Cavalieri Principle, the percentage of an area in the cross-section of a specimen can be considered equivalent to the percentage of its volume in a 3D representation [23]. In the study of concrete, it is easier to use 2D images than 3D, in order to represent information. Jianguo Han used the technique of 2D image analysis to study the characteristics and distribution of aggregates in concrete[8]. The research carried out by Masad *et al.* demonstrated the usefulness of this technique to obtain the percentage of holes in bituminous mixtures using 2D images of X-ray tomographies and contrasting it with the volumetric tests usually used in this type of materials [24]. Scrivener, in one of his investigations, discussed the limitations of the use of 2D images and affirmed that these images correspond to a fraction of total area, directly equivalent to the fraction of volumen [25].

2 MATERIALS

The experimental campaign involved the production of a concrete made with coarse lightweight aggregate using the Fanjul method [26], in order to produce LWAC with a target density of 1700 kg/m^3. This method allows the maximum control in the design of concretes with a pre-set density. Table 2 includes the mix proportions.

CEM I 52.5 R cement with an absolute density of 3176 kg/m^3 was used; expanded clay was used as lightweight aggregate (Arlita Leca produced by Weber, commercially referred as M). Limestone sand was also used as fine aggregate. Prior to mixing lightweight aggregates were presaturated. At the time of mixing, the water content of the lightweight aggregates (54%) and the surface water content (3.7%) were determined in order to make the necessary corrections and maintain a constant effective w/c ratio of 0.6.

Table 2: Mix proportions to produce 1 m^3 of concrete.

	Cement	**Water**	**Fine aggregate**	**LWA**
Wheight (kg)	350	210	991	149
Volume (m^3)	0.110	0.210	0.369	0.309

Table 3: Characteristics of aggregates and the methods/standard used for testing.

Property	Method	Fine	Coarse	LWA
Dry particle density (kg/ m^3)	Fernández-Fanjul el al [27]	2708		482
Bulk density (kg/ m^3)	UNE EN 1097-3	1605		269
24 h Water absortion (%)	UNE EN 1097-6	0.12		36.9
Granulometric fraction (di/Di)	UNE EN 933-1	0/4		6/16

Table 4: Mortar characterization

Age (days)	Density (kg/m^3)	Absortion (%)	Porosity (%)
7	2039	11.25	22.98
28	2049	10.78	22.25
90	2060	10.92	22.52

The main characteristics of aggregates and the methods/standard used for testing are presented in Table 3.

In order to know the physical properties of the mortar that forms part of the lightweight concrete, and knowing its dosage, prismatic mortar probes of 40x40x160 mm (according to UNE EN 196-1) were manufactured to characterize its density, porosity and absorption. The curing of the specimens was in water at a temperature of 20 ± 1°C; their values determined at 7, 28 and 90 days of age (Table 4).

3 METHODOLOGY

In order to achieve the objectives set forth in the previous section, the methodology described below and represented in the diagram of Fig. 1 was adopted.

3.1 Manufacturing and preparation of the concrete specimens

The concrete was manufactured considering the following variables when making the specimens (cylindrical of 150 mm of φ and 300 mm of height): the compaction has been performed using an electric needle vibrator of 18000 rpm/min and a needle diameter of φ25 mm. The specimens were vibrated with six different times (5–10–20–40–80–160 seconds) in a single layer.

After being made and cured in the water at a temperature of 20 ± 1°C, the specimens were saw-cut through its longitudinal axis, their bulk densities were determined by the hydrostatic balance method. Subsequently, their sections were photographed for the subsequent image analysis.

3.2 Experimental phase

The specimen's halves were then saw-cut into four equal parts, resulting in octaves, which had their bulk densities determined. Using the density values of the upper and lower sections, the segregation index was obtained according to the methodology indicated by Ke [10].

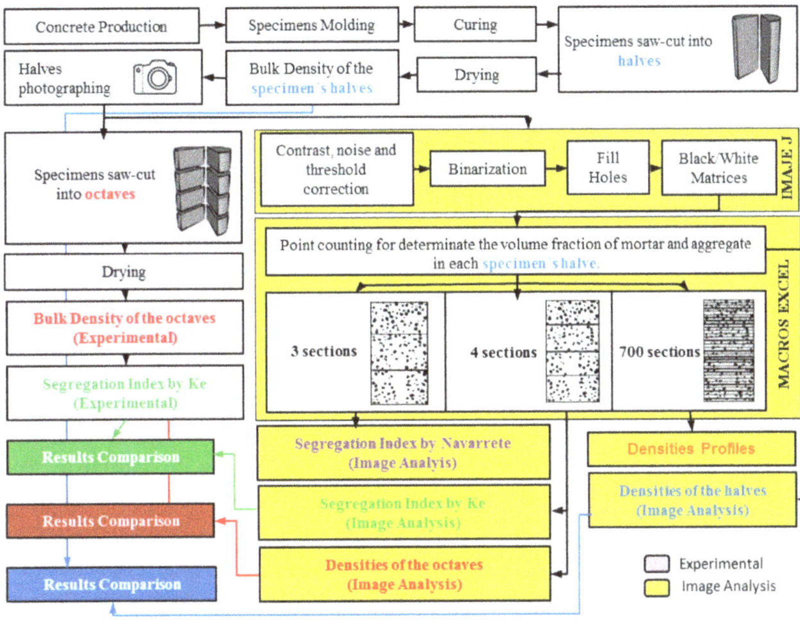

Figure 1: Methodology diagram.

3.2.1 Image processing

The same treatment was performed for all the samples: from the original image the perspective was corrected, seeking to eliminate any errors caused by the unevenness of the camera or the surface where the specimens were located. Once the perspective correction was done, the contrast and threshold were adjusted, the noise was reduced, the image was binarized and the internal voids of the aggregates were filled using ImageJ (Fig. 2).

3.2.2 Binarization

With the binarization what is tried is to classify each pixel of the image, differentiating between the lightweight aggregate and the mortar. A binary code is established where the

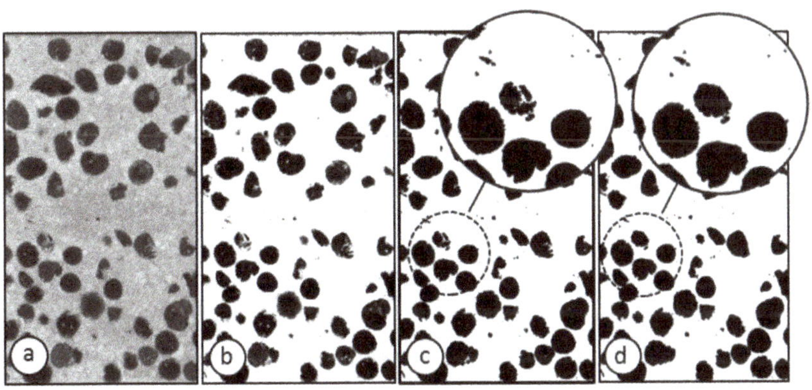

Figure 2: a) Original image. b) Contrast, threshold and noise adjustments. c) Binarization. d) Fill holes.

black color, representing the lightweight aggregate, have a numerical value of 1 and the white colour, representing the mortar matrix, have a numerical value of 0.

3.2.3 Black and White matrices data processing

The black and white matrices, which represent each half of the specimen, were horizontally separated into four parts, which represent an eighth of each of the specimens. In this way, the densities of the upper and lower octaves of each specimen were obtained by image analysis. From the densities obtained for these octaves, using image analysis, the Segregation Index proposed by Ke [8] was calculated.

In addition to the study, the same procedure has been applied separating the sample halves into three equal parts, from where the values for the volumetric fraction of aggregates of the respective regions were determined and used to calculate the Segregation Index proposed by Navarrete *et al.* [9].

To obtain the densities by image analysis, the percentage of each material in each zone was quantified as described above and, since we know the densities of the mortar matrix and of the lightweight aggregates, it is possible to determine the density of the section analyzed by means of the Eq. 4, where N_{mortar} is the percentage of mortar pixels present in the analyzed area, N_{LWA} is the percentage of lightweight aggregate pixels present in the analyzed area, ρ_{mortar} is the bulk density of mortar at 28 days of age (Table 4) and ρ_{LWA} is the dry density of the lightweight aggregates (Table 3).

$$\rho_{secci\ n} = \frac{N_{mortar} * \rho_{mortar} + N_{LWA} * \rho_{LWA}}{N_{mortar} + N_{LWA}} \quad (4)$$

Microsoft Excel was used for processing the data from the Black and White Matrices. Each concrete sample generated one file containing one Spreadsheet arranged in 701 rows and 326 columns, equivalent to the 700x325 pixels of each image and it reference axes. In order to facilitate the processing of the data and for obtaining more reliable results, Macros have been programmed in Visual Basic to speed up the process.

The first step is unifying all the Spreadsheets from each concrete section into a single file using the file *UNIR.xlsm*. All samples must be placed in the same directory before running the first Macro pressing the button *'Combine data in a single file'*. This Macro generates a new file called *'AI_Serie_Matrices.xlsx'* in the same directory of the other samples (Fig. 3).

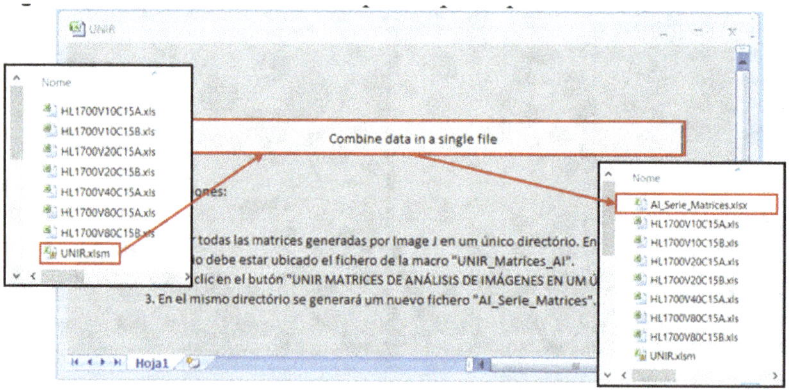

Figure 3: UNIR.xlsm is the tool to unify the samples in a single file (AI_Serie_Matrices.xlsx).

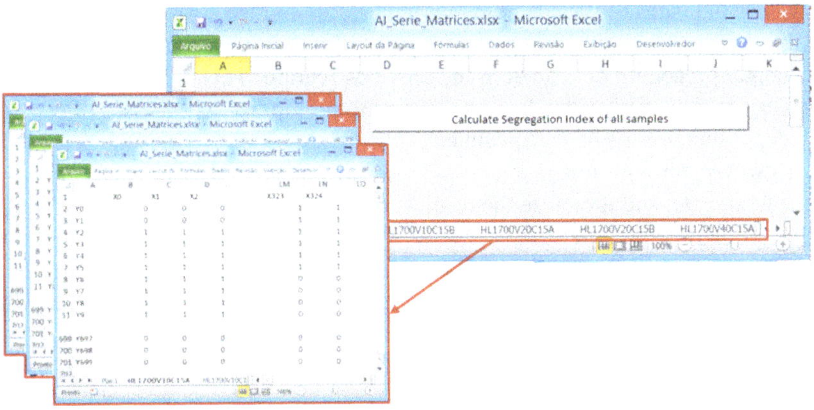

Figure 4: AI_Serie_Matrices.xlsx is the file to run the second Macro by pressing *'Calculate Segregation Index of all samples'*. All the samples are united in this file, one sample in each Spreadsheet.

Opening the file *'AI_Serie_Matrices.xlsx'* and keeping the file *'UNIR.xslm'* running in the background, the second Macro is executed by pressing *'Calculate Segregation Index of all samples'*. Internally, the Macro runs a process to count the points of each region, and based on the input data (densities of mortar and aggregates), combines this information to calculate the densities of each region (Fig. 4), the segregation indexes of Ke *et al.*, Navarrete *et al.* and traces the densities profile. The process is automatically repeated for each sample.

4 RESULTS AND DISCUSSION

4.1 Validation of the image analysis methodology

4.1.1 Density: experimental and image analysis methodology.

Once described the methodology used in the research, this section shows the results obtained experimentally and through the methodology of image analysis. Figure 5, on the left, shows

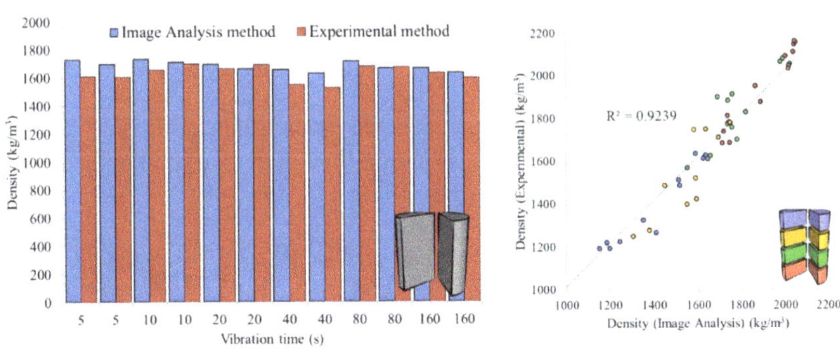

Figure 5: On the left, comparison of the dry densities of concrete specimens compacted with different vibration times and obtained through and by image analysis. On the right, experimental densities, vs image analysis of the eighth eight concrete specimen sections.

the dry density results of the concrete saw-cut halves vibrated with different times. As can be seen, the results obtained by image analysis are slightly higher than those obtained with the experimental procedure described in the previous section. This difference is higher in those with a few seconds of vibration because, during the compaction process, a high percentage of entrained air remains inside the specimen [7] and causes a decrease in the density of the concrete. However, through image analysis, voids are considered as a mortar, which is denser than the aggregate, and therefore the result is a higher density.

In Fig. 5, on the right, the comparison between the densities obtained in the octaves of the concrete specimens is shown. As can be observed, there is a high coincidence between the values of dry concrete density obtained through experimental procedure and the values obtained by image analysis. The coefficient of determination (R^2) is related to the straight line 1:1 and its R^2 value is 0.9239, which, according to the Evans scale [27], shows a very strong correlation between both results.

These results demonstrate that concrete density can be obtained using photographed section (provided component densities are known) allowing for more in-depth concrete section analysis.

As an example, Fig. 6 shows the evolution of the density of the four concrete sections subjected to different compaction times. As can be seen in Fig. 6, the density of the concrete of the upper section (ρ_{top}) decreases with the increase of the vibration time when the entire concrete specimen is saw-cut into four sections. In addition, the density of the concrete of the lower section (ρ_{bottom}) rises by the increase in the time of vibration. All this will lead, as will be seen in the following section, to an increase in the segregation of the concrete with the time of vibration.

Thus, demonstrate and verify that density of the concrete can be obtained by image analysis. Next step is to obtain the segregation indexes proposed by Ke according to eqn. (1) and by López-Navarrete according to eqn. (3).

4.1.2 Segregation index according to Ke: experimental procedure vs image analysis.

As in the previous section, this section aims to verify the possibility of obtaining the segregation index according to eqn. (3) by means of the image analysis of each concrete section comparing them with the results obtained experimentally.

In Fig. 7 shows a very strong correlation ($R^2 = 0.967$) between the I_s values obtained by experimental and by image analysis method. Thus, the results showed the possibility to determine the segregation index according to Ke methodology with the photographed concrete section and without the need to perform experimentally the determination of the densities.

As an example, the evolution of the concretes segregation index has been represented in Fig. 8. As can be seen, the increase in the vibration time increased the segregation index of

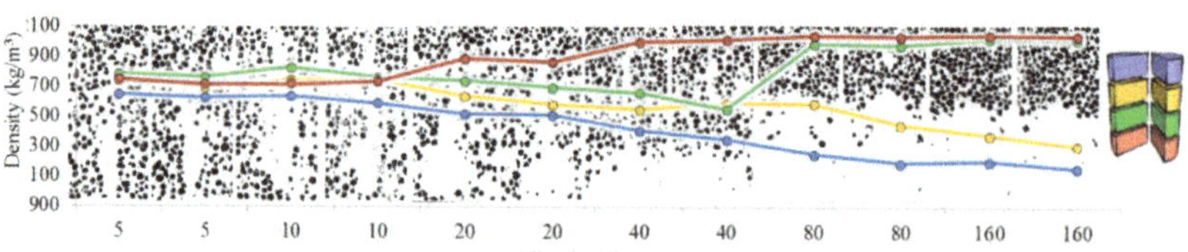

Figure 6: Evolution of the four concrete density sections increasing the vibration time obtained through image analysis.

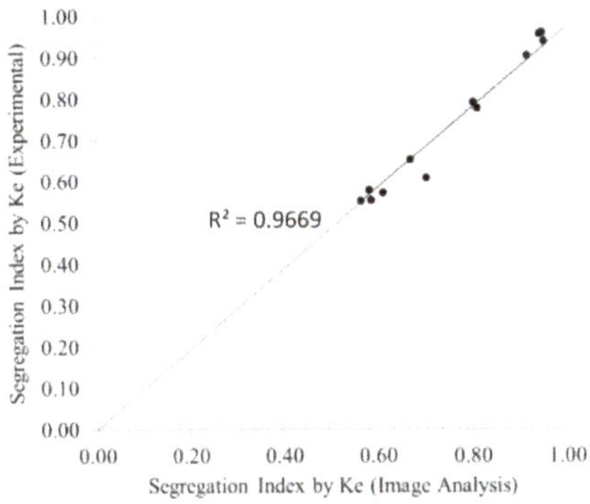

Figure 7: Comparison between segregation indexes (according to Ke) obtained by experimental and image analysis method.

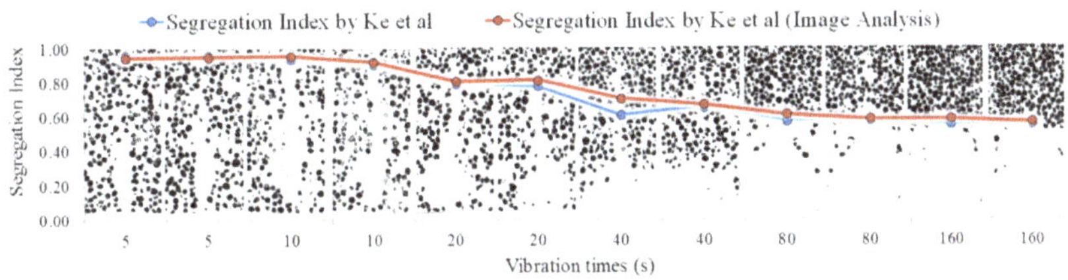

Figure 8: Segregation Indexes (I_s) evolution (Image Analysis).

the concrete. Homogeneous concrete were manufactured with reduced vibration times (5–10 s) because values of I_s obtained were approximately 1. However, as the compaction times increased, segregation became evident. The maximum values of I_s obviously, were presented with 80–160 seconds of vibration, where all the lightweight aggregate was located in the upper zone of the concrete specimens (Fig. 8).

4.2 Other results obtained

4.2.1 Density profile

As demonstrated in the results, the image analysis tool obtains very closely the density values of the concretes studied. Thus, it allows to study in detail the mechanisms of the segregation in concrete.

For that reason, using Visual Basic, an application in Microsoft Excel was developed to obtain the density profile of the concretes. Figure 9 shows the segregation sheets that can be obtained using the application programmed in Visual Basic. In red, the part related to the mortar is plotted and in blue the part related to the lightweight aggregate. Therefore, image analysis based applications can be used to study with a greater degree of detail the segregation of concrete that is impossible or very difficult to carry out experimentally.

4.2.2 Segregation index according to Navarrete *et al.*

In the last section, since the use of the images to determine the density has been demonstrated, the authors wanted to compare the segregation indexes obtained according to Ke and those obtained according to Navarrete *et al.*

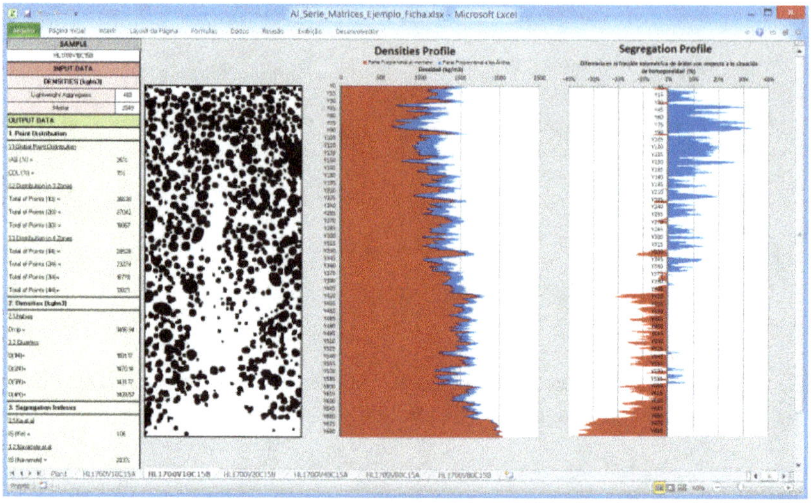

Figure 9: Result sheets. Density profile of the specimen related to lightweight aggregate and mortar parts in each section.

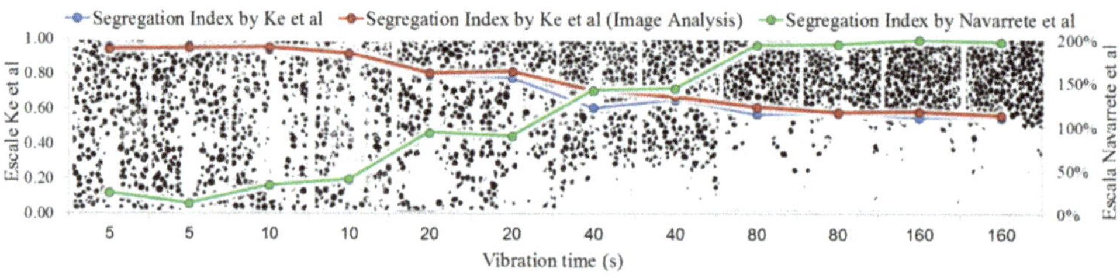

Figure 10: Segregation index (I_s) according to Ke and VI index according to Navarrete.

The scales in both methods are different since a value of 1 or 0% means homogeneous concrete according to Ke or Navarrete method, respectively. In contrast, a value of 0 or 200% means a totally segregated concrete.

Plotted in Fig. 10, there was a similar trend in both methods. It was found homogeneous values with the shorter vibration times (5–10 s) and the maximum segregation with vibration time of 160 s.

5 CONCLUSIONS

In this paper, the main objective has been to determine the density of segregated lightweight concrete using image analysis software and to compare with experimental results.

- The results showed that the concrete density could be obtained using a photographed section (as long as the densities of the constituents are known).
- Segregation indexes of lightweight concretes can be obtained through the photograph of the section. The method allows replacing the experimental method for determining the density of each section using images of the concrete specimens.
- By means of the image analysis, the density profile was obtained using a tool developed in Visual Basic. This tool let to perform a deeper analysis of the segregation phenomenon.

Therefore, this study demonstrates that the image analysis allows a deeper knowledge of the behavior of segregated concrete opening new possibilities for engineers to understand and analyze the mechanisms of segregation in lightweight concrete.

ACKNOWLEDGEMENTS

This research was funded by the University of Alicante (GRE13-03) and (VIGROB-256).

REFERENCES

[1] Hwang, C.L. & Hung, M.F., Durability design and performance of self-consolidating lightweight concrete. *Construction and Building Materials*, **19**(8), pp. 619–626, 2015. https://doi.org/10.1016/j.conbuildmat.2005.01.003

[2] Sari, D. & Pasamehmetoglu, A.G., The effects of gradation and admixture on the pumice lightweight aggregate concrete. *Cement and Concrete Research*, **35**(5), pp. 936–942, 2015. https://doi.org/10.1016/j.cemconres.2004.04.020

[3] Rossignolo, J.A., Agnesini, M.V. & Morais, J.A., Properties of high-performance LWAC for precast structures with Brazilian lightweight aggregates. *Cement and Concrete Composites*, **25**, pp. 77–82, 2003. https://doi.org/10.1016/s0958-9465(01)00046-4

[4] Rossignolo, J.A., *Concreto Leve Estrutural: produção, propriedades, microestrutura e aplicações.*São Paulo: PINI, 2009.

[5] Tattersall, G.H. & Baker, P.H., The effect of vibration on the rheological properties of fresh concrete. *Magazine of Concrete Research*, **40**, pp. 79–89, 1988. https://doi.org/10.1680/macr.1988.40.143.79

[6] Banfill, P.F.G., Teixeira, M.A.O.M. & Craik, R.J.M., Rheology and vibration of fresh concrete: predicting the radius of action of poker vibrators from wave propagation. *Cement and Concrete Research*, **41**(9), pp. 932–941, 2011. https://doi.org/10.1016/j.cemconres.2011.04.011

[7] Aïtcin, P.C. & Flatt, R., *Science and Technology of Concrete Admixtures,* 2016.

[8] Han, J., Wang, K., Wang, X. & Monteiro, P.J.M, 2D image analysis method for evaluating coarse aggregate characteristic and distribution in concrete. *Construction and Building Materials*, **127**, pp. 30–42, 2016. https://doi.org/10.1016/j.conbuildmat.2016.09.120

[9] Navarrete, I. & Lopez, M., Estimating the segregation of concrete based on mixture design and vibratory energy. *Construction and Building Materials*, **122**, pp. 384–390, 2016. https://doi.org/10.1016/j.conbuildmat.2016.06.066

[10] Barbosa, F.S., Farage, M.C.R., Beaucour, A.-L. & Ortola, S., Evaluation of aggregate gradation in lightweight concrete via image processing. *Construction and Building Materials*, **29**, pp. 7–11, 2012. https://doi.org/10.1016/j.conbuildmat.2011.08.081

[11] Yu, Q.L., Spiesz, P. & Brouwers, H.J.H., Ultra-lightweight concrete: conceptual design and performance evaluation. *Cement and Concrete Composites*, **61**, pp. 18–28, 2015. https://doi.org/10.1016/j.cemconcomp.2015.04.012

[12] Larrard, F.D. & Belloc, A., L'influence du granulat sur la resistance a la compression des betons. *Bulletin des Laboratoires des Ponts et Chaussées*, pp. 41–52, 1999.

[13] Ke, Y., Beaucour, A.L., Ortola, S., Dumontet, H. & Cabrillac, R., Comportement Mécanique des Bétons de Granulats Légers: Étude Expérimentale et Modélisation *Rencontres Du Génie Civil Urbain, Costruire. Les Nouveaux* Défis, **24**, 2006.
[14] Gerritse, A., Design considerations for reinforced lightweight concrete. *International Journal of Cement Composites and Lightweight Concrete*, **3**(1), pp. 57–69, 1981. https://doi.org/10.1016/0262-5075(81)90031-2
[15] American Concrete Institute, *213R-14 Guide for Structural Lightweight-Aggregate Concrete*. 2003, Reported by ACI Committee 213.
[16] Ke, Y., Beaucour, A.L., Ortola, S., Dumontet, H. & Cabrillac, R., Influence of volume fraction and characteristics of lightweight aggregates on the mechanical properties of concrete. *Construction and Building Materials*, **23**(8), pp. 2821–2828, 2009. https://doi.org/10.1016/j.conbuildmat.2009.02.038
[17] Ke, Y., *Characterization of the mechanical behavior of lightweight aggregate concretes: experiment and modelling*. Université de Cergy-Pontoise, 2008.
[18] Barbosa, F.S., Beaucour, A.L., Fanage, M.C.R. & Ortola, S., Image processing applied to the analysis of segregation in lightweight aggregate concretes. *Construction and Building Materials*, **25**, pp. 3375–3381, 2011. https://doi.org/10.1016/j.conbuildmat.2011.03.028
[19] Esmaeilkhanian, B., Khayat, K.H., Yahia, A. & Feys, D., Effects of mix design parameters and rheological properties on dynamic stability of self-consolidating concrete. *Cement and Concrete Composites,* **54**, pp. 21–28, 2014. https://doi.org/10.1016/j.cemconcomp.2014.03.001
[20] Jacek Kwasny, S.M., Sonebi, M., Taylor, S.E. & Bai, Y., Influence of the type of coarse lightweight aggregate on properties of semilightweight self-consolidating concrete. *Journal of Materials in Civil Engineering*, **24**(12), pp. 1474–1483, 2012. https://doi.org/10.1061/(asce)mt.1943-5533.0000527
[21] Navarrete, I., *Stratified concrete: understanding its stratification process and modelling its structural behavior*. Pontificia Universidad Catolica de Chile: Santiago de Chile, 2015.
[22] Esmaeilkhanian, B., Feys, D., Khayat, K.H. & Yahia, A., New test method to evaluate dynamic stability of self-consolidating concrete. *ACI Materials Journal*, **111**(13), pp. 299–308, 2014.
[23] Baddeley, A. & Vedel Jensen, E.B., *Stereology for Statisticians*. 2005.
[24] Masad, E., Muhunthan, B., Shashidhar, N. & Harman, T., Internal structure characterization of asphalt concrete using image analysis. *Journal of Computing in Civil Engineering*, **13**(2), pp. 88–95, 1999. https://doi.org/10.1061/(asce)0887-3801(1999)13:2(88)
[25] Scrivener, K.L., Backscattered electron imaging of cementitious microstructures: understanding and quantification. *Cement and Concrete Composites*, **26**, pp. 935–945, 2004. https://doi.org/10.1016/j.cemconcomp.2004.02.029
[26] Fernández-Fanjul, A. & Tenza-Abril, A.J., Méthode FANJUL: Dosage pondéral des bétons légers et lourds. *Annales du Bâtiment et des Travaux Publics*, **5**, pp. 32–50, 2012.
[27] Fernández-Fanjul, A., Tenza-Abril, A.J. & Baeza-Brotons, F., A new methodology for determining particle density and absorption of lightweight, normal-weight and heavy weight aggregates in aqueous medium. *Construction and Building Materials*, **146**, pp. 630–643, 2017. https://doi.org/10.1016/j.conbuildmat.2017.04.052

Author index

www.ingramcontent.com/pod-product-compliance
Lightning Source LLC
LaVergne TN
LVHW062235200926
848484LV00006B/1544